CourseMate Engaging. Trackable. Affordable.

CourseMate brings course concepts to life with interactive ~~study,~~
and exam preparation tools that support Applied CALC.

INCLUDES:
Integrated eBook, interactive teaching and learning tools, and **Engagement Tracker,**
a first-of-its-kind tool that monitors student engagement in the course.

ON THE WEB

APPLIED **CALC** *Are you in?*

ONLINE RESOURCES INCLUDED!

FOR INSTRUCTORS:
• First Day Class Instructions
• Solution Builder
• PowerPoint® Slides
• Instructor Prep Cards
• Student Review Cards
• Tech Cards

FOR STUDENTS:
• Quizzing
• Flashcards
• Glossary
• Interactive eBook
• Solution Videos
• Lecture Videos
• Additional Examples
• Extra Practice Exercises
• Answers & Solutions
• Technology Guides
• and more!

Students sign in at **login.cengagebrain.com**

BROOKS/COLE
CENGAGE Learning™

Applied CALC
Frank C. WIlson

VP/Editor-in-Chief: Michelle Julet

Publisher: Richard Stratton

Senior Development Editor: Laura Wheel

Editorial Assistant: Haeree Chang

Media Editors: Heleny Wong, Philip Lanza

Senior Content Project Manager: Cathy Brooks

Executive Marketing Manager, 4LTR Press:
Robin Lucas

Product Development Manager, 4LTR Press:
Steven E. Joos

Project Manager, 4LTR Press: Kelli Strieby

Senior Art Director: Jill Ort

Senior Print Buyer: Diane Gibbons

Marketing Director: Mandee Eckersley

Marketing Manager: Ashley Pickering

Marketing Communications Manager:
Mary Anne Payumo

Editorial Development and Composition:
Lachina Publishing Services

Production Service:
Lachina Publishing Services

Cover Designer: Hannah Wellman

Cover Image: ©istockphoto.com/Lefthome

Contributors: Linda Meng, Sue Steele

For product information and technology assistance, contact us at
Cengage Learning Customer & Sales Support, 1-800-354-9706
For permission to use material from this text or product,
submit all requests online at **www.cengage.com/permissions**
Further permissions questions can be emailed to
permissionrequest@cengage.com

Library of Congress Control Number: 2010930763

ISBN-13: 978-0-8400-6563-6

ISBN-10: 0-8400-6563-9

Brooks/Cole
20 Channel Center Street
Boston, MA 02210
USA

Cengage Learning is a leading provider of customized learning solutions with office locations around the globe, including Singapore, the United Kingdom, Australia, Mexico, Brazil, and Japan. Locate your local office at:
international.cengage.com/region

Cengage Learning products are represented in Canada by Nelson Education, Ltd.

For your course and learning solutions, visit **www.cengage.com**

Purchase any of our products at your local college store or at our preferred online store **www.cengagebrain.com**

Disclaimer

In this book, I have incorporated real-world data from the financial markets to the medical field. In each case, I have done my best to present the data accurately and interpret the data realistically. However, I do not claim to be an expert in financial, medical, and other similar fields. My interpretations of real-world data and my associated conclusions may not adequately consider all relevant factors. Therefore, readers are encouraged to seek professional advice from experts in the appropriate fields before making decisions related to the topics addressed herein.

Despite the usefulness of mathematical models as representations of real-world data sets, most mathematical models have a certain level of error. It is common for model results to differ from raw data set values. Consequently, conclusions drawn from a mathematical model may differ (sometimes dramatically) from conclusions drawn by looking at raw data sets. Readers are encouraged to interpret model results with this understanding.

—Frank Wilson, Author

Printed in the United States of America
1 2 3 4 5 6 7 14 13 12 11 10

CALC

Brief Contents

Shutterstock

1 Functions and Linear Models   2

2 Nonlinear Models   20

3 The Derivative   46

4 Differentiation Techniques   74

Alan Schein Photography/Corbis

5 Derivative Applications   96

6 The Integral   134

Shutterstock

7 Advanced Integration Techniques and Applications   172

Shutterstock

8 Multivariable Functions and Partial Derivatives   202

U.S. Coast Guard

STUDY
YOUR WAY

At no additional cost, you have access to online learning resources that include **tutorial videos, printable flashcards, quizzes,** and more!

Watch videos that offer step-by-step conceptual explanations and guidance for each chapter in the text.

Along with the printable flashcards and other online resources, you will have a multitude of ways to check your comprehension of key mathematical concepts.

You can find these resources at **login.cengagebrain.com.**

Contents

1 Functions and Linear Models 2

1.1 Functions   2

1.2 Linear Functions   11

2 Nonlinear Models   20

2.1 Quadratic Function Models   20

2.2 Exponential Function Models   35

3 The Derivative   46

3.1 Average Rate of Change   46

3.2 Limits and Instantaneous Rates of Change   51

3.3 The Derivative as a Slope: Graphical Methods   58

3.4 The Derivative as a Function: Algebraic Method   66

3.5 Interpreting the Derivative   70

4 Differentiation Techniques   74

4.1 Basic Derivative Rules   74

4.2 The Product and Quotient Rules   79

4.3 The Chain Rule   82

4.4 Exponential and Logarithmic Rules   88

4.5 Implicit Differentiation   92

5 Derivative Applications   96

5.1 Maxima and Minima   96

5.2 Applications of Maxima and Minima   107

5.3 Concavity and the Second Derivative   116

5.4 Related Rates   129

6 The Integral   134

6.1 Indefinite Integrals   134

6.2 Integration by Substitution   141

6.3 Using Sums to Approximate Area   145

6.4 The Definite Integral   156

6.5 The Fundamental Theorem of Calculus   164

7 Advanced Integration Techniques and Applications   172

7.1 Integration by Parts   172

7.2 Area Between Two Curves   178

7.3 Differential Equations and Applications   185

7.4 Differential Equations: Limited Growth and Logistic Growth Models   192

8 Multivariable Functions and Partial Derivatives   202

8.1 Multivariable Functions   202

8.2 Partial Derivatives   211

8.3 Multivariable Maxima and Minima   221

8.4 Constrained Maxima and Minima and Applications   231

Index   241

Shutterstock

Shutterstock

LEARN YOUR WAY

With **Applied CALC**, you have a multitude of study aids at your fingertips.

The Student Solutions Manual contains worked-out solutions to every odd-numbered exercise, to further reinfore your understanding of mathematical concepts in **Applied CALC**.

In addition to the Student Solutions Manual, Cengage Learning's **CourseMate** offers exercises and questions that correspond to every section and chapter in the text for extra practice.

For access to these study aids, sign in at **login.cengagebrain.com**.

Chapter 1

Functions and
Linear Models

Mathematical functions are a powerful tool used to model real-world phenomena. Whether simple or complex, functions give us a way to forecast expected results. Remarkably, anything that has a constant rate of change may be accurately modeled with a linear function. For example, the cost of filling your car's gas tank is a linear function of the number of gallons purchased.

1.1 **Functions**

1.2 **Linear Functions**

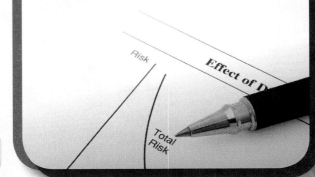

1.1 Functions

Our society is a complex system of relationships between people, places, and things. Many of these relationships are interconnected. In mathematics, we often model the relationship between two or more interdependent quantities by using a **function**. In this section, we will show how to distinguish between functions and non-functions and will practice using function notation. We will also demonstrate how to use technology to draw a function graph and discuss how to find the domain of a function.

> **Definition:** Function
> A function is a rule that associates each
> input with *exactly one* output.

Shutterstock

Often the rule is represented in a table of data with the inputs on the left-hand side and the outputs on the right-hand side. For example, the amount of money we pay to fill up our gas tank is a function of the number of gallons pumped (see Table 1.1).

In this case, the input to the function is *gallons pumped* and the output of the function is *total fuel cost*. Fuel cost is a function of the number of gallons pumped because each input has exactly one output.

Similarly, the weekly wage of a service station employee is a function of the number of hours worked. The *number of hours worked* is the input to the function, and the *weekly wage* is the output of the function. Since weekly wage is a function of the number of hours worked, an employee who works 40 hours expects to be paid the same wage each time she works that amount of time.

Table 1.1

Gallons Pumped	Total Fuel Cost
10	$27.99
15	$41.99
20	$55.98

EXAMPLE 1 Determining If a Table of Data Represents a Function

➡ As reported by the U.S. Census Bureau, the population rank of Massachusetts in selected years is shown in the table (see Table 1.2). Is the year a function of the population rank of Massachusetts?

Table 1.2

Massachusetts Population Rank	Year
13	2006
13	2007
14	2008
15	2009

same input { 13 / 13 }   different outputs { 2006 / 2007 }

SOLUTION

According to the definition of a function, each input must have exactly one output. The input value 13 has two different outputs: 2006 and 2007. Since the input 13 has more than one output, the year is *not* a function of the population rank of Massachusetts.

Function Notation

When we encounter functions in real life, they are often expressed in words. To make functions easier to work with, we typically use symbolic notation to represent the relationship between the input and the output. Let's return to the fuel cost table introduced previously (Table 1.3).

Table 1.3

Gallons Pumped	Total Fuel Cost
10	$27.99
15	$41.99
20	$55.98

Observe that the fuel cost is equal to $2.799 times the number of gallons pumped. We represent this symbolically as $C(g) = 2.799g$, where $C(g)$ represents the total fuel cost when g gallons are pumped. ($C(g)$ is read "C of g.") The letter C is used to represent the name of the rule, and the letter g in the parentheses indicates that the rule works with different values of g (see Figure 1.1).

We call the output variable of a function the **dependent variable** because the value of the output variable *depends* on the value of the input variable. The input

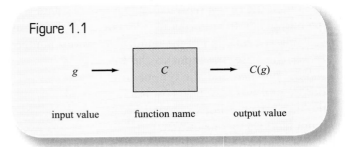

Figure 1.1

$$g \longrightarrow \boxed{C} \longrightarrow C(g)$$

input value   function name   output value

variable is called the **independent variable.** (One way to remember the meaning of the terms is to observe that both *input* and *independent* begin with *in*.) From the table, we see that

$$C(10) = 27.99$$
$$C(15) = 41.99$$
$$C(20) = 55.98$$

For this function, the independent variable took on the values 10, 15, and 20, and the dependent variable assumed the values 27.99, 41.99, and 55.98.

EXAMPLE 2 Determining a Linear Model from a Verbal Description

➡ An electronics store employee earns $8.50 per hour. Write an equation for the employee's earnings as a function of the hours worked. Then calculate the amount of money the employee earns (in dollars) by working 30 hours.

SOLUTION

Since the employee earns $8.50 for each hour worked, the employee's total earnings are equal to $8.50 times the total number of hours worked. That is,

$$E(h) = 8.50h$$

where E is the employee's earnings (in dollars) and h is the number of hours worked. To calculate the amount of money earned by working 30 hours, we evaluate this function at $h = 30$.

Shutterstock

$$E(30) = 8.5(30)$$
$$= 255$$

The employee earns $255 for 30 hours of work.

Function notation is extremely versatile. Suppose we are given the function $f(x) = x^2 - 2x + 1$. We may evaluate the function using either numerical values or nonnumerical values. For example,

$$f(2) = (2)^2 - 2(2) + 1 \qquad f(\Delta) = (\Delta)^2 - 2(\Delta) + 1$$
$$= 4 - 4 + 1$$
$$= 1$$

$$f(a + 2) = (a + 2)^2 - 2(a + 2) + 1$$
$$= (a^2 + 4a + 4) - 2a - 4 + 1$$
$$= a^2 + 2a + 1$$

In each case, we replaced the value of x in the function $f(x) = x^2 - 2x + 1$ with the quantity in the parentheses. Whether the independent variable value was 2, Δ, or $a + 2$, the process was the same.

EXAMPLE 3 Evaluating a Function Using Function Notation

Evaluate the function $s(t) = t^3 + 4t$ at $t = 3$, $t = \Delta$, and $t = a^2$.

SOLUTION

$$s(3) = (3)^3 + 4(3) \qquad s(\Delta) = (\Delta)^3 + 4(\Delta)$$
$$= 27 + 12$$
$$= 39$$

$$s(a^2) = (a^2)^3 + 4(a^2)$$
$$= a^6 + 4a^2$$

Graphs of Functions

Functions are represented visually by plotting points on a Cartesian coordinate system (see Figure 1.2). The horizontal axis shows the value of the independent variable (in this case, x), and the vertical axis shows the value of the dependent variable (in this case, y).

When using the coordinate system, y is typically used in place of the function notation $f(x)$. That is, $y = f(x)$. This is true even if the function has a name other than f.

The point of intersection of the horizontal and vertical axes is referred to as the **origin** and is represented by the ordered pair $(0, 0)$. To graph an ordered pair (a, b), we move from the origin $|a|$ units horizontally and $|b|$

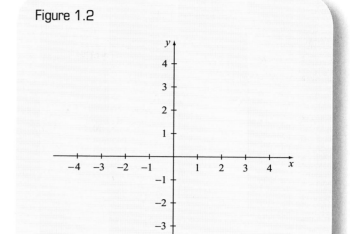

Figure 1.2

units vertically and draw a point. If $a > 0$, we move to the right. If $a < 0$, we move to the left. Similarly, if $b > 0$, we move up, and if $b < 0$, we move down. For example, consider the table of values with its associated interpretation in Table 1.4 and the graph in Figure 1.3.

When we are given the equation of a function, we can generate a table of values and then plot the corresponding points. Once we have drawn a sufficient

Table 1.4

x	y	Horizontal	Vertical
−3	2	left 3	up 2
−1	−4	left 1	down 4
1	−2	right 1	down 2
2	4	right 2	up 4

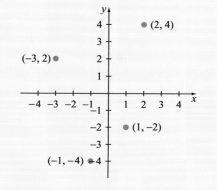

Figure 1.3

number of points to be able to determine the basic shape of the graph, we typically connect the points with a smooth curve. For example, the function $y = x^3 - 9x$ has the table of values and graph shown in Figure 1.4.

EXAMPLE 4 Estimating Function Values from a Graph

➡ Estimate $f(-3)$ and $f(2)$ using the graph of $f(x) = x^3 - 16x$ shown in Figure 1.5.

SOLUTION

It appears from Figure 1.6 that $f(-3) \approx 20$ and $f(2) \approx -25$. Calculating these values with the algebraic equation, we see that

$$f(-3) = (-3)^3 - 16(-3) \qquad f(2) = (2)^3 - 16(2)$$
$$= -27 + 48 \qquad \text{and} \qquad = 8 - 32$$
$$= 21 \qquad\qquad\qquad = -24$$

One drawback of using a graph to determine the values of a function is that it is difficult to be precise. For this reason, algebraic methods are typically preferred when precision is important.

Not all data sets represent functions. If any value of the independent variable is associated with more than one value of the dependent variable, the table of data and its associated graph will not represent a function. For example, consider $y = \pm\sqrt{2x}$ (see Figure 1.7).

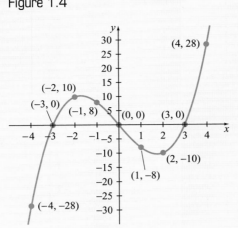

Figure 1.4

x	y
−4	−28
−3	0
−2	10
−1	8
0	0
1	−8
2	−10
3	0
4	28

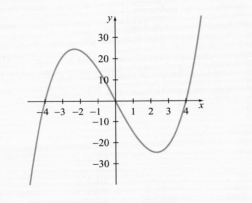

Figure 1.5

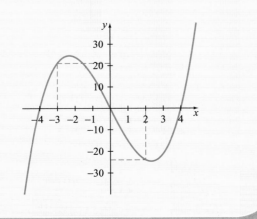

Figure 1.6

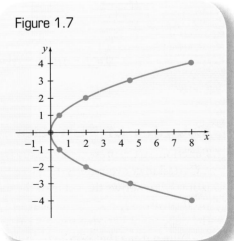

Figure 1.7

x	y
0	0
0.5	−1
0.5	1
2	−2
2	2
4.5	−3
4.5	3
8	−4
8	4

Each positive value of x is associated with two different values of y. We can easily determine this from the graph by observing that a vertical line drawn through any positive value of x will cross the graph twice. This observation leads us to the Vertical Line Test.

Vertical Line Test
If every vertical line drawn on a graph intersects the graph in at most one place, then the graph is the graph of a function. Otherwise, the graph is not the graph of a function.

EXAMPLE 5 Determining If a Graph Represents a Function

The graph of $y = x^2$ is shown in Figure 1.8. Does the graph represent a function?

Figure 1.8

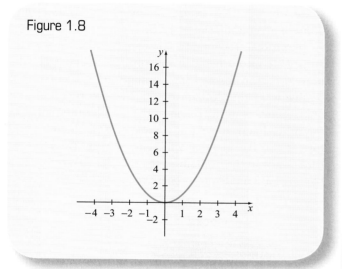

SOLUTION

Each of the vertical lines drawn crosses the graph in exactly one place, as shown in Figure 1.9. Therefore, the graph passes the Vertical Line Test and y is a function of x.

Figure 1.9

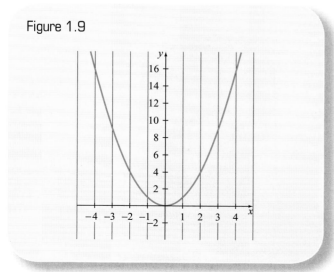

EXAMPLE 6 Determining If a Graph Represents a Function

The graph of $y = \pm\sqrt{x}$ is shown in Figure 1.10. Does the graph represent a function?

Figure 1.10

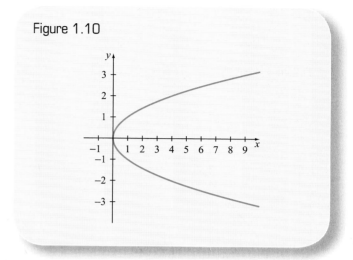

SOLUTION

Notice that many of the vertical lines in Figure 1.11 cross the graph in two places. The graph does not pass the Vertical Line Test, so y is not a function of x. Each positive value of x corresponds with two values of y.

Figure 1.11

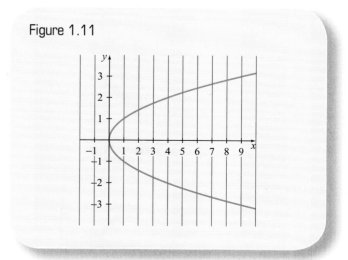

The graphs shown thus far have been continuous. That is, each graph could be drawn without lifting the pencil. However, some function graphs are *discontinuous*. A discontinuous graph has a break in the graph. The break in the graph is referred to as a **discontinuity**.

EXAMPLE 7 Determining If a Discontinuous Graph Represents a Function

The graph of $h(x) = \begin{cases} x \text{ if } x < 1 \\ 2 \text{ if } x > 1 \end{cases}$ is shown in Figure 1.12. Determine if the graph represents a function.

Figure 1.12

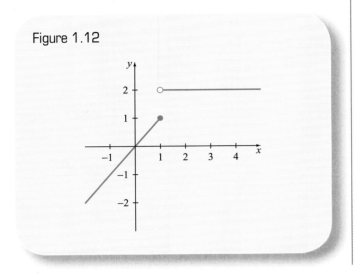

SOLUTION

The graph is discontinuous at $x = 1$. The value of the function at $x = 1$ is 1. (The open dot at indicates that the point $(1, 2)$ is not part of the graph.) Although the graph is discontinuous, each value of the independent variable is associated with exactly one value of the dependent variable. Therefore, the graph represents a function.

Domain and Range

We are often interested in the set of all possible values of the independent variable and the set of all possible values of the dependent variable.

It is easy to remember the meaning of the terms if we observe that *input, independent variable,* and *domain* all contain the word *in.*

Consider a common kitchen blender. If we put an orange into a blender and turn the blender on, the orange is transformed into orange juice. We say that *orange* is in the domain of the blender function and *orange juice* is in the range of the blender function. On the other hand, if we put a rock into the blender and turn it on, the blender will self-destruct. We say that *rock* is not in the domain of the blender function, since blenders are unable to process rocks.

Oranges

In domain ✔

Rocks

Not in domain ✗

Domain and Range

The set of all possible values of the independent variable of a function is called the **domain**. The set of all possible values of the dependent variable of a function is called the **range**.

Finding the domain of a function frequently involves solving an equation or an inequality. Recall that *solving an equation* means finding the value of the variable that makes the equation a true statement.

The domain of most frequently used mathematical functions is the set of all real numbers. However, there are three common situations in which the domain of a function is restricted to a subset of the real numbers. They are the following:

1. A zero in the denominator

2. A negative value under a square root symbol (radical)

3. The context of a word problem

Let's look at an example for each of the situations.

EXAMPLE 8 Determining the Domain of a Function

 What is the domain of $g(x) = \dfrac{3x - 1}{2x + 6}$?

SOLUTION

We know that the value in the denominator must be nonzero, since division by zero is undefined. That is, $2x + 6 \neq 0$. To find the value of x that must be excluded from the domain, we must solve the following equation:

$$2x + 6 = 0$$
$$2x + 6 - 6 = 0 - 6 \qquad \text{Subtract 6 from both sides}$$
$$2x = -6$$
$$\frac{2x}{2} = \frac{-6}{2} \qquad \text{Divide both sides by 2}$$
$$x = -3$$

The domain of the function is all real numbers except -3. We may rewrite the function equation with the domain restriction as follows:

$$g(x) = \frac{3x - 1}{2x + 6} \qquad x \neq -3$$

EXAMPLE 9 Determining the Domain of a Function

 What is the domain of $y = \sqrt{x - 3}$?

SOLUTION

We know that the value underneath the radical must be nonnegative, since the square root of a negative number is undefined in the real number system. Therefore, $x - 3 \geq 0$. Solving for x, we get

$$x - 3 \geq 0$$
$$x - 3 + 3 \geq 0 + 3 \qquad \text{Add 3 to both sides}$$
$$x \geq 3$$

The domain of the function is all real numbers greater than or equal to 3.

EXAMPLE 10 Determining the Domain of a Function

The revenue R from the sale of x gallons of gasoline is given by the equation

$$R(x) = 2.839x$$

What is the domain of the function?

SOLUTION

In the context of this problem, it doesn't make sense to talk about selling a negative number of gallons of gas. So $x \geq 0$. The domain of the function is all nonnegative real numbers.

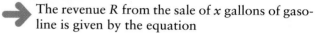

1.1 Exercises

In Exercises 1–3, determine whether the output is a function of the input.

1. $W(a) =$ your weight in pounds when you were a years old

2.

Time of Day	Temperature (°F)
11:00 A.M.	68
1:00 P.M.	73
3:00 P.M.	75

3.

Fish Caught	Salmon in Catch
169	24
182	32
182	47

In Exercises 4–7, calculate the value of the function at the designated input and interpret the result.

4. $C(x) = 39.95x$ at $x = 4$, where $C(x)$ is the cost of buying x pairs of shoes

5. $H(t) = -16t^2 + 120$ at $t = 2$, where $H(t)$ is the height of a cliff diver above the water t seconds after he jumped from a 120-foot cliff

6. *Earnings per Share* Find $S(t)$ at $t = 8$, where $S(t)$ is the annual net sales of Apple Computer Corporation t years after 2000.

Year	Net Sales (in millions of dollars)
5	13,931
6	19,315
7	24,578
8	37,491
9	42,905

Source: www.apple.com

7. *Stock Price* Find $P(t)$ at $t = 4$, where $P(t)$ is the stock price of a Procter & Gamble Company at the end of the week and t is the number of weeks after March 8, 2010.

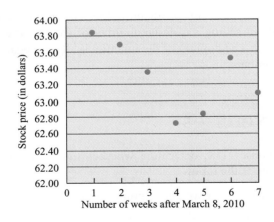

Source: Procter & Gamble Company

In Exercises 8–10, determine whether the graphs represent functions by applying the Vertical Line Test.

8.

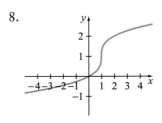

9.

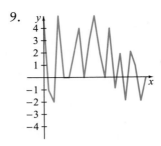

10.

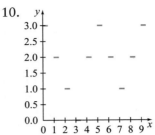

In Exercises 11 and 12, graph the function on your graphing calculator, using the specified viewing window. Note that $a \le x \le b$ means $x\ min = a$ and $x\ max = b$. Similarly, $c \le y \le d$ means $y\ min = c$ and $y\ max = d$.

11. $y = -x^2 - 4$; $-3 \le x \le 3$, $-10 \le y \le 1$

12. $y = -2x^2 + 1$; $-2 \le x \le 2$, $-3 \le y \le 2$

In Exercises 13 and 14, use the graph to estimate the value of the function at the indicated x-value. Then calculate the exact value algebraically.

13. $y = 5|x| - x^2$; $x = 1$

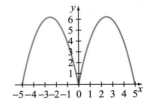

14. $y = -\dfrac{4(x^2 - 4)}{x^3 + 2x^2 + 3x + 6}$; $x = -2$

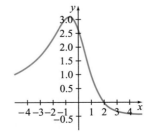

In Exercises 15–20, determine the domain of the function.

15. $f(p) = p^2 - 2p + 1$

16. $h(r) = \dfrac{r^2 - 1}{r^2 + 1}$

17. $g(t) = \dfrac{3t - 3}{4t - 4}$

18. $f(a) = a^3 - \sqrt{a + 1}$

19. $f(x) = \dfrac{\sqrt{2x + 6}}{x^2 + 3}$

20. $P(n)$ = the profit earned from the sale of n bags of candy

1.2 Linear Functions

The cost of filling our car's gas tank, the number of calories we consume by eating a few bags of fruit snacks, and the amount of sales tax we pay when we buy new clothes are examples of linear functions. In this section, we will show how to calculate the slope of a linear function. We will discuss how to interpret the physical and graphical meaning of slope, *x*-intercept, and *y*-intercept. We will also show three different ways to write a linear equation and demonstrate how to find the equation of a linear function from a data set.

Definition: Graph of a Linear Function
The line passing through any two points (x_1, y_1) and (x_2, y_2) with $x_1 \neq x_2$ is referred to as the **graph of a linear function**.

Linear functions are characterized by a constant rate of change. That is, increasing a domain value by one unit will always change the corresponding range value by a constant amount. The converse is also true. Any table of data with a constant rate of change represents a linear function. The constant rate of change is referred to as the **slope** of the linear function.

Definition: Slope
The slope of a linear function is the change in the output that occurs when the input is increased by one unit. The slope m may be calculated by dividing the difference of two outputs by the difference in the corresponding inputs. That is,

$$m = \frac{y_2 - y_1}{x_2 - x_1}$$

where (x_1, y_1) and (x_2, y_2) are data points of the linear function.

According to the package labeling, the number of calories C in n 51-gram servings of Skittles Sour candy is shown in Table 1.5.

Table 1.5

Servings (n)	Calories (C)
1	200
2	400
3	600

Source: Skittles package labeling

Notice that the calorie count increases by 200 calories for each additional serving. Since the dependent variable C is changing at a constant rate (200 calories per serving), the data may be modeled by a linear function. In this case, the linear function is $C = 200n$.

To calculate the slope of the function, we may use any two data points. Using the data points $(x_1, y_1) = (1, 200)$ and $(x_2, y_2) = (3, 600)$, the slope is

$$m = \frac{600 - 200}{3 - 1} \frac{\text{calories}}{\text{servings}}$$

$$= \frac{400}{2} \frac{\text{calories}}{\text{servings}}$$

$$= 200 \,\text{calories per serving}$$

A one-serving increase in the input results in a 200-calorie increase in the output.

Will we get the same result if we use different points? Let's check using (1, 200) and (2, 400) (see Figure 1.13).

$$m = \frac{200 - 400 \text{ calories}}{1 - 2 \text{ servings}}$$

$$= \frac{-200 \text{ calories}}{-1 \text{ serving}}$$

$$= 200 \text{ calories per serving}$$

The result is the same! With linear functions, we may use any two points to calculate the slope.

Not all lines are functions. Vertical lines fail to pass the Vertical Line Test, so they are not functions. As shown in Example 1, vertical lines have an undefined slope.

EXAMPLE 1 Determining the Slope of a Line from Two Points on the Line

➡ What is the slope of the line going through (2, 4) and (2, 8) (Figure 1.14)?

SOLUTION

$$m = \frac{8 - 4}{2 - 2}$$

$$= \frac{4}{0}$$

$$= \text{undefined}$$

Since division by zero is not defined, the line has an undefined slope. Vertical lines are the only lines that have an undefined slope.

EXAMPLE 2 Determining the Slope of a Line from Two Points on the Line

➡ What is the slope of the line going through (2, 4) and (5, 4) in Figure 1.15?

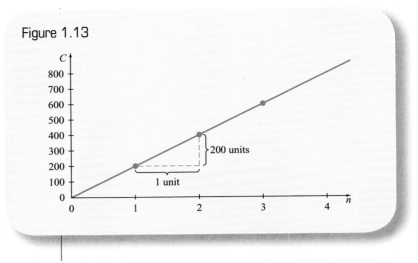

Figure 1.13

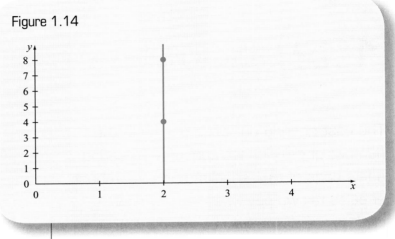

Figure 1.14

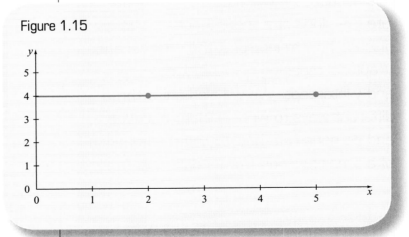

Figure 1.15

SOLUTION

$$m = \frac{4 - 4}{5 - 2}$$

$$= \frac{0}{3}$$

$$= 0$$

Any line with a zero slope is a horizontal line.

The absolute value of the slope is referred to as the **magnitude** of the slope. In a general sense, slope is a measure of steepness: The greater the magnitude of the slope, the greater the steepness of the line. The graph of a line with a negative slope falls as the independent variable increases. The graph of a line with a positive slope rises as the independent variable increases.

EXAMPLE 3 Determining the Sign and Magnitude of a Line's Slope from a Graph

Determine from the graph in Figure 1.16 which lines have a negative slope and which lines have a positive slope. Then identify the line whose slope has the greatest magnitude.

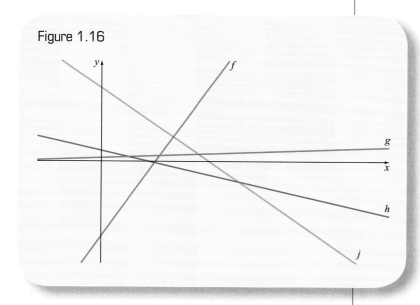

Figure 1.16

SOLUTION

Since lines h and j fall as the value of x increases, they have negative slopes. Since lines f and g rise as the value of x increases, they have positive slopes.

The steepest line is the line whose slope has the greatest magnitude. Although both f and j are steep, line f is steeper. Therefore, f has the slope with the greatest magnitude.

Intercepts

In discussing linear functions, it is often useful to talk about where the line crosses the x- and y-axes. Knowing these **intercepts** helps us determine the equation of the line.

> **Definition:** *y*-Intercept
> The *y*-intercept is the point on the graph where the function intersects the *y*-axis. It occurs when the value of the independent variable is 0. It is formally written as an ordered pair (0, *b*), but *b* itself is often called the *y*-intercept.

EXAMPLE 4 Finding the y-intercept of the Graph of a Linear Function

What is the y-intercept of the linear function $y = 3x + 5$?

SOLUTION

At the y-intercept, the x-coordinate is 0.

$$y = 3(0) + 5 \qquad \text{Substitute 0 for } x$$

$$= 5$$

So the y-intercept of the function is (0, 5).

> **Definition:** *x*-Intercept
> The *x*-intercept is the point on the graph where the function intersects the *x*-axis. It occurs when the value of the dependent variable is 0. It is formally written as an ordered pair (*a*, 0), but *a* itself is often called the *x*-intercept.

EXAMPLE 5 Finding the x-intercept of the Graph of a Linear Function

What is the x-intercept of the linear function $y = 3x + 5$?

SOLUTION

At the x-intercept, the y-coordinate is 0.

$$y = 3x + 5$$

$$0 = 3x + 5 \qquad \text{Substitute 0 for } y$$

$$-3x = 5$$

$$x = -\frac{5}{3}$$

So the x-intercept of the function is $\left(-\frac{5}{3}, 0\right)$.

EXAMPLE 6 Determining the Slope and Intercepts of the Graph of a Linear Function

Determine the slope, the y-intercept, and the x-intercept of the linear function from its graph (Figure 1.17).

Figure 1.17

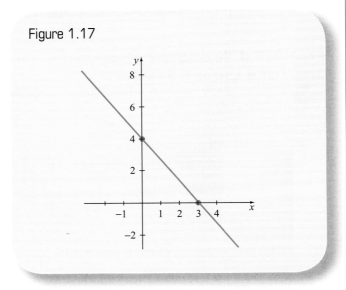

SOLUTION

Since the graph of the function crosses the y-axis at $(0, 6)$, the y-intercept is $(0, 6)$. Since the graph crosses the x-axis at $(3, 0)$, the x-intercept is $(3, 0)$.

Since the line is falling as x increases, the slope will be negative. The slope of the line is

$$m = \frac{y_2 - y_1}{x_2 - x_1}$$

$$= \frac{6 - 0}{0 - 3}$$

$$= \frac{6}{-3}$$

$$= -2$$

Linear Equations

The graph of any line may be represented by a linear equation. Vertical and horizontal lines have the simplest linear equations.

Equations of Vertical and Horizontal Lines

- The equation of a vertical line passing through a point (a, b) is $x = a$.

- The equation of a horizontal line passing through a point (a, b) is $y = b$.

If we know the slope and y-intercept of a linear function, we are able to easily determine the slope-intercept form of the linear equation.

Slope-Intercept Form of a Line

A linear function with slope m and y-intercept $(0, b)$ has the equation

$$y = mx + b$$

EXAMPLE 7 Determining the Slope-Intercept Form of a Line

What is the slope-intercept form of the linear function with slope 5 and y-intercept $(0, 4)$?

SOLUTION

Since $m = 5$ and $b = 4$,

$$y = 5x + 4$$

is the slope-intercept form of the line.

EXAMPLE 8 Finding a Linear Model from a Verbal Description

For breakfast, we decide to eat an apple containing 5.7 grams of dietary fiber and a number of servings of Cheerios, each containing 3 grams of dietary fiber. (**Source:** Cheerios box label) Write the equation for our dietary fiber intake as a function of number of servings of cereal.

SOLUTION

We know that if we don't eat any cereal, we will consume 5.7 grams of dietary fiber (from the apple). So the y-intercept of the fiber function is $(0, 5.7)$. Since each serving of Cheerios contains the same amount of fiber, we know that the function is linear. The slope of the function is 3 grams per serving. So the equation of the fiber function is

$$F(n) = 3n + 5.7$$

where $F(n)$ is the amount of dietary fiber (in grams) and n is the number of servings of cereal.

To check our work, we directly calculate the number of grams of dietary fiber in a breakfast containing two servings of cereal and one apple.

$$\text{Fiber} = 3 + 3 + 5.7 = 11.7 \text{ grams}$$

Using the fiber function formula, we get

$$F(2) = 3(2) + 5.7$$
$$= 11.7 \text{ grams}$$

The results are the same, so we are confident that our formula is correct.

EXAMPLE 9 Determining a Linear Model from a Table of Data

The amount of sales tax paid on a clothing purchase in Seattle is a function of the sales price of the clothes, as shown in Table 1.6.

If the function is linear, write the equation for the sales tax as a function of the sales price.

Table 1.6

Sales Price (p)	Tax (T)
$20.00	$1.72
$30.00	$2.58
$40.00	$3.44

Source: www.cityofseattle.net

SOLUTION

We must first determine if the function is linear. Since each $10 increase in sales price increases the sales tax by a constant $0.86, we conclude that the function is linear. The slope of the function is

$$m = \frac{2.58 - 1.72 \text{ dollars}}{30 - 20 \text{ dollars}}$$

$$= \frac{0.86 \text{ dollars}}{10 \text{ dollars}}$$

$$= 0.086 \text{ tax dollar per sales price dollar}$$

(In other words, for each dollar increase in the sales price, the sales tax increases by 8.6 cents.) Since a sale of $0 results in $0 sales tax, we know that the y-intercept is $(0, 0)$. The equation of the function is

$$T = 0.086p$$

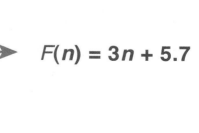

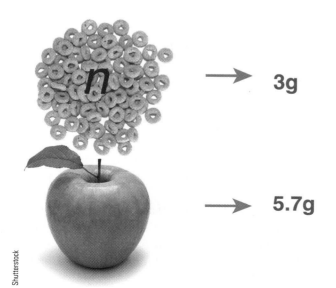

n → **3g**

→ **5.7g**

$F(n) = 3n + 5.7$

To check our work, we substitute the point (40, 3.44) into the equation.

$$3.44 = 0.086(40) \qquad p = 40 \text{ and } T = 3.44$$

$$3.44 = 3.44$$

The statement is true, so we are confident that our equation is correct.

EXAMPLE 10 Finding a Linear Model from the Graph of a Linear Function

The graph in Figure 1.18 shows the balance of a checking account as a function of the number of ATM withdrawals from the account.

(a) Identify the x-intercept and y-intercept and interpret their physical meaning.
(b) Write the linear equation for the function.

Figure 1.18

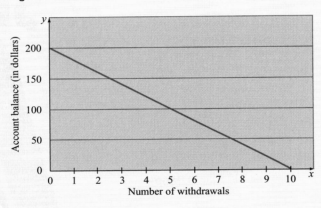

SOLUTION

(a) The x-intercept is (10, 0). When there have been 10 withdrawals, the account balance is $0. The y-intercept is (0, 200). When there have been no withdrawals, the account balance is $200.

(b) The slope of the function is

$$m = \frac{200 - 0 \,\text{dollars}}{0 - 10 \,\text{withdrawals}}$$

$$= -\$20 \,\text{per withdrawal}$$

So the slope-intercept form of the linear equation is

$$B = -20w + 200$$

where B is the account balance and w is the number of withdrawals.

Finding the Equation of a Line

As demonstrated in Example 10, to find the slope-intercept form of a line from two points, we need to calculate the slope and the y-intercept of the function. To do this, we proceed as follows:

1. Calculate the slope.

2. Write the function in slope-intercept form, substituting the slope for m.

3. Select one of the points and substitute the output value for y and the input value for x.

4. Solve for b.

5. Write the function in slope-intercept form, substituting the y-intercept for b.

EXAMPLE 11 Finding the Slope-Intercept Form of a Line from Two Points

Find the equation of the line passing through the points (3, 5) and (7, 1).

SOLUTION

$$m = \frac{5 - 1}{3 - 7}$$

$$= \frac{4}{-4}$$

$$= -1$$

So the slope is -1. Substituting in the slope and the point (3, 5), we get

$$y = -1 \cdot x + b$$

$$5 = -1(3) + b$$

$$b = 8$$

The y-intercept is (0, 8).

The slope-intercept form of the line is $y = -1 \cdot x + 8$ or $y = -x + 8$.

Other Forms of Linear Equations

There are two additional forms of linear equations that are commonly used: standard form and point-slope form.

The standard form of a linear equation is extremely useful when working with systems of equations.

The point-slope form of a line is especially useful when we know a line's slope and the coordinates of a point on the line.

Standard Form of a Line

A linear equation may be written as

$$Ax + By = C$$

where *A*, *B*, and *C* are real numbers. If $A = 0$, the graph of the equation is a horizontal line. If $B = 0$, the graph of the equation is a vertical line. In the equation, *A* and *B* cannot both be zero.

Point-Slope Form of a Line

A linear function written as

$$y - y_1 = m(x - x_1)$$

has slope *m* and passes through the point (x_1, y_1).

EXAMPLE 12 Finding a Linear Model from a Verbal Description

Based on data from 2000 to 2007, per capita consumption of whole milk has been decreasing by approximately 0.233 gallons per year. In 2007, the per capita consumption of whole milk was 6.4 gallons. (**Source:** *Statistical Abstract of the United States, 2010,* Table 210) Find an equation for the per capita milk consumption as a function of years since 2000.

SOLUTION

The slope of the line is $m = -0.233$. The point $(7, 6.4)$ lies on the line, since 2007 is 7 years after 2000. The point-slope form of the line is

$$y - 6.4 = -0.233(t - 7)$$

If preferred, the equation may be rewritten in slope-intercept form,

$$y = -0.233(t - 7) + 6.4$$
$$y = -0.233t + 1.631 + 6.4$$
$$y = -0.233t + 8.031$$

or standard form,

$$0.233t + y = 8.031$$

Graphing Linear Functions

To graph a linear function, we first generate a table of values by substituting different values of *x* into the equation and calculating the corresponding value of *y*. For example, if we are given the linear equation $y = 4x - 8$, we may choose to evaluate the function at $x = -1$, $x = 0$, $x = 1$, $x = 2$, and $x = 3$, as shown in Table 1.7.

Table 1.7

x	y
−1	−12
0	−8
1	−4
2	0
3	4

We then plot the points and connect them with a straight line, as shown in Figure 1.19. (Although we plotted multiple points, only two points are necessary to determine the line.)

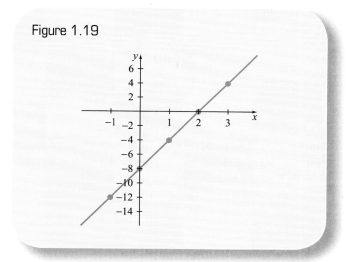

Figure 1.19

EXAMPLE 13 Graphing a Linear Function

Graph the function $y = 2x - 3$.

SOLUTION

From the equation, we see that the *y*-intercept is $(0, -3)$. We need to find one more point. Evaluating the function at $x = 2$ yields $y = 1$, so $(2, 1)$ is a point on the line. We plot each point and connect the points with a straight line, as shown in Figure 1.20.

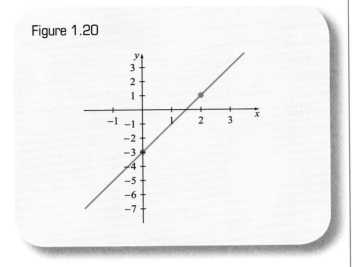

Figure 1.20

Recall that the standard form of a line is $Ax + By = C$. Graphing a line from standard form is remarkably easy. The y-intercept of the graph occurs when the x-value is equal to zero. Similarly, the x-intercept of the graph occurs when the y-value is equal to zero. Using these facts, we can quickly determine the x- and y-intercepts of the function. To find the x-intercept, we set $y = 0$.

$$Ax + By = C$$
$$Ax + B(0) = C$$
$$Ax = C$$
$$x = \frac{C}{A}$$

Notice that the x-coordinate of the x-intercept is the constant term divided by the coefficient on the x-term.

To find the y-intercept, we set $x = 0$.

$$Ax + By = C$$
$$A(0) + By = C$$
$$By = C$$
$$y = \frac{C}{B}$$

Notice that the y-coordinate of the y-intercept is the constant term divided by the coefficient on the y-term. Using this procedure to find intercepts will allow us to graph linear equations in standard form quickly by hand.

EXAMPLE 14 Graphing a Linear Function

Graph the linear function $2x + y = 4$.

SOLUTION

The x-intercept is found by dividing the constant term by the coefficient on the x-term.

$$x = \frac{4}{2}$$
$$= 2$$

The point $(2, 0)$ is the x-intercept.

The y-intercept is found by dividing the constant term by the coefficient on the y-term.

$$y = \frac{4}{1}$$
$$= 4$$

The point $(0, 4)$ is the y-intercept. We graph the x- and y-intercepts and then draw the line through the intercepts, as shown in Figure 1.21.

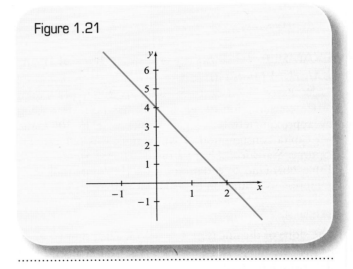

Figure 1.21

1.2 Exercises

In Exercises 1–3, calculate the slope of the linear function passing through the points.

1. $(2, 5)$ and $(4, 3)$

2. $(1.2, 3.4)$ and $(2.7, 3.1)$

3. $(2, 2)$ and $(5, 2)$

In Exercises 4–6, find the x-intercept and y-intercept of the linear function.

4. $y = 5x + 10$

$$0 = 5x + 10$$
$$-10 \qquad -10$$
$$-10 = 5$$

5. $y = 2x + 11$

6. $3x - y = 4$

In Exercises 7–9, write the equation of the linear function passing through the points in slope-intercept form, point-slope form, and standard form.

7. $(2, 5)$ and $(4, 3)$

8. $(1.2, 3.4)$ and $(2.7, 3.1)$

9. $(-2, 2)$ and $(5, 2)$

In Exercises 10–13, graph the line.

10. $y = 4x - 2$

11. $y - 4 = 0.5(x - 2)$

12. $2x - 3y = 5$

13. $y = -\dfrac{2}{3}x + \dfrac{4}{3}$

In Exercises 14 and 15, use the graph to determine the equation of the line.

14.

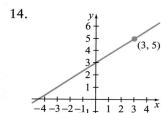

15.

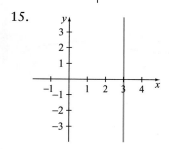

In Exercises 16–18, determine if the table of data may be represented by a linear function. If so, calculate the slope and interpret its real-life significance.

16. *Two-Year Median Household Income*

Two-Year Period	U.S. Two-Year Median Household Income (in terms of year 2008 dollars)
2005–2006	51,283
2007–2008	51,233

Source: www.census.gov

17. *Take-Home Pay*

Months (since March 2010)	Take-Home Pay (dollars)
0	4121
1	4073

Source: Employee pay stubs

18. *Solid Waste Disposal*

Clean Wood (pounds)	Cost to Dispose of Clean Wood at Enumclaw Transfer Station
500	$18.75
700	$26.25
900	$33.75
1000	$37.50

Source: www.dnr.metrokc.gov

19. *Nutrition* The Recommended Daily Allowance for dietary fiber is 25 grams for a person on a 2,000-calorie-per-day diet. A $\frac{3}{4}$-cup serving of General Mills Wheaties cereal contains 2.1 grams of dietary fiber. (**Source:** Package labeling) A 1-cup serving of 2 percent milk fortified with vitamin A doesn't contain any dietary fiber. A large (8 to $8\frac{7}{8}$ inches long) banana contains 3.3 grams of dietary fiber. (**Source:** www.nutri-facts.com) Find the linear equation that models the data. How many servings of Wheaties with milk would you have to eat along with a large banana to consume 8 grams of dietary fiber? (Round up to the nearest number of servings.)

20. What is the equation of the line that passes through $(4, 7)$ and $(4, -2)$?

Chapter 2

Nonlinear
Models

Mathematical functions are commonly used to model real-world data. Although every model has its limitations, models are often used to forecast expected results. Selecting which mathematical model to use is relatively easy once you become familiar with the basic types of mathematical functions. Remarkably, these basic functions may be used to effectively model many real-world data sets. For example, based on data from 1996 to 2009, the number of Walmart Supercenters in the United States may be effectively modeled by a quadratic function.

2.1 Quadratic Function Models

2.2 Exponential Function Models

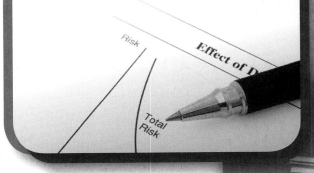

2.1 Quadratic Function Models

Walmart is one of the most widely recognized retailers in the country. Despite the economic downturn that occurred in the late 2000s, Walmart continued to increase the number of Walmart Supercenters in the United States. (**Source:** www.walmartstores.com) The number of Supercenters increased dramatically between 1996 and 2007, as shown in Figure 2.1.

How many Walmart Supercenters will there be in the United States in 2013? Nobody knows; however, using a mathematical model, we can predict what may happen.

In this section, we will discuss the identifying features of quadratic functions. We will then demonstrate how to use algebraic methods and quadratic regression to model data sets whose graphs open upward or downward over their domain. Let's begin by looking at polynomials.

Figure 2.1

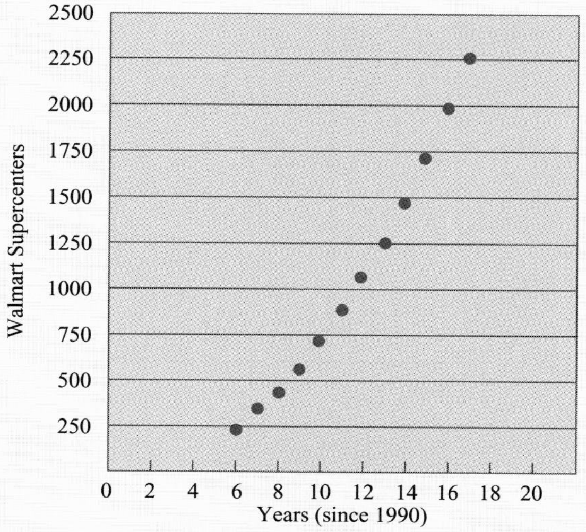

A polynomial is a function that is the sum of terms of the form ax^n, where a is a real number and n is a nonnegative integer. For example, each of the following functions is a polynomial.

$$f(x) = 6x^4 - 3x^3 + 5x^2 - 2x + 9$$

$$g(x) = 4x^2 + 2x$$

$$h(x) = -1.3x + 9.7$$

$$j(x) = x^2 - 2x + 6$$

Can a constant term like 9 be written in the form ax^n? Yes! Since $x^0 = 1$ for nonzero x, $9x^0 = 9$. Consequently, a constant function $s(x) = 9$ is also a polynomial.

The **degree of a polynomial** is the value of its largest exponent. For example, the degree of the polynomial $f(x) = 6x^4 - 3x^3 + 5x^2 - 2x + 9$ is 4, since 4 is the largest exponent. Since the equation of any line may be written as $y = ax^1 + b$, lines are polynomials of degree 1. We worked with first-degree polynomials (lines) in the preceding chapter. In this section, we are interested in polynomials of degree 2. Polynomials of degree 2 are called **quadratic functions**.

> **Quadratic Function**
> A polynomial function of the form
> $f(x) = ax^2 + bx + c$ with $a \neq 0$ is called
> a **quadratic function**. The graph of a
> quadratic function is a **parabola**.

When a parabola opens upward, we say that it is "concave up." When the parabola opens downward, we say that it is "concave down." The steepness of the sides of the parabola and its concavity are controlled by the value of a, the coefficient on the x^2 term in its equation. If $a > 0$, the graph is concave up. If $a < 0$, the graph is concave down. As the magnitude of a increases, the steepness of the graph increases. (For $a > 0$, the magnitude of a is a. For $a < 0$, the magnitude of a is $-a$.) Consider the graphs in Figures 2.2 and 2.3. These graphs have the same values for b and c ($b = -2$ and $c = 3$) but differing values of a.

In Figure 2.2, since a is the coefficient on the x^2 term, $a = 1$. The magnitude of a is 1. Since $a > 0$, the graph is concave up.

In Figure 2.3, since a is the coefficient on the x^2 term, $a = 2$. The magnitude of a is 2. Increasing the magnitude of a increased the graph's steepness.

Figure 2.2

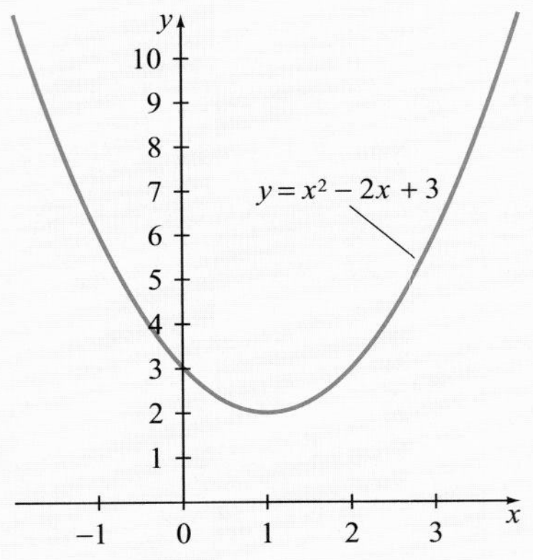

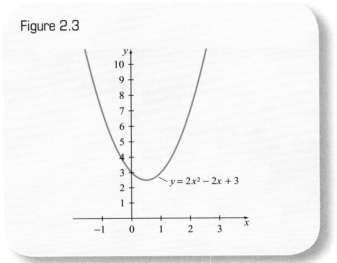

Figure 2.3

Consider the graphs in Figures 2.4 and 2.5. These graphs have the same values for b and c ($b = 4$ and $c = 0$) but differing values of a.

Figure 2.4

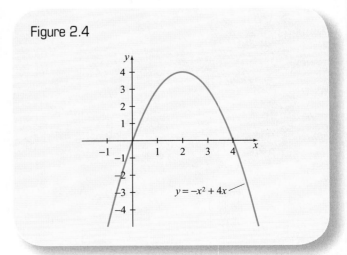

Figure 2.5

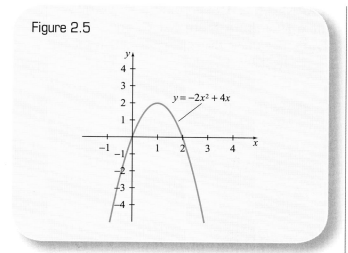

y-value of a concave down parabola occurs at the vertex (see Figure 2.6b).

Figure 2.6b Concave Down Parabola

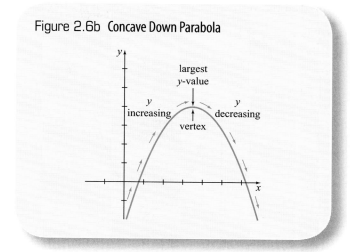

In Figure 2.4, since *a* is the coefficient on the x^2 term, $a = -1$. The magnitude of *a* is 1. Since $a < 0$, the graph is concave down.

In Figure 2.5, since *a* is the coefficient on the x^2 term, $a = -2$. The magnitude of *a* is 2. Increasing the magnitude of *a* from 1 to 2 increased the graph's steepness.

The coefficient *b* in the equation is the slope of the graph at the *y*-intercept. It is also called the *initial rate of change*. The constant term *c* in the equation indicates that the point $(0, c)$ is the *y*-intercept of the graph.

Recall that the graph of a function is said to be increasing if the value of *y* gets bigger as the value of *x* increases. Similarly, the graph of a function is said to be decreasing if the value of *y* gets smaller as the value of *x* increases. The vertex of a parabola is the point on the graph of a quadratic function where the curve changes from decreasing to increasing (or vice versa). The minimum *y*-value of a concave up parabola occurs at the vertex (see Figure 2.6a). Similarly, the maximum

For parabolas with *x*-intercepts, the *x*-coordinate of the vertex always lies halfway between the *x*-intercepts. Recall that as a result of the quadratic formula, we know that the *x*-coordinate of the vertex is $x = \dfrac{-b}{2a}$. The *y*-coordinate of the vertex is obtained by evaluating the quadratic function at this *x*-value. For example, for the quadratic function $y = x^2 - 2x + 3$, we know that $a = 1$ and $b = -2$. The *x*-coordinate of the vertex is

$$
\begin{aligned}
x &= \frac{-b}{2a} \\
&= \frac{-(-2)}{2(1)} \\
&= \frac{2}{2} \\
&= 1
\end{aligned}
$$

Evaluating the function at $x = 1$ yields

$$
\begin{aligned}
y &= x^2 - 2x + 3 \\
&= (1)^2 - 2(1) + 3 \\
&= 1 - 2 + 3 \\
&= 2
\end{aligned}
$$

Therefore, the vertex of the function is $(1, 2)$.

Parabolas are symmetrical. That is, if we draw a vertical line through the vertex of the parabola, the portion of the graph on the left of the line is the mirror image of the portion of the graph on the right of the

Figure 2.6a Concave Up Parabola

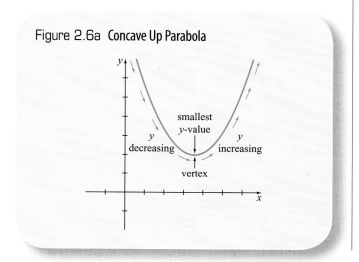

line. The line is referred to as the **axis of symmetry** (see Figure 2.7).

Since the axis of symmetry is a vertical line passing through the vertex, the equation of the axis of symmetry is $x = \dfrac{-b}{2a}$.

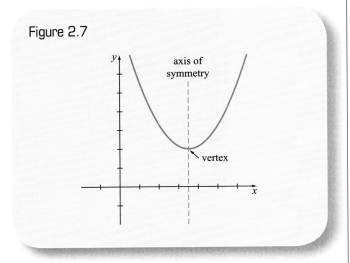

Figure 2.7

EXAMPLE 1 Describing the Graph of a Quadratic Function from Its Equation

Determine the concavity, the y-intercept, and the vertex of the quadratic function $y = 3x^2 + 6x - 1$.

SOLUTION

We have $a = 3$, $b = 6$, and $c = -1$. Since $a > 0$, the parabola is concave up. Since $c = -1$, the y-intercept is $(0, -1)$. The x-coordinate of the vertex is given by

$$x = \frac{-b}{2a}$$
$$= \frac{-6}{2(3)}$$
$$= \frac{-6}{6}$$
$$= -1$$

The y-coordinate of the vertex is obtained by evaluating the function at $x = -1$.

$$y = 3x^2 + 6x - 1$$
$$= 3(-1)^2 + 6(-1) - 1$$
$$= 3 - 6 - 1$$
$$= -4$$

The vertex of the parabola is $(-1, -4)$.

EXAMPLE 2 Determining the Equation of a Parabola from Its Graph

Determine the equation of the parabola shown in Figure 2.8.

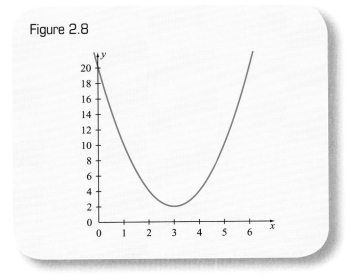

Figure 2.8

SOLUTION

The y-intercept is $(0, 20)$, so $c = 20$. The vertex is $(3, 2)$. Since the x-coordinate of the vertex is $\dfrac{-b}{2a}$, we know that

$$\frac{-b}{2a} = 3$$
$$-b = 6a$$
$$b = -6a$$

Since $f(x) = ax^2 + bx + c$, we have

$$f(x) = ax^2 + (-6a)x + (20) \qquad \text{Since } b = -6a$$
$$= ax^2 - 6ax + 20$$

The vertex is $(3, 2)$, so $f(3) = 2$. Therefore,

$$f(x) = ax^2 - 6ax + 20$$
$$2 = a(3)^2 - 6a(3) + 20 \qquad \text{Substitute in } (3, 2)$$
$$2 = 9a - 18a + 20$$
$$-18 = -9a$$
$$a = 2$$

Since $b = -6a$, $b = -12$. The equation of the parabola is $f(x) = 2x^2 - 12x + 20$.

We can check our work by substituting in a different point from the graph. The parabola passes through the point $(4, 4)$. Consequently, this point should satisfy the equation $f(x) = 2x^2 - 12x + 20$.

$$f(x) = 2x^2 - 12x + 20$$

$$4 \overset{?}{=} 2(4)^2 - 12(4) + 20 \qquad \text{Substitute in } (4, 4)$$

$$4 \overset{?}{=} 2(16) - 48 + 20$$

$$4 \overset{?}{=} 32 - 48 + 20$$

$$4 = 4$$

Recognizing the relationship between the graph of a parabola and its corresponding quadratic function provides a quick way to evaluate the accuracy of a quadratic model. Quadratic models may be determined algebraically or by using quadratic regression. We will demonstrate both methods in the next several examples.

Let's return to the Walmart Supercenter data introduced at the beginning of the section. At first glance, the Walmart data don't look at all like a parabola; however, we do observe that the data appear to be increasing at an ever-increasing rate (concave up). Plotting the quadratic equation $f(t) = 9.27t^2 - 30.8t + 97.0$ together with the data set, we observe that, in fact, a quadratic equation fits the data very well (Figure 2.9). (**Source:** www.walmartstore.com) This equation is found by performing quadratic regression on the data, a process that we cover on the Chapter 2 Tech Card.

To make the coefficients of the quadratic function model relatively small, we **aligned the data.** We let $t = 0$ in 1990, $t = 1$ in 1991, and so on. Doing quadratic regression on the aligned data yielded the equation $f(t) = 9.27t^2 - 30.8t + 97.0$ with the coefficient of determination $r^2 = 0.9999$. Recall that the coefficient of determination is a measure of how well the model fits the data. The closer r^2 is to 1, the better the model fits the data.

Using the model, we predict how many Walmart Supercenters there will be administered in 2013. In 2013, $t = 23$.

$$f(23) = 9.27(23)^2 - 30.8(23) + 97.0$$
$$= 4292$$

We estimate that there will be 4,292 Walmart Supercenters in 2013.

A quadratic function model for a data set may be generated by using the quadratic regression feature on a graphing calculator, as demonstrated on your Chapter 2 Tech Card. However, the fact that the calculator can create a quadratic model does not guarantee that the model will be a good fit for the data.

Although the quadratic model fits the Walmart Supercenter data very well from 1996 to 2007, we must be cautious in using it to predict future behavior. (Predicting the output value for an input value outside the interval of the input data is called **extrapolation.**) For this data set, we may feel reasonably comfortable with an estimate 2 or 3 years beyond the last data point; however, we would doubt the accuracy of the model 100 years beyond 2007. For example, in 2107 ($t = 117$), the estimated number of Walmart Supercenters is 123,390. This figure is nearly 55 times the number of Supercenters in 2007!

We also need to consider the effect of national economic conditions on Walmart's growth. Although Walmart continued to add Supercenters during the national economic downturn in 2008 and 2009, the company did so at a slower pace. The quadratic model predicted the number of Supercenters in 2008 and 2009 to be 2,546 and 2,858, respectively. However, the actual number of stores was 2,447 in 2008 and 2,612 in 2009. (**Source:** www.walmart stores.com) This illustrates that mathematical models are often imperfect in predicting future data.

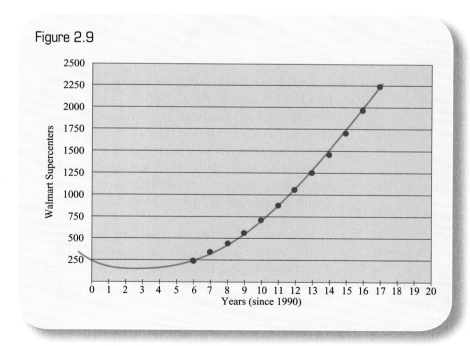

Figure 2.9

Walmart Supercenters vs. Years (since 1990)

It is often possible to model a data set with more than one mathematical model. When selecting a model, we should consider the following:

1. The graphical fit of the model to the data

2. The correlation coefficient (r) or the coefficient of determination (r^2)

3. The known behavior of the thing being modeled

Recall that the closer the correlation coefficient is to 1 or -1, the better the model fits the data. Similarly, the closer the coefficient of determination is to 1, the better the model fits the data.

EXAMPLE 3 Using Quadratic Regression to Forecast Prescription Drug Sales

➡️ Retail prescription drug sales in the United States increased from 2000 to 2008 as shown in Table 2.1.

Table 2.1

Years (since 2000) (t)	Retail Sales (billions of dollars) S(t)
0	145.6
1	161.3
2	182.7
3	204.2
4	216.7
5	226.1
6	243.2
7	249.2
8	253.6

Source: *Statistical Abstract of the United States, 2010,* Table 153

Model the data using a quadratic function. Then use the model to predict retail prescription drug sales in 2012 and 2014.

SOLUTION

We observe from the scatter plot that the data appear concave down (Figure 2.10).

A quadratic function may fit the data well. We use quadratic regression to find the quadratic model that best fits the data.

Figure 2.10

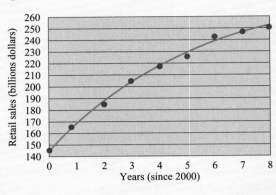

Based on data from 2000 to 2008, retail prescription drug sales in the United States may be modeled by

$$S(t) = -1.088t^2 + 22.68t + 143.1 \text{ billion dollars}$$

where t is the number of years since 2000.

The coefficient of determination ($r^2 = 0.9958$) is extremely close to 1. The graph also appears to pass near each data point. The model appears to fit the data well.

In 2012, $t = 12$, and in 2014, $t = 14$. Evaluating the function at each t-value, we get

$$S(12) = -1.088(12)^2 + 22.68(12) + 143.1$$

$$= 258.6$$

$$S(14) = -1.088(14)^2 + 22.68(14) + 143.1$$

$$= 247.4$$

According to the model, prescription drug sales will be $258.6 billion in 2012 and $247.4 billion in 2014. Although the 2012 estimate seems reasonable, the 2014 estimate (which is less than the 2012 estimate) does not seem reasonable because we expect prescription drug spending to continually increase. This example illustrates the importance of making sure mathematical results are reasonable in the real world context.

You may ask, "Is there a way to find a quadratic model without using quadratic regression?" There is. The model may not be the model of best fit, but it may still model the data effectively. In Example 4, we repeat Example 3 using an algebraic method to find a quadratic model.

EXAMPLE 4 Using Algebraic Methods to Model Prescription Drug Sales

Retail prescription drug sales in the United States increased from 2000 to 2008, as shown in Table 2.2.

Table 2.2

Years (since 2000) (t)	Retail Sales (billions of dollars) S(t)
0	145.6
1	161.3
2	182.7
3	204.2
4	216.7
5	226.1
6	243.2
7	249.2
8	253.6

Source: *Statistical Abstract of the United States, 2010,* Table 153

Model the data using a quadratic function. Then use the model to predict retail prescription drug sales in 2012 and 2014.

SOLUTION

Given any three data points, we can find a quadratic function that passes through the points, provided that the points define a nonlinear function. We will pick the points $(0, 145.6), (4, 216.7),$ and $(8, 253.6)$ from the table. A quadratic function is of the form $S(t) = at^2 + bt + c$. Each of the points must satisfy this equation.

$145.6 = a(0)^2 + b(0) + c$ — Substitute $t = 0, S(t) = 145.6$

$c = 145.6$

$216.7 = a(4)^2 + b(4) + c$ — Substitute $t = 4, S(t) = 216.7$

$216.7 = 16a + b(4) + 145.6$ — Since $c = 145.6$

$71.1 = 16a + 4b$

$253.6 = a(8)^2 + b(8) + c$ — Substitute $t = 8, S(t) = 253.6$

$253.6 = 64a + 8b + 145.6$ — Since $c = 145.6$

$108 = 64a + 8b$

We can now find the values of a and b by solving the system of equations.

$$71.1 = 16a + 4b$$
$$108 = 64a + 8b$$

$$\begin{bmatrix} 16 & 4 & | & 71.1 \\ 64 & 8 & | & 108 \end{bmatrix}$$

$$\begin{bmatrix} 16 & 4 & | & 71.1 \\ 0 & 8 & | & 176.4 \end{bmatrix} \quad 4R_1 - R_2$$

$$\begin{bmatrix} 32 & 0 & | & -34.2 \\ 0 & 8 & | & 176.4 \end{bmatrix} \quad 2R_1 - R_2$$

$32a = -34.2 \qquad 8b = 176.4$

$a = -1.06875 \qquad b = 22.05$

A quadratic model for the data is

$$S(t) = -1.069t^2 + 22.05t + 145.6$$

Graphing this model with the data shows that it fits the data relatively well (Figure 2.11).

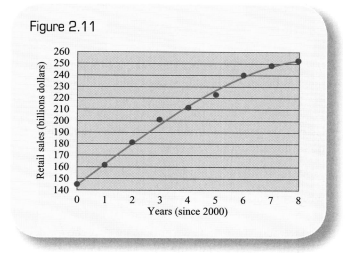

Figure 2.11

In 2012, $t = 12$, and in 2014, $t = 14$. Evaluating the function at each t-value, we get

$$S(12) = -1.069(12)^2 + 22.05(12) + 145.6$$
$$\approx 256.3 \text{ billion}$$
$$S(14) = -1.069(14)^2 + 22.05(14) + 145.6$$
$$\approx 244.8 \text{ billion}$$

These estimates are close to the estimates from Example 3. In Example 3, we estimated that prescription drug sales were $258.6 billion in 2012 and $247.4 billion in 2014. As was the case with Example 3, the 2014 estimate seems unreasonable since it is less than the 2012 estimate.

EXAMPLE 5 Using Quadratic Regression to Forecast Medicare Costs

Because of their costly medical needs, many elderly Americans rely on Medicare to pay their medical expenses. The number of people enrolled in Medicare has risen substantially since 2000, as shown in Table 2.3.

Table 2.3

Years (since 2000) (t)	Medicare Enrollees (in millions) $M(t)$
0	39.7
3	41.2
4	41.9
5	42.6
6	43.4
7	44.3
8	45.2

Source: *Statistical Abstract of the United States, 2010,* Table 142

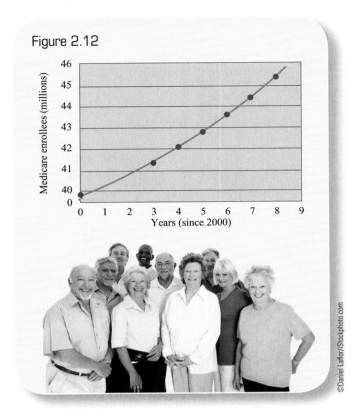

Figure 2.12

Model the data with a quadratic function and forecast the number of Medicare enrollees there will be in 2013.

SOLUTION

Using quadratic regression, we determine that the quadratic model of best fit is

$$M(t) = 0.0360t^2 + 0.402t + 39.7 \text{ million people}$$

where t is the number of years since 2000.

Since the coefficient of determination ($r^2 = 0.9999$) is extremely close to 1, we anticipate that the model fits the data well. Graphing the data and the model together yields Figure 2.12.

The graph passes near each data point. The model appears to fit the data well. Zooming out, however, we see that the graph decreases from about 42 million to roughly 39.5 million enrollees between 1985 and 1994 (Figure 2.13). This doesn't make sense since we know that the number of enrollees has increased continually. Consequently, we choose to limit the use of our model to a practical domain of $t = -6$ to $t = 14$. For these domain values, we expect the results will make sense.

We use the model to predict the number of Medicare enrollees in 2013. In 2013, $t = 13$.

$$M(13) = 0.0360(13)^2 + 0.402(13) + 39.7$$

$$= 51.01$$

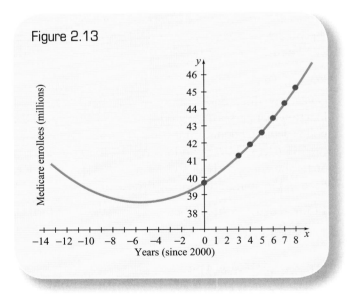

Figure 2.13

We estimate that in 2013, there will be approximately 51 million Medicare enrollees.

EXAMPLE 6 Using Algebraic Methods to Find a Quadratic Model for Net Sales

Based on the data in Table 2.4, find a quadratic model for the net sales of the Kellogg Company algebraically.

Table 2.4

Years (since 2004) (t)	Kellogg Company Net Sales (millions of dollars) R(t)
0	9614
1	10,177
2	10,907
3	11,776
4	12,822

Source: Kellogg Company 2009 Annual Report, p. 5

SOLUTION

We can use any three points to find a quadratic model. We choose to use the first, middle, and last points. Since we are given the y-intercept, $(0, 9614)$, we know that $c = 9614$. We have

$$R(t) = at^2 + bt + c$$
$$= at^2 + bt + 9614$$

$$R(2) = a(2)^2 + b(2) + 9614$$
$$10{,}907 = 4a + 2b + 9614$$
$$4a + 2b = 1293$$

$$R(4) = a(4)^2 + b(4) + 9614$$
$$12{,}822 = 16a + 4b + 9614$$
$$16a + 4b = 3208$$
$$4a + b = 802 \qquad \text{Divide both sides by 4}$$

We must solve the system of equations

$$4a + 2b = 1293$$
$$4a + b = 802$$

We will solve the system using the substitution method. Solving the first equation for $4a$ yields $4a = -2b + 1293$. Substituting this result into the second equation $4a + b = 802$ yields

$$(-2b + 1293) + b = 802 \qquad \text{Since } 4a = -2b + 1293$$
$$-b + 1293 = 802$$
$$-b = -491$$
$$b = 491$$

Since $4a = -2b + 1293$,

$$4a = -2(491) + 1293$$
$$4a = 311$$
$$a = 77.75$$

The quadratic function that models the revenue of the Kellogg Company is $R(t) = 77.75t^2 + 491t + 9614$, where t is the number of years since the end of 2004 and $R(t)$ is the revenue from sales in millions of dollars.

In each of the preceding examples, the quadratic model fit the data well. This is not always the case, as demonstrated in Example 7.

EXAMPLE 7 Determining When a Quadratic Model Should Not Be Used

→ The per capita consumption of low-fat ice cream is shown in Table 2.5.

Table 2.5

Years (since 1980) (t)	Ice cream (pounds) I(t)
0	7.1
5	6.9
10	7.7
15	7.4
20	7.3
21	7.3
22	6.5
23	7.5
24	7.3
25	6.7
26	6.9
27	7.0

Source: *Statistical Abstract of the United States, 2010,* Table 212

Explain why you do or do not believe that a quadratic function will model the data set well.

SOLUTION

We first draw the scatter plot of the data set (Figure 2.14).

Recall that a parabola changes from increasing to decreasing (or vice versa) exactly once. This scatter plot changes from increasing to decreasing and decreasing to increasing multiple times. This causes us to doubt that a quadratic model will fit the data well. In short, the scatter plot doesn't look like a parabola or a portion of a parabola.

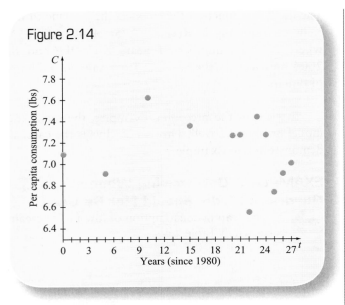

Figure 2.14

The quadratic model that best fits the data is $C(t) = -0.00258t^2 + 0.0647t + 6.99$ and is shown in Figure 2.15.

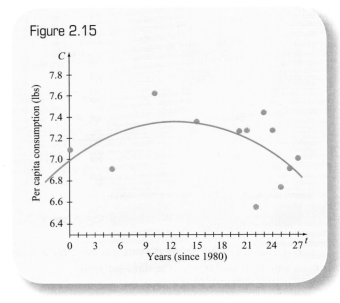

Figure 2.15

The coefficient of determination ($r^2 = 0.2413$) is not at all close to 1. Additionally, we can see visually that the model does not fit the data well.

Once a model has been created, we can often use it to make educated guesses about what may happen in the future. Forecasting the future is a key function for many businesses.

EXAMPLE 8 Using Quadratic Regression to Forecast the Number of Alternative Fuel Vehicles

The number of alternative fuel vehicles has increased significantly since 2003 (Table 2.6).

Table 2.6

Alternative Fuel Vehicles in Use in the United States	
Years (since 2003) (t)	Vehicles (thousands) (F)
0	534
1	565
2	592
3	635
4	696

Source: *Statistical Abstract of the United States, 2010,* Table 1061

Model the number of alternative vehicles in use with a quadratic function and forecast the year in which the number of vehicles will reach 1 million.

SOLUTION

Using quadratic regression, we determine the model to be

$$A(t) = 5.43t^2 + 17.7t + 536 \text{ thousand vehicles}$$

where t is the number of years since the end of 2003. Since 1 million is the same as 1,000 thousand, we want to know at what value of t does $A(t) = 1000$. This problem may be solved algebraically or graphically using technology. We will solve the problem using technology and allow you to use the method of your choice in the exercises.

GRAPHICAL SOLUTION

We graph the function $A(t) = 5.43t^2 + 17.7t + 536$ and the horizontal line $y = 1000$ simultaneously (Figure 2.16). $A(t) = 1000$ at the point at which these two functions intersect.

Using the intersect feature on the graphing calculator, we determine that the point of intersection is (7.757, 1000). Since $t = 7$ is the end of 2011, $t = 7.757$ is well into 2012. About 75 percent of the way through 2012, the number of alternative fuel vehicles in use is predicted to reach 1 million.

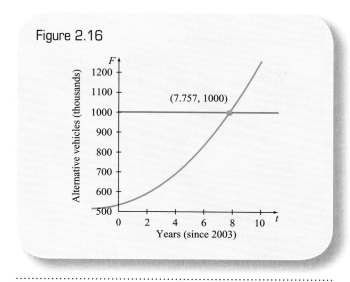

Figure 2.16

(7.757, 1000)

Alternative vehicles (thousands)

Years (since 2003)

2.1 Exercises

In Exercises 1–5, determine the concavity, y-intercept, and vertex of the quadratic equation.

1. $y = x^2 - 2x + 1$

2. $f(x) = -2x^2 + 4$

3. $g(x) = 3x^2 + 3x$

4. $h(t) = -1.2t^2 + 2.4t + 4.5$

5. $f(t) = 2.8t^2 - 1.4t + 2.1$

In Exercises 6–10, determine the equation of the parabola from the graph.

6.

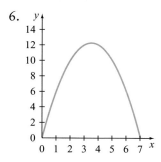

7.

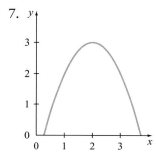

8.

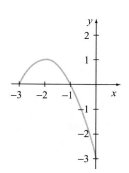

9.

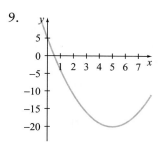

10.

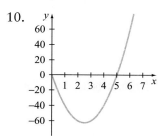

In Exercises 11–15, do the following:
   (a) Find a quadratic model for the data algebraically.
   (b) Graph the model together with the scatter plot.
   (c) Explain why you do or do not believe that the function is a good model.

11. **Medicare and Prescriptions**

Medicare and Prescriptions by Year

Years (since 1995) (t)	Medicare Enrollees (millions) (m)	Prescriptions (millions) (p)
0	37.6	2125
5	39.7	2865
6	40.1	3009
7	40.5	3139
8	41.1	3215
9	41.7	3274

Source: *Statistical Abstract of the United States, 2006,* Tables 126, 132

For the model, use Medicare Enrollees as the independent variable and Prescriptions as the dependent variable. That is, model Prescriptions as a function of Medicare Enrollees.

12. *United States Population*

U.S. Population

Years (since 1990)	U.S. Population (millions)
0	250
2	257
4	263
6	270
8	276
10	282
12	288
14	294
16	298
18	304

Source: *Statistical Abstract of the United States, 2006* and *2010*, Tables 2 and 7

13. *Walmart Net Sales*

Walmart Net Sales

Year (since 1996)	Net Sales (billions of dollars) (S)
0	89
1	100
2	112
3	131
4	156
5	181
6	204
7	230
8	256
9	285
10	312

Source: Walmart Annual Report, 2006, p.18–19

14. *Consumer Spending*

Consumer Spending on Books

Years (since 2004) (x)	Adult Trade Book Spending (millions of dollars) (b)
0	14,952.1
1	15,532.4
2	15,937.2
3	16,370.1
4	16,670.0
5	16,985.4

Source: *Statistical Abstract of the United States, 2007,* Table 1119

15. *Oil Production*

Oil Production Deficit

Years (since 1985) (t)	U.S. Oil Field Production Minus Net Imports of Oil (millions of barrels) (b)
0	2180
2	1396
4	698
6	639
8	58
10	−199
11	−342
12	−608
13	−856
14	−997
15	−1176
16	−1280
17	−1236
18	−1450
19	−1699

Source: *Statistical Abstract of the United States, 2007,* Table 881

In Exercises 16–26, do the following:
  (a) Draw a scatter plot for the data.
  (b) Explain why a quadratic function may or may not fit the data.

If it appears that a quadratic function may fit the data:
  (c) Use quadratic regression to find the quadratic function that best fits the data.
  (d) Graph the model together with the scatter plot.

(e) Explain why you do or do not believe that the function is a good model.
(f) Answer any additional questions that may be given.

16. *Company Net Worth*

United Services Automobile Association Net Worth

Years (since 1999) (x)	USAA Net Worth (billions of dollars) (w)
0	6.9
1	7.1
2	7.4
3	7.9
4	9.1
5	10.1
6	11.2
7	13.1

Source: USAA 2006 Report to Members

According to the model, what will be the net worth of USAA in 2012?

17. *Football Player Salaries*

NFL Player Salaries

Years (since 2000) (x)	NFL Player Average Salary (thousands of dollars) (s)
0	787
1	986
2	1180
3	1259
4	1331
5	1400

Source: *Statistical Abstract of the United States, 2007,* Table 1228

18. *Electronics Industry*

Computer and Electronic Products Industry Employees

Years (since 2000) (x)	Employees (thousands) (E)
0	1820
2	1507
3	1355
4	1323
5	1320

Source: *Statistical Abstract of the United States, 2007,* Table 980

19. *Theme Park Tickets*

Disneyland Theme Park Tickets—Adults

Days in Park (d)	2007 Park Hopper Bonus Ticket Cost (dollars) (T)
1	83
2	122
3	159
4	179
5	189

Source: www.disneyland.com

According to the model, what is the predicted price for a six-day ticket?

20. *Theme Park Tickets*

Disneyland Theme Park Tickets—Children

Days in Park (d)	2007 Park Hopper Bonus Ticket Cost (dollars) (T)
1	73
2	102
3	129
4	149
5	159

Source: www.disneyland.com

21. *Company Net Income*

Johnson & Johnson Net Income

Years (since 1990) (t)	Net Income (millions of dollars) (I)
0	1195
1	1441
2	1572
3	1786
4	1998
5	2418
6	2958
7	3385
8	3798
9	4348
10	4998
11	5899

Source: *Johnson & Johnson Annual Report, 2001,* p. 18

22. *Movies*

Motion Picture Screens and Movie Attendance by Year

Year (since 2000) (*t*)	Movie Attendance (millions) (*m*)	Motion Picture Screens (*S*)
0	1421	38,000
1	1487	37,000
2	1639	36,000
3	1574	37,000
4	1536	37,000

Source: *Statistical Abstract of the United States, 2006,* Table 1234

Use movie attendance as the independent variable and movie screens as the dependent variable. That is, model movie screens as a function of movie attendance.

23. *Malaria in Ghana*

Clinical Malaria Cases Reported in Children under 5 Years

Years (since 1998) (*t*)	Malaria Cases in Children (*C*)
0	576,563
1	1,126,807
2	1,303,685
3	1,316,724
4	966,923

Source: http://www.afro.who.int/malaria/country-profile/ghana.pdf

24. *Coffee Sales*

Starbucks Corporation Sales

Years (since 09/93) (*t*)	Income from Sales (millions of dollars) (*S*)
0	163.5
1	284.9
2	465.2
3	696.5
4	966.9
5	1308.7
6	1680.1
7	2169.2
8	2649.0
9	3288.9

Source: moneycentral.msn.com

According to the model, when did Starbucks Corporation sales reach $5,000 million?

25. *Poultry Pricing*

Average Retail Price of Fresh Whole Chicken

Years (since 1985) (*t*)	Price (dollars per pound) (*P*)
0	0.78
5	0.86
6	0.86
7	0.88
8	0.91
9	0.90
10	0.94
11	1.00
12	1.00
13	1.06
14	1.05
15	1.08

Source: *Statistical Abstract of the United States, 2001,* Table 706, p. 468

26. *College Enrollment*

Private College Enrollment

Years (since 1980) (*t*)	Students (in thousands) (*S*)
0	2640
2	2730
4	2765
6	2790
8	2894
10	2974
12	3103
14	3145
16	3247
18	3373

Source: *Statistical Abstract of the United States, 2001,* Table 205

2.2 Exponential Function Models

United Arab Emirates (UAE) is the fastest-growing country in the world. According to The World Factbook, published by the Central Intelligence Agency, the population of UAE is increasing by 3.689 percent per year. This means that for every 100,000 people in the population, 3,689 new people are added during the course of a year. Anything that changes at a constant percentage rate may be represented with an exponential function.

In this section, we will demonstrate how exponential functions can be used to model increasing data sets such as the population of UAE. We will develop exponential models from tables of data and verbal descriptions. In addition, we will show what an exponential function graph looks like.

The countries with the largest percentage growth rates are shown in Table 2.7.

In 2009, 4.80 million people lived in the United Arab Emirates. Assuming that UAE will continue to grow at a rate of 3.689 percent per year, the population of UAE can be modeled by $U(t) = 4.80(1.03689)^t$ (Figure 2.17).

Notice that the independent variable, t, appears as an exponent. The function U is called an **exponential function**.

Table 2.7

Top Ten Fastest Growing Countries		
Rank	Country	Percentage Growth (percent)
1	United Arab Emirates	3.689
2	Burundi	3.69688
3	Niger	3.68
4	Kuwait	3.55
5	Gaza Strip	3.35
6	Mayotte	3.32
7	Congo, Democratic Republic of the	3.21
8	Ethiopia	3.21
9	Oman	3.14
10	Burkina Faso	3.10

Source: CIA The World Factbook

Figure 2.17

Exponential Function

If a and b are real numbers with $a \neq 0, b > 0$, and $b \neq 1$, then the function

$$y = ab^x$$

is called an **exponential function**. The value b is called the **base** of the exponential function.

Why must b be positive? Consider the function $y = (-1)^x$. For integer values, y oscillates between -1 and 1 (Table 2.8). However, the function is undefined for numerous noninteger values of x (Table 2.9).

Table 2.8

x	y
2	1
1	−1
0	1
1	−1
2	1

Table 2.9

x	y
0.5	Undefined
0.3	Undefined
0.1	Undefined
1.4	1
1.5	Undefined

On the other hand, if $b > 0$, then the value of b^x is defined for *all* integer and noninteger values of x.

Why don't we allow b to be 1? If $b = 1$, then

$$y = ab^x$$
$$y = a(1)^x$$
$$y = a \qquad \text{Since } (1)^x = 1 \text{ for all } x$$

The graph of the function $y = a$ is a horizontal line and does not exhibit the same graphical behavior as all other functions of the form $y = ab^x$ with $b > 0$. By eliminating the case of $b = 1$, we are able to talk about a family of like functions.

Exponential functions are used frequently to model growth and decay situations, such as growth in population, depreciation of a vehicle, or growth in a retirement account. When the growth or decay of a quantity is modeled by an exponential function $y = ab^x$, the independent variable is frequently time. The beginning value of the quantity at time $x = 0$ is referred to as the **initial value** of the function and, as demonstrated next, is equal to a.

$$y = ab^0$$
$$= a(1) \qquad \text{Since } b^0 = 1 \text{ for all nonzero real numbers } b$$
$$= a$$

Graphically speaking, the constant a in $y = ab^x$ corresponds with the y-intercept of the exponential graph. In most real-life applications, a will be positive. The initial value of the UAE population function $U(t) = 4.80(1.03689)^t$ is 4.80. That is, the model states that the population was 4.80 million in the year $t = 0$ (2009).

If the function is increasing, the base b of $y = ab^x$ is referred to as the **growth factor** of the function. If the function is decreasing, the base b of $y = ab^x$ is referred to as the **decay factor** of the function. Changing the value of x by one unit changes y by a factor of b. The growth factor of the UAE population function $U(t) = 4.80(1.03689)^t$ is 1.03689, since $b = 1.03689$. That is, the population of UAE is increasing by a factor of 1.03689 annually. Next year's population is forecast to be 1.03689 times this year's population.

Recall that the x-intercept of a function occurs when $y = 0$. For what values of x does $ab^x = 0$? Let's consider the exponential function $y = 2^x$. We'll generate a table of values and plot a few points to get an idea about what is happening graphically (Figure 2.18).

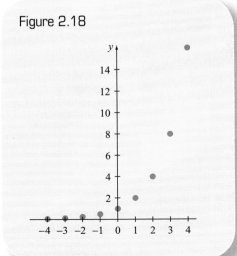

Figure 2.18

x	$y = 2^x$
−4	$\frac{1}{16}$
−3	$\frac{1}{8}$
−2	$\frac{1}{4}$
−1	$\frac{1}{2}$
0	1
1	2
2	4
3	8
4	16

Observe that as x increases, y also increases. Also, observe that y is positive for all values of x. That is, *there are no x-intercepts*. This remains true even if we pick a negative number with a larger magnitude, say $x = -33$.

$$2^{-33} = \frac{1}{2^{33}}$$

$$2^{-33} = \frac{1}{8,589,934,592}$$

We call the line $y = 0$ a **horizontal asymptote** of the function $y = 2^x$. For sufficiently small values of x, the graph of $y = 2^x$ approaches the graph of the line $y = 0$. All exponential functions of the form $y = ab^x$ have a horizontal asymptote at $y = 0$.

Exponential Function Graphs

Exponential function graphs will take on one of the four basic shapes specified in Table 2.10. Each graph has a horizontal asymptote at $y = 0$ and a y-intercept at $(0, a)$.

EXAMPLE 1 Comparing Exponential Graphs

Compare and contrast the graphs of $f(x) = 3(2)^x$ and $g(x) = 4(0.5)^x$. (You may sketch the graphs by hand by plotting several points, or you may use technology to graph the functions.)

SOLUTION

Both f and g are exponential functions. In the function equation of f, $a = 3$ and $b = 2$. The graph of f is concave up, since $a > 0$, and is increasing, since $b > 1$. The graph of f has a y-intercept at $(0, 3)$.

In the function equation of g, $a = 4$ and $b = 0.5$. The graph of g is concave up, since $a > 0$, and is decreasing, since $b < 1$. The graph of g has a y-intercept at

Table 2.10

Exponential Function Graphs: $y = ab^x$				
Value of a	Value of b	Concavity of Graph	Increasing/Decreasing	Sample Graph
$a > 0$	$b > 1$	Concave up	Increasing	
$a > 0$	$0 < b < 1$	Concave up	Decreasing	
$a < 0$	$b > 1$	Concave down	Decreasing	
$a < 0$	$0 < b < 1$	Concave down	Increasing	

(0, 4). The graphs of both functions have a horizontal asymptote at $y = 0$ (Figure 2.19).

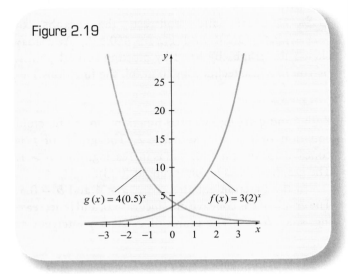

Figure 2.19

$g(x) = 4(0.5)^x$ $\qquad f(x) = 3(2)^x$

Properties of Exponents

As seen earlier in this section, it is often necessary to apply the properties of exponents when working with exponential functions. Since these properties are typically covered in depth in an algebra course, we will only summarize them here.

Properties of Exponents

If b, m, and n are real numbers with $b > 0$, the following properties hold.

Property	Example
1. $b^{-n} = \dfrac{1}{b^n}$	1. $2^{-3} = \dfrac{1}{2^3}$
2. $b^m \cdot b^n = b^{m+n}$	2. $3^2 \cdot 3^4 = 3^{2+4} = 3^6$
3. $\dfrac{b^m}{b^n} = b^{m-n}$	3. $\dfrac{5^6}{5^4} = 5^{6-4} = 5^2$
4. $b^{mn} = (b^m)^n = (b^n)^m$	4. $6^{2 \cdot 3} = (6^2)^3 = (6^3)^2$

Finding an Exponential Function from a Table

We can easily determine if a table of values models an exponential function by calculating the ratio of consecutive outputs for evenly spaced inputs. The ratio will be constant if the table of values models an exponential function. Consider Table 2.11.

Table 2.11

x	y
0	6
2	24
4	96
6	384

The domain values are equally spaced (two units apart). Calculating the ratios of the consecutive range values, we get

$$\frac{24}{6} = 4 \qquad \frac{96}{24} = 4 \qquad \frac{384}{96} = 4$$

In each case, the ratio was 4. Therefore, the table of values models an exponential function. (This test only works if the domain values are equally spaced.) But what is the equation of the function?

We can find the equation of the function algebraically. We know that an exponential function must be of the form $y = ab^x$. Substituting the point (0, 6) into the equation and solving, we get

$$y = ab^x$$
$$6 = ab^0 \qquad \text{Substitute } x = 0 \text{ and } y = 6$$
$$6 = a$$

Since $a = 6$, $y = 6b^x$. Substituting the point (2, 24) into the equation $y = 6b^x$, we get

$$y = 6b^x$$
$$24 = 6b^2 \qquad \text{Substitute } x = 2 \text{ and } y = 24$$
$$4 = b^2$$
$$b = 2 \qquad b \neq -2 \text{ since the base of an exponential}$$
$$\text{function must be positive}$$

Therefore, the exponential function that models the table data is $y = 6(2)^x$.

EXAMPLE 2 Finding an Exponential Equation from a Table

Find the equation of the exponential function modeled by Table 2.12.

Table 2.12

x	y
1	21
2	63
3	189
4	567

SOLUTION

We know that an exponential function must be of the form $y = ab^x$. Substituting the point $(2, 63)$ into the equation, we get

$$y = ab^x$$

$$63 = ab^2 \qquad \text{Substitute } x = 2 \text{ and } y = 63$$

Substituting the point $(1, 21)$ into the equation $y = ab^x$, we get

$$y = ab^x$$

$$21 = ab^1 \qquad \text{Substitute } x = 1 \text{ and } y = 21$$

We can eliminate the a variable by dividing the first equation by the second equation. (*Note:* What we are really doing is dividing both sides of the first equation by the same nonzero quantity, expressed in two different forms.)

$$\frac{63}{21} = \frac{ab^2}{ab^1}$$

$$3 = \frac{a}{a} \cdot b^{2-1} \qquad \text{Since } \frac{b^m}{b^n} = b^{m-n}$$

$$3 = 1 \cdot b$$

$$b = 3$$

We may then substitute the value of b into either equation to find a.

$$21 = ab^1$$

$$21 = a \cdot 3^1 \qquad \text{Substitute } b = 3$$

$$\frac{21}{3} = a$$

$$a = 7$$

The exponential function that models the table data is $y = 7(3)^x$.

When using the method of dividing one equation by the other, computations will tend to be easier if we divide the equation with the largest exponent by the equation with the smallest exponent.

The technique shown in Example 2 may be generalized for all exponential functions. Using the result from the generalized solution will allow us to determine the exponential equation more quickly.

Consider an exponential function $y = ab^x$ whose graph goes through the points (x_1, y_1) and (x_2, y_2). We have

$$y = ab^x$$

$$y_1 = ab^{x_1} \qquad \text{Substitute } x = x_1 \text{ and } y = y_1$$

and

$$y = ab^x$$

$$y_2 = ab^{x_2} \qquad \text{Substitute } x = x_2 \text{ and } y = y_2$$

Dividing the second equation by the first equation yields a quick way to calculate b.

$$\frac{y_2}{y_1} = \frac{ab^{x_2}}{ab^{x_1}}$$

$$\frac{y_2}{y_1} = \frac{a}{a} b^{x_2 - x_1}$$

$$\frac{y_2}{y_1} = b^{x_2 - x_1}$$

$$\left(\frac{y_2}{y_1}\right)^{\frac{1}{x_2 - x_1}} = \left(b^{x_2 - x_1}\right)^{\frac{1}{x_2 - x_1}}$$

$$\left(\frac{y_2}{y_1}\right)^{\frac{1}{x_2 - x_1}} = b^{\frac{x_2 - x_1}{x_2 - x_1}}$$

$$b = \left(\frac{y_2}{y_1}\right)^{\frac{1}{x_2 - x_1}}$$

Using the points $(1, 21)$ and $(3, 189)$ from Example 2, we get

$$b = \left(\frac{y_2}{y_1}\right)^{\frac{1}{x_2 - x_1}}$$

$$b = \left(\frac{189}{21}\right)^{\frac{1}{3-1}} \qquad \text{Substitute } x_1 = 0, x_2 = 2, y_1 = 21, \text{ and } y_2 = 189$$

$$= 9^{1/2}$$

$$= \sqrt{9} \qquad \text{Recall that for } b \geq 0, b^{1/2} = \sqrt{b}$$

$$= 3$$

This method is especially useful when checking to see if a data table with unequally spaced inputs is an exponential function. If it is an exponential function, the value of b will be constant regardless of which two points we substitute into the formula.

Using Exponential Regression to Model Data

Data sets with near-constant ratios of change may be modeled using the exponential regression feature on our graphing calculator. It is often helpful to do a scatter plot of the data first to see if the graph looks like an exponential function.

EXAMPLE 3 Using Exponential Regression to Model the Population of Akron, Ohio

The population of Akron, Ohio, is shown in Table 2.13. Model the population with an exponential function and forecast the population of Akron in 2014.

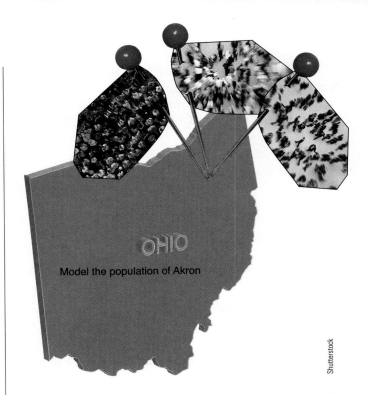

Model the population of Akron

Shutterstock

Table 2.13

Years (since 1970) (x)	Population (thousands) (y)
0	275
10	237
20	223
30	217
39	208

Source: U.S. Census Bureau

SOLUTION

We first draw a scatter plot so that we can visually predict whether an exponential function will fit the data well (Figure 2.20). (Note that because we have zoomed in on the data, the "x-axis" is given by $y = 200$ instead of $y = 0$.)

Since the scatter plot is decreasing and concave up, an exponential function may fit the data well. Using techniques given on the Chapter 2 Tech Card, we can calculate an exponential model for the data set (Figure 2.21).

Figure 2.20

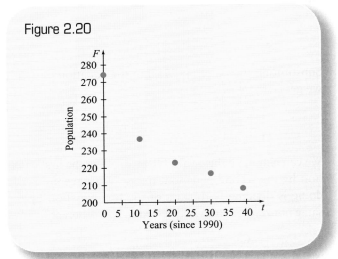

Figure 2.21

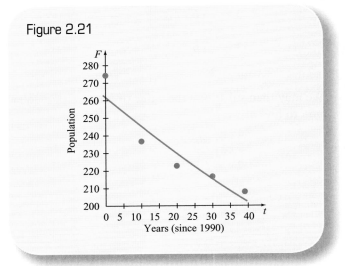

The model $y = 263.2(0.9934)^x$ is a terrible fit for the data. What happened?

Recall that an exponential function of the form $y = ab^x$ has a horizontal asymptote at $y = 0$. From the scatter plot of the data, it appears that the population is approaching a constant value of $y = 200$ instead of $y = 0$. (This value is not precise. We are *guessing* what the horizontal asymptote would be based on the scatter plot of the data set.) If we subtract 200 from each of the *y*-values, we get a new data set with a horizontal asymptote of $y = 0$. This is referred to as an **aligned data set** (Table 2.14).

Table 2.14

Year (*x*)	Population – 200 (thousands) (*y*)
0	75
10	37
20	23
30	17
39	8

Using exponential regression to find a model for the aligned data set yields $y = 70.46(0.9479)^x$, a function that fits the aligned data well (Figure 2.22).

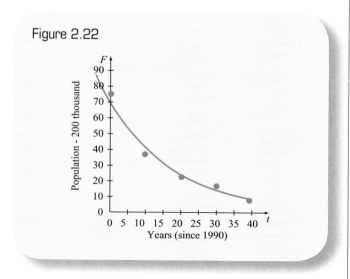

Figure 2.22

Subtracting 200 from each of the *y*-values of the original data set had the graphical implication of moving the data down 200 units. To return the aligned data model to the position of the original data, we need to move the model up 200 units. To do this, we add back the 200.

$$P(x) = 70.46(0.9479)^x + 200 \qquad \text{Add 200 to shift the data graph upward}$$

The model created from the aligned data fits those data much better than the unaligned data model (Figure 2.23).

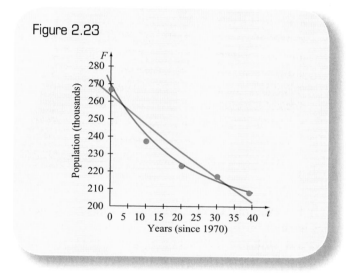

Figure 2.23

To estimate the population of Akron, Ohio, in 2014, we evaluate $P(x) = 70.46(0.9479)^t + 200$ at $x = 44$.

$$P(x) = 70.46(0.9479)^{44} + 200$$

$$\approx 206.7$$

We forecast that the population of Akron, Ohio, in 2014 will be 206,700.

The methods of aligning data and of performing exponential regression are detailed on the Chapter 2 Tech Card in the back of your book.

Finding an Exponential Function from a Verbal Description

Much of the information we encounter in the media is given to us in terms of percentages. Consider these typical news headlines: "Gasoline Prices Flare Up 9 Percent" and "Home Values Increase by 6 Percent." Anything that increases or decreases at a constant percentage rate may be modeled with an exponential function. The value of the growth factor b is $1 + r$, where r is the decimal form of the percentage (e.g., 5 percent = 0.05). Note that if $r < 0$, then $b < 1$. A negative rate of growth is referred to as **depreciation**, while a positive rate of growth is referred to as **appreciation**.

Annual Percentage Growth Rate or Decay Rate

Let $y = ab^t$ model the amount of a quantity at time t years. For positive values of r, the annual percentage growth rate r of the quantity y is given by $r = b - 1$. The annual growth factor is $b = r + 1$. For negative values of r, the annual percentage decay rate r of the quantity y is given by $r = b - 1$. The annual decay factor is $b = r + 1$.

For example, Detroit, Michigan, was one of the cities hardest hit by the economic downturn that started around 2006. In 2009, home values in Detroit were falling at a rate of 18.8 percent per year. (**Source:** www.zillow.com) In other words, $r = -0.188$. The corresponding decay factor is $b = 1 + (-0.188) = 0.812$. The average price of a Detroit home at the end of 2009 was $91,000. To forecast the value of a Detroit home at the end of 2010, we multiply the home value by the decay factor. The predicted value is ($91,000)(0.812) = $73,892.

EXAMPLE 4 Using an Exponential Model to Forecast a Car's Value

New cars typically lose 50 percent of their value in the first 3 years. Many cars lose more than 25 percent of their value in the first year alone. (**Source:** Runzheimer International) In 2007, a new Honda Accord VP Sedan had a manufacturer's suggested retail price of $20,020. In 2010, the 2007 Accord was valued at $13,580. (**Source:** www.kbb.com) Determine the rate of depreciation of the 2007 Accord and, assuming the depreciation rate remains constant, predict the value of the car in 2014.

SOLUTION

Since the car is depreciating at a constant percentage rate, an exponential function may be used to model the data. The initial value of the car is $20,020, so

$a = 20{,}020$. We use the given data to calculate the decay factor.

$$b = \left(\frac{13{,}580}{20{,}020}\right)^{\frac{1}{3-0}}$$

$$= 0.8786$$

The exponential function is

$$V(t) = 20{,}020(0.8786)^t$$

where t is the age of the car in years.

We want to know the value of the car in 2014, so $t = 7$.

$$V(t) = 20{,}020(0.8786)^7$$

$$\approx 8091 \qquad \text{Values may vary depending upon rounding of intermediate values.}$$

We estimate that the car will be valued at $8,091 when it is 7 years old.

EXAMPLE 5 Using an Exponential Function to Model Inflation

Inflation (rising prices) causes money to lose its buying power. In the United States, inflation hovers around 3 percent annually.

(a) If a candy bar costs $0.99 today, what will it cost 10 years from now?

(b) When will a candy bar cost $2.00?

SOLUTION

Since we're assuming a constant percentage increase in the price of the candy bar, an exponential model should be used. The initial value is $0.99, so $a = 0.99$. The price of the candy bar is increasing by 3 percent annually, so $r = 0.03$. The cost of the candy bar is given by

$$C(t) = 0.99(1 + 0.03)^t \text{ dollars}$$

$$= 0.99(1.03)^t \text{ dollars}$$

where t is the number of years from today.

(a) Evaluating the function at $t = 10$, we get

$$C(10) = 0.99(1.03)^{10}$$

$$= 0.99(1.344)$$

$$= 1.33$$

Ten years from now we estimate that the candy bar will cost $1.33.

(b) We want to know when $C(t) = 2$.

$$2 = 0.99(1.03)^t$$

$$2.02 = 1.03^t$$

$$C(t) = 0.99(1.03)^t$$

99¢

$2.00

18

12

6

Shutterstock

We can solve the problem graphically by breaking the equation $2.02 = 1.03^t$ into two separate functions: $y_1 = 2.02$ and $y_2 = 1.03^t$. By graphing these functions simultaneously, we can determine the point of intersection of the two functions using the graphing calculator.

We generate the graph in Figure 2.24 and determine the point of intersection.

The functions intersect when $t = 23.79$. In about 24 years, we estimate that the cost of a candy bar will be $2.00.

We also could have used the guess-and-check method to approximate the result by evaluating $C(t) = 0.99(1.03)^t$ at various values of t. Since

$$C(23) = 0.99(1.03)^{23} = 1.97 \quad \text{and} \quad C(24) = 0.99(1.03)^{24} = 2.03$$

we determine that the price reaches $2.00 between the 23rd and 24th years.

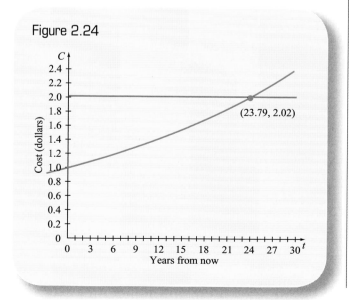

Figure 2.24

2.2 Exercises

In Exercises 1–5, do the following:
  (a) Determine if the graph of the function is increasing or decreasing.
  (b) Determine if the graph of the function is concave up or concave down.
  (c) Identify the coordinate of the y-intercept.
  (d) Graph the function to verify your conclusions.

1. $y = 4(0.25)^x$

2. $y = 0.5(2)^x$

3. $y = 0.4(5)^x$

4. $y = -1.2(2.3)^x$

5. $y = 3(0.9)^x$

In Exercises 6–10, use algebraic methods to find the equation of the exponential function that fits the data in the table.

6.

x	y
0	2
1	6
2	18
3	54

7.

x	y
1	10
2	20
3	40
4	80

8.

x	y
2	16
4	4
6	1
8	0.25

9.

x	y
1	1
2	4
4	64
5	256

10.

x	y
0	256
5	8
7	2
10	0.25

In Exercises 11–13, use exponential regression to model the data in the table. Use the model to predict the value of the function when $t = 30$, and interpret the real-world meaning of the result.

The Consumer Price Index is used to measure the increase in prices over time. In each table, the index is assumed to have the value 100 in the year 1984.

11. *Dental Prices*

Price of Dental Services

Years (since 1980) (t)	Price Index (I)
0	78.9
5	114.2
10	155.8
15	206.8
20	258.5
25	324.0
28	376.9

Source: *Statistical Abstract of the United States, 2010,* Table 135

12. *Food Prices*

Price of Food (Excluding Alcohol)

Years (since 1990) (t)	Price Index (I)
0	81.9
2	84.8
4	87.6
6	92.2
8	96.0
10	100.0
12	104.9
14	110.2
16	115.4
17	119.8

Source: *Statistical Abstract of the United States, 2010,* Table 723

13. *Purchased Meals and Beverages*

Purchased Meals and Beverages

Years (since 1990) (t)	Price Index (I)
0	78.6
2	83.4
4	86.3
6	90.3
8	95.1
10	100.0
12	105.9
14	111.5
16	118.6
17	123.0

Source: *Statistical Abstract of the United States, 2010,* Table 723

In Exercises 14–18, find the exponential function that fits the verbal description.

14. *Salaries* Instructors' salaries are $45,000 per year and are expected to increase by 3.5 percent annually. What will instructors' salaries be 5 years from now?

15. *Savings Account* A savings account balance is currently $235 and is earning 2.32 percent per year. When will the balance reach $250?

16. *Television Price* The cost of a 26-inch LCD HD television in 2010 was $389.99. (Source: www.bestbuy.com) Television prices are expected to decrease by 16 percent per year. How much is the LCD HD television expected to cost in 2013?

17. *Concert Admission* Reserved seating at a Black Eyed Peas concert in Baltimore, Maryland, cost $94.70 in August 2010. (**Source:** www.ticketmaster.com) Suppose that due to the growing popularity of the band, the concert ticket price is expected to increase by 28 percent per year. What is the expected price of a Black Eyed Peas ticket in August 2011?

18. *Tuition* Tuition is currently $2,024 per year and is increasing by 12 percent annually. In how many years from now will tuition reach $3,567?

In Exercises 19–21, find the solution by solving the equation graphically.

19. $(1.2)^x = 5$

20. $(1.05)^x = 2$

21. $(0.3)^x = 0.09d$

The Derivative

It is impossible to determine how quickly a person is running from a single photograph, since speed is calculated as a change in distance over a change in time. Nevertheless, we may estimate a person's speed at a particular instant in time by determining the distance traveled over a small interval of time (e.g., one second). A runner's speed may be classified as a rate of change in distance. A key component of calculus is the study of rates of change.

3.1 Average Rate of Change

3.2 Limits and Instantaneous Rates of Change

3.3 The Derivative as a Slope: Graphical Methods

3.4 The Derivative as a Function: Algebraic Method

3.5 Interpreting the Derivative

3.1 Average Rate of Change

Colleges and universities periodically raise their tuition rates in order to cover rising staffing and facilities costs. As a result, it is often difficult for students to know how much money they should save to cover future tuition costs. By calculating the *average rate of change* in the tuition price over a period of years, we can estimate projected increases in tuition costs. In this section, we will demonstrate how to calculate the average rate of

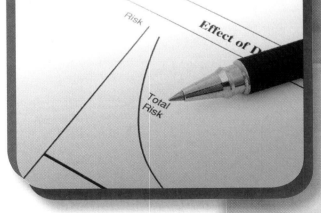

change in the value of a function over a specified interval [a, b]. (The interval notation [a, b] is equivalent to $a \leq x \leq b$.)

In calculating the difference quotient, we answer the question, "Over the interval [a, b], on average, how much does a one-unit increase in the x-value change the y-value of the function?"

The Difference Quotient:
An Average Rate of Change

The **average rate of change** of a function $y = f(x)$ over an interval [a, b] is

$$\frac{f(b) - f(a)}{b - a}$$

This expression is referred to as the **difference quotient.** For a linear function, the difference quotient gives the slope of the line.

EXAMPLE 1 Calculating an Average Rate of Change from a Table

The annual cost of tuition and fees for full-time, resident, undergraduate students majoring in the Arts and Sciences at the University of Pittsburgh is shown in Table 3.1.

Table 3.1

Years (since 2004–2005) t	Annual Tuition and Fees (dollars) $f(t)$	Change in Tuition from Prior Year (dollars)
0	10,130	
1	10,736	606
2	11,368	632
3	12,106	738
4	12,832	726
5	13,344	512

Source: University of Pittsburgh

What is the average rate of change in the tuition and fees from the 2004–2005 academic year to the 2009–2010 academic year? Rounded to the nearest dollar, what do you estimate the 2012–2013 tuition and fees will be?

SOLUTION

The average rate of change may be calculated by using the difference quotient formula. For the period 2004–2005 to 2009–2010, the interval $[a, b] = [0, 5]$. The average rate of change of the tuition is

$$\frac{f(b) - f(a)}{b - a} = \frac{f(5) - f(0)}{5 - 0}$$

$$= \frac{13{,}344 - 10{,}130 \, \text{dollars}}{5 \, \text{years}}$$

$$= \frac{3214 \, \text{dollars}}{5 \, \text{years}}$$

$$= 642.8 \, \text{dollars per year}$$

Over the five-year period between 2004–2005 and 2009–2010, tuition increased by $3,214. Although the annual increase varied from year to year, the average annual increase was $642.80.

The 2009–2010 tuition and fees cost was $13,344. We predict that tuition and fees will increase by $642.80 per year in subsequent years. To predict the 2012–2013 tuition and fees cost, we repeatedly increase the annual cost by $642.80.

2010–2011:	$13{,}344 + 642.80 = 13{,}986.80$
2011–2012:	$13{,}986.80 + 642.80 = 14{,}629.60$
2012–2013:	$14{,}629.60 + 642.80 = 15{,}272.40$

We estimate that the 2012–2013 tuition and fees cost will be $15,272.40. This value may be calculated more quickly as follows: $13{,}344 + 3(642.80) = 15{,}272.40$.

When determining the meaning of an average rate of change in a real-life problem, it is essential to find the units of measurement of the result. Fortunately, the units are easily determined. **The units of the rate of change are the units of the output divided by the units of the input.** In Example 1, the units of the output were *dollars* and the units of the input were *years.* Consequently, the units of the average rate of change were *dollars* divided by *years,* or *dollars per year.*

EXAMPLE 2 Calculating an Average Rate of Change from an Equation

Based on data from 1990–2009, the population of Washington state may be modeled by the function $P(t) = 4.933(1.017)^t$, where P is the population in millions of people and t is the number of years since 1990. (**Source:** Modeled from data at www.census.gov) According to the model, what is the average rate of change in the population between 1990 and 2010?

SOLUTION

We observe that in the model, t is the number of years since 1990. So for the model, $t = 0$ represents 1990 and $t = 20$ represents 2010. The average rate of change in the population over the interval $[0, 20]$ is

$$\frac{P(20) - P(0)}{20 - 0} = \frac{6.911 - 4.933 \, \text{million people}}{20 \, \text{years}}$$

$$= 0.0989 \, \text{million people per year}$$

$$= 98{,}900 \, \text{people per year}$$

Between 1990 and 2010, the population of Washington state increased by an average of 98,900 people per year.

Graphical Interpretation of the Difference Quotient

A line connecting any two points on a graph is referred to as a **secant line.** Graphically speaking, the difference quotient for a function $y = f(x)$ is the slope of the secant line connecting $(a, f(a))$ and $(b, f(b))$ (Figure 3.1).

Figure 3.1 Secant line slope $= \dfrac{f(b) - f(a)}{b - a}$

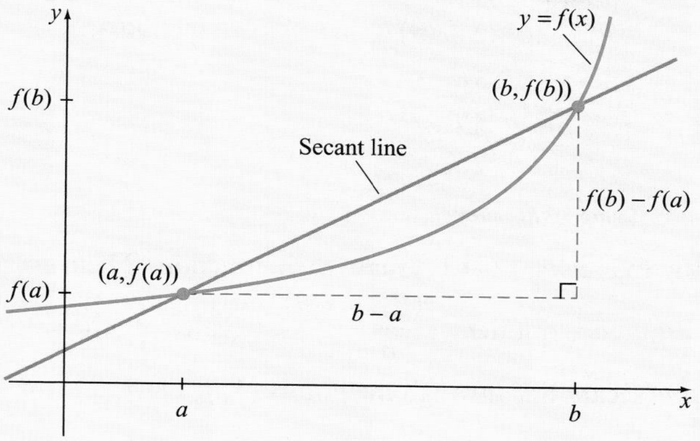

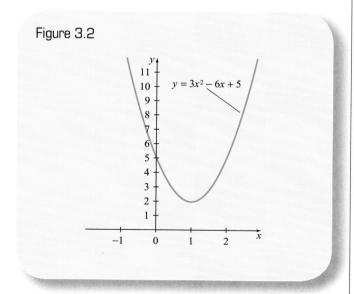

Figure 3.3

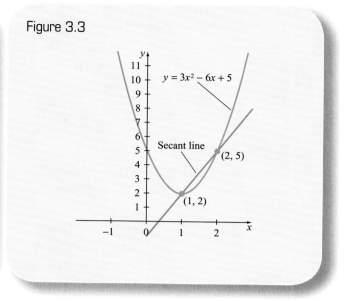

EXAMPLE 3 Finding the Slope of a Secant Line

➤ The graph of the function $f(x) = 3x^2 - 6x + 5$ is shown in Figure 3.2. Calculate the slope of the secant line of f that passes through $(1, 2)$ and $(2, 5)$.

Figure 3.2

SOLUTION

We first plot the points $(1, 2)$ and $(2, 5)$ and draw the line connecting them (Figure 3.3).

This line is the secant line of the graph between $x = 1$ and $x = 2$. The slope of the secant line is given by

$$m = \frac{f(b) - f(a)}{b - a}$$

$$= \frac{f(2) - f(1)}{2 - 1}$$

$$= \frac{5 - 2}{1}$$

$$= 3$$

The slope of the secant line is 3. That is, on the interval $[1, 2]$, a one-unit increase in x results in a three-unit increase in y, on average.

3.1 Exercises

In Exercises 1–5, calculate the average rate of change of the function over the given interval.

1. $f(x) = 2x - 5$ over the interval $[3, 5]$

2. $v(x) = -x^3$ over the interval $[-1, 1]$

3. $v(m) = m^2 - m$ over the interval $[-3, 4]$

4. $z = \dfrac{\ln(x)}{x}$ over the interval $[1, 5]$

5. $q(x) = \sqrt{x + 2}$ over the interval $[0, 6]$

In Exercises 6–10, calculate the average rate of change in the designated quantity over the given interval(s).

6. *Air Temperature* Temperature between 11:00 A.M. and 3:00 P.M.

Time of Day	Temperature (°F)
11:00 A.M.	68
1:00 P.M.	73
3:00 P.M.	75

7. *Dow Jones Industrial Average* Dow Jones Industrial Average between 2000 and 2007 and between 2002 and 2008.

Year	Dow Jones Industrial Average Closing Value at the End of the Year (points)
2000	10,787
2002	8342
2007	13,265
2008	8776

Source: www.census.gov

8. *Reading Scores* A third-grade student's reading score between the first and third quarter.

Quarter	Reading Score (words per minute)
1	69
2	107
3	129

Source: Author's data

9. *Newspaper Subscriptions* Daily newspaper subscriptions as the number of cable TV subscribers increased from 50.5 million (in 1990) to 67.7 million (in 2000).

Cable TV Subscribers (millions)	Daily Newspaper Circulation (millions)
50.5	62.3
60.9	58.2
66.7	56.0
67.7	55.8

Source: www.census.gov

10. *Poverty Level* Percentage of people below poverty level when the unemployment rate decreased from 5.6 percent (in 1990) to 4.0 percent (in 2000).

Unemployment Rate (percentage)	People below Poverty Level (percentage)
5.6	13.5
4.2	11.8
4.0	11.3

Source: www.census.gov

In Exercises 11–13, graph each function. Then use the difference quotient, $\dfrac{f(b) - f(a)}{b - a}$, to calculate the slope of the secant line through the points $(1, f(1))$ and $(3, f(3))$.

11. $f(x) = 2^x$

12. $f(x) = 5$

13. $f(x) = (x - 2)^2$

In Exercises 14 and 15, use the difference quotient, $\dfrac{f(b) - f(a)}{b - a}$, to calculate the slope of the secant line through the points $(1, f(1))$ and $(3, f(3))$ for the given graph of f.

14.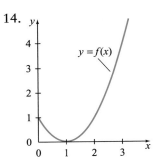

15.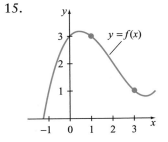

In Exercises 16–20, calculate the average rate of change in the state populations (in people per year) between 2000 and 2009.

16. *Population*

Montana

Date	Population (in thousands)
2000	902.2
2009	975.0

Source: www.census.gov

17. *Population*

Massachusetts

Date	Population (in thousands)
2000	6349.1
2009	6593.6

Source: www.census.gov

18. *Population*

Missouri

Date	Population (in thousands)
2000	5595.2
2009	5987.6

Source: www.census.gov

19. *Population*

West Virginia

Date	Population (in thousands)
2000	1808.3
2009	1819.8

Source: www.census.gov

20. *Population*

Pennsylvania

Date	Population (in thousands)
2000	12,281.1
2009	12,604.8

Source: www.census.gov

3.2 Limits and Instantaneous Rates of Change

In the 2008 Olympic games in Beijing, Usain Bolt set the world record for the 100-meter dash with a time of 9.69 seconds. How fast was he running (in meters per second) when the accompanying picture was taken?

From a single photo, it is impossible for us to determine his speed. However, if we knew how long it took him to reach various checkpoints during the race, we could approximate his speed at the finish line. In this section, we will demonstrate how to use the difference

Nicolas Asfouri/AFP/Getty Images

quotient to estimate and the derivative to calculate an *instantaneous rate of change.*

Whereas an average rate of change is calculated over an interval $[a, b]$, an instantaneous rate of change is calculated at a single value a. For example, the average highway speed of a person over a 200-mile trip may have been 59 miles per hour. However, when he passed a state patrol car exactly 124 miles into the trip, he was speeding at 84 miles per hour. His *average* speed on the interval $[0, 200]$ was 59 miles per hour; however, his *instantaneous* speed at $d = 124$ was 84 miles per hour.

EXAMPLE 1 Estimating an Instantaneous Rate of Change

Suppose that a runner in the 100-meter dash recorded the times shown in Table 3.2.

Table 3.2

Time (seconds) (t)	Total Distance Traveled (meters) ($D(t)$)
0	0
4.85	50
8.70	90
9.18	95
9.60	99
9.69	100

Estimate his speed when he crossed the finish line.

SOLUTION

His average speed over various distances may be calculated using the difference quotient, $\dfrac{D(b) - D(a)}{b - a}$.

His average speed over the 100-meter distance was

$$\text{Average speed} = \frac{100 - 0 \text{ meters}}{9.69 - 0 \text{ second}}$$

$$= 10.32 \text{ meters per second}$$

His average speed over the last 50 meters was

$$\text{Average speed} = \frac{100 - 50 \text{ meters}}{9.69 - 4.85 \text{ seconds}}$$

$$= \frac{50 \text{ meters}}{4.84 \text{ seconds}}$$

$$= 10.33 \text{ meters per second}$$

His average speed over the last 10 meters was

$$\text{Average speed} = \frac{100 - 90 \text{ meters}}{9.69 - 8.70 \text{ seconds}}$$

$$= \frac{10 \text{ meters}}{0.99 \text{ seconds}}$$

$$= 10.10 \text{ meters per second}$$

His average speed over the last 5 meters was

$$\text{Average speed} = \frac{100 - 95 \text{ meters}}{9.69 - 9.18 \text{ seconds}}$$

$$= \frac{5 \text{ meters}}{0.51 \text{ second}}$$

$$= 9.80 \text{ meters per second}$$

And his average speed over the last meter was

$$\text{Average speed} = \frac{100 - 99 \text{ meters}}{9.69 - 9.60 \text{ seconds}}$$

$$= \frac{1 \text{ meter}}{0.09 \text{ second}}$$

$$= 11.11 \text{ meters per second}$$

Although each calculation yielded a different result, all of these difference quotients estimate the runner's finish-line speed. Which of the estimates do you think is most accurate?

The last calculation best estimates his finish-line speed because it measures the change in distance over the smallest interval of time: 0.1 second. (Reducing the time interval to an even smaller amount of time, say 0.01 second, would further improve the estimate.) We estimate that the runner's speed when he crossed the finish line was 11.11 meters per second.

To estimate the *instantaneous rate of change* of a function $y = f(x)$ at a point $(a, f(a))$, we calculate the average rate of change of the function over a very small interval $[a, b]$. If we let the variable h represent the distance between $x = a$ and $x = b$, then $b = a + h$. Consequently, the difference quotient may be rewritten as

$$\frac{f(b) - f(a)}{b - a} = \frac{f(a + h) - f(a)}{(a + h) - a}$$

$$= \frac{f(a + h) - f(a)}{h}$$

The Difference Quotient as an Estimate of an Instantaneous Rate of Change

The *instantaneous rate of change* of a function $y = f(x)$ at a point $(a, f(a))$ may be *estimated* by calculating the **difference quotient of f at a**,

$$\frac{f(a + h) - f(a)}{h}$$

using an h arbitrarily close to 0. (If $h = 0$, the difference quotient is undefined.)

EXAMPLE 2 Estimating an Instantaneous Rate of Change

→ Based on data from 2004–2005 and projections for 2006–2009, the amount of money spent by consumers on books may be modeled by

$$b(x) = 41.01x^3 - 416.3x^2 + 2999x + 49{,}180$$
million dollars

where x is the number of years since 2004. (**Source:** Modeled from *Statistical Abstract of the United States, 2007*, Table 1119) According to the model, how quickly was consumer spending on books increasing in 2010?

SOLUTION

In 2010, $t = 6$. Using the difference quotient

$$\frac{S(6 + h) - S(6)}{h}$$

and selecting increasingly small values of h, we generate Table 3.3.

Table 3.3

h	$\dfrac{S(6 + h) - S(6)}{h}$
1.000	2795
0.100	2465
0.010	2436
0.001	2433

We conclude that in 2010 consumer spending on books is increasing by $2,433 million per year.

Limits

In Examples 1 and 2, we estimated the instantaneous rate of change at a point by calculating the average rate of change over a "short" interval by picking "small" values of h. The terms *short* and *small* are vague. Numerically, what does "small" mean? Mathematicians struggled with this dilemma for years before developing the concept of the **limit**. We will explore the limit concept graphically before giving a formal definition.

Consider the graph of the function $f(x) = -x^2 + 4$ over the interval $[-3, 3]$ (Figure 3.4).

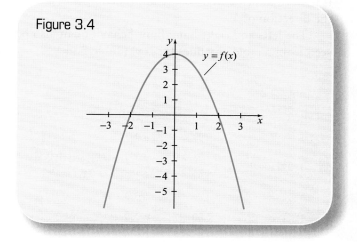

Figure 3.4

We ask the question, "As x gets close to 2, to what value does $f(x)$ get close?"

Observe from the graph of f that as the value of x moves from 0 to 2, the value of $f(x)$ moves from 4 to 0. We represent this behavior symbolically with the notation

$$\lim_{x \to 2^-} f(x) = 0$$

which is read, "the limit of $f(x)$ as x approaches 2 *from the left* is 0." This is commonly referred to as a **left-hand limit** because we approach $x = 2$ through values to the *left* of 2.

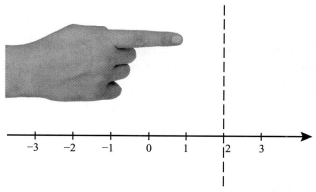

Observe from the graph of f that as the value of x moves from 3 to 2, the value of $f(x)$ moves from -5 to 0. We write

$$\lim_{x \to 2^+} f(x) = 0$$

Shutterstock

which is read, "the limit of $f(x)$ as x approaches 2 *from the right* is 0." This is commonly referred to as a **right-hand limit** because we approach $x = 2$ through values to the *right* of 2.

On a number line, $-\infty$ lies to the left of $+\infty$. For the left-hand limit, $\lim_{x \to a^-} f(x)$, the minus sign is used to indicate that we are approaching $x = a$ from the direction of $-\infty$. For the right-hand limit, $\lim_{x \to a^+} f(x)$, the plus sign is used to indicate that we are approaching $x = a$ from the direction of $+\infty$.

The left- and right-hand limit behavior can also be seen from a table of values for $f(x)$ (Table 3.4).

Table 3.4

	x	f(x)	
	0.00	4.000	
↓	1.00	3.000	f(x) gets close to
Left of x = 2	1.90	0.390	0 as x nears 2
↓	1.99	0.040	
	2.00	**0.000**	
↑	2.01	−0.040	
Right of x = 2	2.10	−0.410	f(x) gets close to
↑	3.00	−5.000	0 as x nears 2

When the left- and right-hand limits of $f(x)$ approach the same finite value, we say that "the limit exists." In this case, the left- and right-hand limits of $f(x)$ neared the same value ($y = 0$) as x approached 2. We say that "the limit of $f(x)$ as x approaches 2 is 0" and write

$$\lim_{x \to 2} f(x) = 0$$

Sometimes the left- and right-hand limits of a function at a point are not equal. Consider the graph of the piecewise function

$$f(x) = \begin{cases} -x^2 + 2 & x \le 1 \\ x + 1 & x > 1 \end{cases}$$

on the interval $[-3, 3]$ (see Figure 3.5).

We ask the question, "As x gets close to 1, to what value does $f(x)$ get close?"

Observe from the graph that as the value of x moves from 0 to 1, the value of $f(x)$ moves from 2 to 1. We write

$$\lim_{x \to 1^-} f(x) = 1$$

Figure 3.5

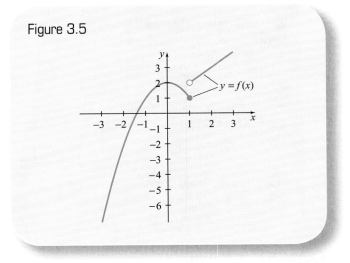

That is, the left-hand limit of $f(x)$ as x approaches 1 is 1. Similarly, as the value of x moves from 2 to 1, the value of $f(x)$ moves from 3 to 2. We write

$$\lim_{x \to 1^+} f(x) = 2$$

That is, the right-hand limit of $f(x)$ as x approaches 1 is 2. Since for this function, the left- and right-hand limits are not equal, we say that "the limit of $f(x)$ as x approaches 1 does not exist" or, simply, "the limit does not exist." This can also be seen from a table of values for $f(x)$ (Table 3.5).

Table 3.5

	x	f(x)	
↓	0.00	2.000	
Left of x = 1	0.90	1.190	f(x) gets close to
↓	0.99	1.020	1 as x nears 1
	1.00	**1.000**	
↑	1.01	2.010	
Right of x = 1	1.10	2.100	f(x) gets close to
↑	2.00	3.000	2 as x nears 1
	3.00	4.000	

One of the most powerful features of limits is that the limit of a function may exist at a point where the function itself is undefined. This feature will be used often when we introduce the limit definition of the derivative later in this section.

The Limit of a Function

If $f(x)$ is defined for all values of x near c, then

$$\lim_{x \to c} f(x) = L$$

means that as x approaches c, $f(x)$ approaches L.

We say that the limit exists if

1. L is a finite number and
2. Approaching c from the left or right yields the same value of L.

The theory surrounding limits is rich and worthy of study; however, in this text, we are most interested in the limit of the difference quotient as h approaches zero. That is,

$$\lim_{h \to 0} \frac{f(a + h) - f(a)}{h}$$

Recall that the difference quotient represents the average rate of change of $f(x)$ over the interval $[a, a + h]$. When we place the limit on the difference quotient, we are symbolically asking, "As the distance between the two x-values (a and $a + h$) gets smaller, what happens to the average rate of change of $f(x)$ on the interval $[a, a + h]$?" The limit of the difference quotient as h approaches zero (if the limit exists) is the **instantaneous rate of change in $f(x)$ at $x = a$**. Note that even though the difference quotient is undefined when $h = 0$, the limit may still exist.

EXAMPLE 3 Calculating an Instantaneous Rate of Change

Let $f(x) = x^2$. What is the instantaneous rate of change of $f(x)$ when $x = 3$?

SOLUTION

We can calculate the instantaneous rate of change by taking the limit of the difference quotient as h approaches zero. The instantaneous rate of change of $f(x)$ at $x = 3$ is given by

$$\lim_{h \to 0} \frac{f(3 + h) - f(3)}{h}$$

$$= \lim_{h \to 0} \frac{(3 + h)^2 - (3)^2}{h} \quad \text{Since } f(x) = x^2$$

$$= \lim_{h \to 0} \frac{(9 + 6h + h^2) - 9}{h}$$

$$= \lim_{h \to 0} \frac{6h + h^2}{h}$$

$$= \lim_{h \to 0} \frac{h(6 + h)}{h}$$

$$= \lim_{h \to 0} (6 + h) \text{ for } h \neq 0 \quad \text{Since } \frac{h}{h} = 1 \text{ for } h \neq 0$$

As h nears zero, what happens to the value of $6 + h$? Let's pick values of h near zero (Table 3.6).

Table 3.6

h	$6 + h, h \neq 0$
-0.100	5.900
-0.010	5.990
-0.001	5.999
0.000	Undefined
0.001	6.001
0.010	6.010
0.100	6.100

As seen from the table, even though the difference quotient is undefined when $h = 0$, the value of the simplified difference quotient, $6 + h$, gets close to 6 as h approaches zero. In fact, by picking sufficiently small values of h, we can get as close to 6 as we would like. So the instantaneous rate of change of $f(x)$ when $x = 3$ is 6.

Observe that we can attain the same result by plugging in $h = 0$ after canceling out the h in the denominator of the difference quotient. That is,

$$= \lim_{h \to 0} (6 + h) = 6 + 0$$

$$= 6$$

Throughout the rest of this chapter, we will substitute in $h = 0$ after eliminating the h in the denominator of the difference quotient. This process will simplify our computations while still giving the correct result.

The limit of the difference quotient as h approaches zero is used widely throughout calculus and is called the **derivative**.

The Derivative of a Function at a Point

The **derivative** of a function $y = f(x)$ at a point $(a, f(a))$ is

$$f'(a) = \lim_{h \to 0} \frac{f(a + h) - f(a)}{h}$$

$f'(a)$ is read "f prime of a" and is the *instantaneous rate of change* of the function f at the point $(a, f(a))$.

EXAMPLE 4 Calculating the Derivative of a Function at a Point

➡ Given $f(x) = 3x + 1$, find $f'(2)$.

SOLUTION

$$f'(2) = \lim_{h \to 0} \frac{f(2 + h) - f(2)}{h}$$

$$= \lim_{h \to 0} \frac{(3(2 + h) + 1) - (3(2) + 1)}{h} \quad \text{Since } f(2 + h) = (3(2 + h) + 1) \text{ and } f(2) = 3(2) + 1$$

$$= \lim_{h \to 0} \frac{(6 + 3h + 1) - (7)}{h}$$

$$= \lim_{h \to 0} \frac{(3h + 7) - (7)}{h}$$

$$= \lim_{h \to 0} \frac{3h}{h}$$

$$= \lim_{h \to 0} 3 \quad \text{Since } \frac{h}{h} = 1$$

$$= 3$$

In this case, the difference quotient turned out to be a constant value of 3, so taking the limit of the difference quotient as h approached zero did not alter the value of the difference quotient.

For linear functions, the slope of the line is the instantaneous rate of change of the function at any value of x. Consequently, the derivative of a linear function will always be a constant value that is equal to the slope of the line.

EXAMPLE 5 Finding and Interpreting the Meaning of the Derivative of a Function at a Point

➡ The population of Washington state may be modeled by the function

$$P(t) = 4.933(1.017)^t \text{ million people}$$

where t is the number of years since 1990. (**Source:** Modeled from www.census.gov data) Find and interpret the meaning of $P'(25)$.

SOLUTION

Since t is the number of years since 1990, $t = 25$ is the year in 2015. $P'(25)$ is the instantaneous rate of change in the population in 2015, given in millions of people per year.

$$P'(25) = \lim_{h \to 0} \frac{P(25 + h) - P(25)}{h}$$

$$= \lim_{h \to 0} \frac{(4.933(1.017)^{25+h}) - (4.933(1.017)^{25})}{h}$$

$$= \lim_{h \to 0} \frac{4.933((1.017)^{25+h} - (1.017)^{25})}{h} \quad \text{Factor out 4.933}$$

$$= \lim_{h \to 0} \frac{4.933((1.017)^{25}(1.017)^h - (1.017)^{25})}{h}$$

Since $(1.017)^{25+h} = (1.017)^{25}(1.017)^h$

$$= \lim_{h \to 0} \frac{4.933(1.017)^{25}((1.017)^h - 1)}{h} \quad \text{Factor out } (1.017)^{25}$$

$$= \lim_{h \to 0} \frac{7.519((1.017)^h - 1)}{h}$$

Unlike in Example 3, we are unable to eliminate the h in the denominator algebraically and calculate the exact value of $P'(25)$. Nevertheless, by picking a small value for h (say $h = 0.001$), we can estimate the instantaneous rate of change.

$$\frac{7.519((1.017)^{0.001} - 1)}{0.001} = 0.1267 \text{ million people per year}$$

$$= 126.7 \text{ thousand people per year}$$

$$\approx 127 \text{ thousand people per year}$$

According to the model, the population of Washington will be increasing by approximately 127,000 people per year in 2015.

Although we were unable to obtain the exact value of the derivative of the exponential function in Example 5, we will develop the theory in later sections that will allow us to calculate the exact value of the derivative of an exponential function.

3.2 Exercises

In Exercises 1–5, use the difference quotient $\frac{f(a + h) - f(a)}{h}$ (with $h = 0.1, h = 0.01$, and $h = 0.001$) to estimate the instantaneous rate of change of the function at the given input value.

1. $f(x) = x^2; x = 2$

2. $s(t) = -16t^2 + 64; t = 2$

3. $w(t) = 4t + 2; t = 5$

4. $P(t) = 5; t = 25$

5. $P(r) = 500(1 + r)^2; r = 0.07$

In Exercises 6–10, use the derivative to calculate the instantaneous rate of change of the function at the given input value. (In each exercise, you can eliminate the h algebraically.) Compare your answers to the solutions of Exercises 1–5.

6. $f(x) = x^2; x = 2$

7. $s(t) = -16t^2 + 64; t = 2$

8. $w(t) = 4t + 2; t = 5$

9. $P(t) = 5; t = 25$

10. $P(r) = 500(1 + r)^2; r = 0.07$

In Exercises 11–15, use the difference quotient (with $h = 0.1$, $h = 0.01$, and $h = 0.001$) to estimate the instantaneous rate of change of the function at the given input value. You may find it helpful to apply the techniques on the Chapter 3 Tech Card.

11. $f(x) = 2x^{-3}; x = 3$

12. $P(t) = 230(0.9)^t; t = 25$

13. $P(r) = 500(1 + r)^{10}; r = 0.07$

14. $y = \ln(x); x = 2$

15. $g(x) = e^{3x}; x = 1$

In Exercises 16–20, determine the instantaneous rate of change of the function at the indicated input value. (You may find it helpful to apply the techniques on the Chapter 3 Tech Card.) Then explain the real-life meaning of the result.

16. *Yogurt Production* Based on data from 1997–2005, the amount of yogurt produced in the United States annually may be modeled by

    $y(x) = 14.99x^2 + 62.14x + 1555$ million pounds

    where x is the number of years since 1997. (**Source:** Modeled from *Statistical Abstract of the United States, 2007*, Table 846) Find and interpret the meaning of $S'(10)$.

17. *Heart Disease Death Rate* Based on data from 1980–2003, the age-adjusted death rate due to heart disease may be modeled by

    $R(t) = -7.597t + 407.4$ deaths per 100,000 people

where t is the number of years since 1980. (**Source:** *Statistical Abstract of the United States, 2006*, Table 106) Find and interpret the meaning of $D'(55)$.

18. *DVD Player Sales* Based on data from 1997–2004, the number of DVD players sold may be modeled by

    $D(p) = 77,100(0.989)^p$ thousand DVD players

    where p is the average price per DVD player in dollars. (**Source:** Modeled from Consumer Electronics Association data) Find and interpret the meaning of $D'(100)$.

19. *Theme Park Tickets* Based on 2007 ticket prices, the cost of a child's Disney Park Hopper Bonus Ticket may be modeled by

    $T(d) = -3.214d^2 + 41.19d + 34.2$ dollars

    where d is the number of days that the ticket authorizes entrance into Disneyland and Disney California Adventure. (**Source:** www.disneyland .com) Find and interpret the meaning of $T'(4)$.

20. *United States Populations* Based on data from 1995–2005, the population of the United States may be modeled by

    $U(t) = 298,213(1.009)^t$ thousand people

    where t is the number of years since 2005. (**Source:** *World Health Statistics 2006*, World Health Organization) Find and interpret the meaning of $U'(10)$.

Exercises 21 and 22 deal with the velocity of a free-falling object on earth. The vertical position of a free-falling object may be modeled by $s(t) = -16t^2 + v_0 t + s_0$ feet, where v_0 is the velocity of the object and s_0 is the vertical position of the object at time $t = 0$ seconds.

21. *Velocity of a Dropped Object* A can of soda is dropped from a diving board 40 feet above the bottom of an empty pool. How fast is the can traveling when it reaches the bottom of the pool?

22. *Velocity of a Ball* A small rubber ball is thrown into the air by a child at a velocity of 20 feet per second. The child releases the ball 4 feet above the ground. What is the velocity of the ball after 1 second?

3.3 The Derivative as a Slope: Graphical Methods

Based on data from 1990–2003, the amount of money spent on prescription drugs (per capita) may be modeled by

$$P(t) = 2.889t^2 - 2.613t + 158.7 \text{ dollars}$$

where t is the number of years since 1990. (**Source:** *Statistical Abstract of the United States, 2006,* Table 121) To help project future drug costs, an insurance company wants to know what the average annual increase in the amount of money spent on prescription drugs was from 1990 to 2010 and at what rate the drug spending will be increasing at the end of 2010. We will answer these questions by calculating the average and instantaneous rates of change in the prescription drug spending.

In Sections 3.1 and 3.2, you learned how to calculate the average rate of change of a function over an interval and the instantaneous rate of change of a function at a point. In this section, we will revisit these concepts from a graphical standpoint. We will also demonstrate how to use tangent-line approximations to estimate the value of a function.

As shown in Section 3.1, the difference quotient formula, $\dfrac{f(a + h) - f(a)}{h}$, gives the average rate of change in the value of the function between the points $(a, f(a))$ and $(a + h, f(a + h))$. If we let $a = c - h$, then the difference quotient formula becomes $\dfrac{f(c) - f(c - h)}{h}$. We will use this modified form of the difference quotient in our exploration of prescription drug spending, since we will be approaching $t = c$ through values to the left of

$t = c$. Recall from Section 3.1 that, graphically speaking, the difference quotient is the **slope of the secant line** connecting the two points on the graph of a function. According to the prescription drug spending model, $C(0) = 158.7$ and $C(20) = 1262$. We're interested in the slope of the secant line between $(0, 158.7)$ and $(20, 1262)$. The slope of this line represents the average rate of change in the prescription drug spending between 1990 and 2010 (Figure 3.6).

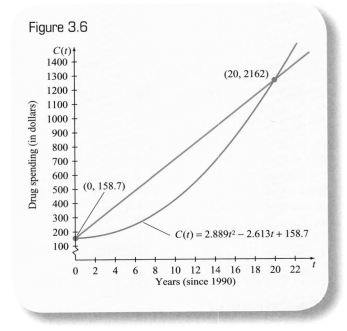

Figure 3.6

The prescription drug spending graph secant line between $(0, 158.7)$ and $(20, 1262)$ has the slope

$$m = \frac{C(20) - C(0) \text{ dollars}}{20 \text{ years}}$$

$$= \frac{1262 - 158.7 \text{ dollars}}{20 \text{ years}}$$

$$= 55.17 \text{ dollars per year}$$

Between 1990 and 2010, the per capita spending on prescription drugs increased by $55.17 per year. Does this mean that from 2010 to 2011, the spending will increase by about $55.17 ? No. Looking at the graph of the model, we notice that the prescription drug spending is rising at an increasing rate as time progresses (the steeper the graph, the greater the magnitude of the rate of change). We can approximate the instantaneous rate of change at the end of 2010 ($t = 20$) by calculating the slope of a secant line through $(20, 1262)$ and a "nearby" point as shown in Figure 3.7.

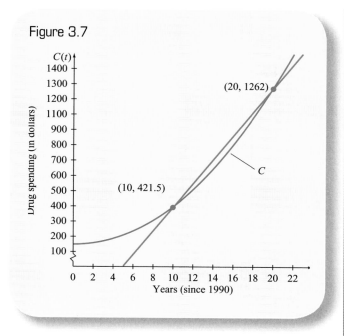

Figure 3.7

The prescription drug spending graph secant line between $(10, 421.5)$ and $(20, 1262)$ has the slope

$$m = \frac{C(20) - C(10) \text{ dollars}}{10 \text{ years}}$$

$$= \frac{1262 - 421.5 \text{ dollars}}{10 \text{ years}}$$

$$= 84.05 \text{ dollars per year}$$

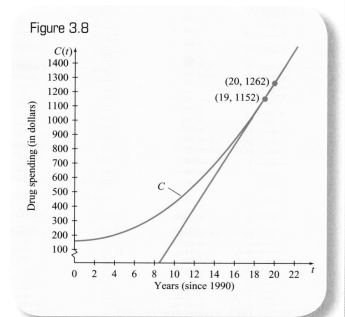

Figure 3.8

As shown in Figure 3.8, the prescription drug spending graph secant line between $(19, 1152)$ and $(20, 1262)$ has the slope

$$m = \frac{C(20) - C(19) \text{ dollars}}{20 - 19 \text{ years}}$$

$$= \frac{1262 - 1152 \text{ dollars}}{1 \text{ year}}$$

$$= 110.0 \text{ dollars per year}$$

By zooming in (Figure 3.9), we can see that between $t = 19$ and $t = 20$, the secant line and the graph of the function are extremely close together.

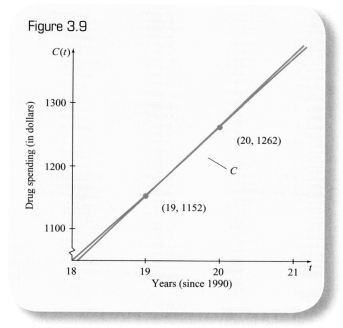

Figure 3.9

The prescription drug spending graph secant line between $(19.9, 1251)$ and $(920, 1262)$ has the slope

$$m = \frac{C(20) - C(19.9) \text{ dollars}}{20 - 19.9 \text{ years}}$$

$$= \frac{1262 - 1251 \text{ dollars}}{0.1 \text{ years}}$$

$$= 110.0 \text{ dollars per year}$$

Through $(19.9, 1251)$ and $(20, 1262)$, the slope of the secant line is $110.0 per year.

Again zooming in, we see that the secant line and the graph of the function are nearly identical between $t = 19.9$ and $t = 20$ (Figure 3.10). In fact, we are unable to visually distinguish between the two.

Figure 3.10

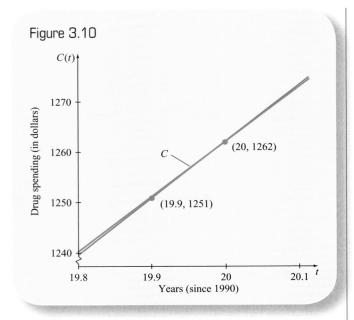

Figure 3.11

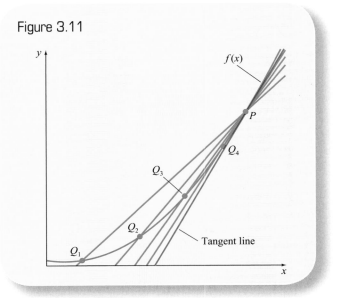

As t gets closer and closer to 20, the slope of the secant line will approach $C'(20)$. This occurs because $\frac{C(20) - C(20 - h)}{h} \approx C'(20)$ for small values of h.

Since h is the horizontal distance between the points, h becomes increasingly small as t nears 20. Using the methods covered in Section 3.2, we determine algebraically that $C'(20) = 112.9$. At the end of 2010, per capita prescription drug spending was increasing at a rate of roughly \$113 per year. In other words, we anticipate that per capita prescription drug spending increased by *approximately* \$113 between 2010 and 2011.

As the preceding example demonstrates, to find the instantaneous rate of change of a function $y = f(x)$ at a point $P = (a, f(a))$, we can find the limit of the slope of the secant line through P and a nearby point Q as the point Q gets closer and closer to P. Graphically speaking, we select values $Q_1, Q_2, Q_3, \ldots$, with each consecutive value of Q_i being a point on the curve that is closer to the point P than the one before it. The limit of the slope of the secant lines (imagine the points P and Q finally coinciding) is the line *tangent* to the curve at the point $P = (a, f(a))$ (Figure 3.11).

The **tangent line** to a graph f at a point $(a, f(a))$ is the line that passes through $(a, f(a))$ and has slope $f'(a)$.

The Graphical Meaning of the Derivative

The derivative of a function f at a point $(a, f(a))$ is the **slope of the tangent line** to the graph of f at that point. The slope of the tangent line at a point is also referred to as the **slope of the curve** at that point.

EXAMPLE 1 Finding the Equation of a Tangent Line

Find the equation of the tangent line to the graph of $f(x) = x^2$ that passes through $(2, 4)$. Then graph the tangent line and the graph of f.

SOLUTION

The slope of the tangent line is $f'(2)$.

$$f'(2) = \lim_{h \to 0} \frac{f(2 + h) - f(2)}{h}$$

$$= \lim_{h \to 0} \frac{(2 + h)^2 - 2^2}{h} \qquad \text{Since } f(x) = x^2$$

$$= \lim_{h \to 0} \frac{4 + 4h + h^2 - 4}{h}$$

$$= \lim_{h \to 0} \frac{4h + h^2}{h}$$

$$= \lim_{h \to 0} \frac{h(4 + h)}{h}$$

$$= \lim_{h \to 0}(4 + h)$$
$$= 4 + 0$$
$$= 4$$

The slope of the tangent line is 4 at the point $(2, 4)$. Using the slope-intercept form of a line, we have $y = 4x + b$. Substituting in the point $(2, 4)$, we get

$$4 = 4(2) + b$$
$$4 = 8 + b$$
$$b = -4$$

The equation of the tangent line is $y = 4x - 4$.

We generate a table of values for $y = 4x - 4$ and $f(x) = x^2$. Then we graph the results (Figure 3.12).

Figure 3.12

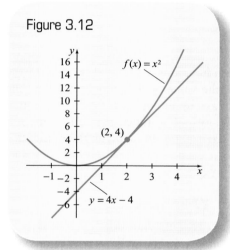

x	y	f(x)
−1	−8	1
0	−4	0
1	0	1
2	4	4
3	8	9
4	12	16

Tangent-Line Approximations

In Example 1 you may have noticed that the tangent line lies very near to the graph of f for values of x near a. In fact, if we zoom in to the region immediately surrounding $(a, f(a))$, the graph of f and the tangent line to the graph of f at $(a, f(a))$ appear nearly identical. For values of x near 2, the tangent-line y-value is a good approximation of the actual function value (see Figure 3.13).

Because it is frequently easier to calculate the values of the tangent line than the values of the function, sometimes the tangent line is used to estimate the value of the function. For example, suppose we wanted to estimate $f(1.9)$ given $f(x) = x^2$. Since $x = 1.9$ is near $x = 2$, we may use the equation of the tangent line, $y = 4x - 4$, to estimate $f(1.9)$. That is, $f(1.9) \approx 4(1.9) - 4 = 3.6$. The actual value is

$$f(1.9) = (1.9)^2$$
$$= 3.61$$

Figure 3.13

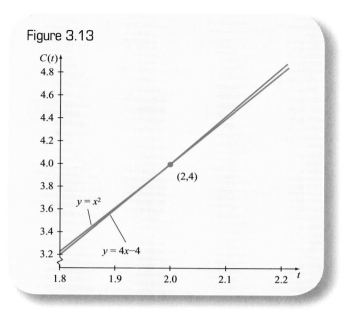

Our tangent-line estimate $(y = 3.6)$ was remarkably close to the actual value of the function $(f(1.9) = 3.61)$.

EXAMPLE 2 Using a Tangent-Line Approximation for Apple iPod Net Sales

Based on data from 2005 to 2009, Apple iPod net sales may be modeled by

$$P(t) = -0.58t^2 + 3.2t + 4.7 \text{ billion dollars}$$

where t is the number of years since the end of 2005. (**Source:** Model based on data from Apple Computer Corporation) According to the model, iPod net sales were \$8.22 billion in 2009.

Use the model to predict how quickly iPod net sales were changing at the end of 2009. Then use a tangent line to estimate iPod net sales at the end of 2010. Compare the estimate to the actual value predicted by the model.

SOLUTION

Since $t = 4$ corresponds with the year 2009, the instantaneous rate of change in iPod net sales at the end of 2009 is given by

$$P'(4) = \lim_{h \to 0} \frac{P(4 + h) - P(4) \text{ million dollars}}{h \text{ years}}$$

We already know that $P(4) = 8.22$. We'll calculate $P(4 + h)$ and then substitute the simplified value into the derivative formula.

$$P(4 + h) = -0.58(4 + h)^2 + 3.2(4 + h) + 4.7$$
$$= -0.58(16 + 8h + h^2) + 12.8 + 3.2h + 4.7$$
$$= -9.28 - 4.64h - 0.58h^2 + 3.2h + 17.5$$
$$= -0.58h^2 - 1.44h + 8.22$$

$$P'(4) = \lim_{h \to 0} \frac{P(4 + h) - P(4)}{h}$$
$$= \lim_{h \to 0} \frac{(-0.58h^2 - 1.44h + 8.22) - 8.22}{h}$$
$$= \lim_{h \to 0} \frac{-0.58h^2 - 1.44h}{h}$$
$$= \lim_{h \to 0} \frac{h(-0.58h - 1.44)}{h}$$
$$= \lim_{h \to 0} (-0.58h - 1.44)$$
$$= -0.58(0) - 1.44$$
$$= -1.44 \ \frac{\text{billion dollars}}{\text{year}}$$

At the end of 2009, iPod net sales were decreasing at a rate of $1.44 billion per year. That is, between 2009 and 2010, the iPod net sales were expected to decrease by *approximately* $1.44 billion.

The point-slope form of the tangent line is $y - y_1 = -1.44(x - x_1)$. Using the point $(4, 8.22)$, we determine

$$y - 8.22 = -1.44(x - 4)$$

This is the tangent-line equation. At $t = 5$, we have

$$y - 8.22 = -1.44(5 - 4)$$
$$y = 8.22 - 1.44$$
$$= 6.78$$

Using the tangent-line equation, we estimate that iPod net sales are about $6.8 billion in 2010. According to the model, the actual number of iPod net sales was somewhat less.

$$P(5) = -0.58(5)^2 + 3.2(5) + 4.7$$
$$= 6.2$$

The model indicates that iPod net sales were $6.2 billion in 2010.

Why was there a discrepancy between the two estimates in Example 2? Look at the graph of the model and the tangent line (Figure 3.14).

Figure 3.14

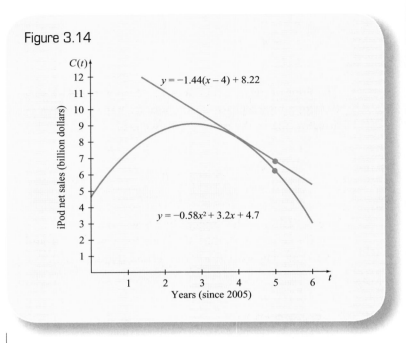

Although both functions were equal at $t = 4$, by the time t reached 5, the value of the model fell below the tangent-line estimate by about $0.6 billion. Although tangent-line estimates of a function's value are not exact, they are often good enough for their intended purpose.

Numerical Derivatives

Often we encounter real-life data in tables or charts. Is it possible to calculate a derivative from a table of data? We'll investigate this question by looking at a table of data for $f(x) = x^2$ (see Table 3.7). For this function, $f'(2) = 4$.

Table 3.7

x	f(x)
0	0
1	1
2	4
3	9
4	16

We can estimate $f'(2)$ by calculating the slope of the secant line through points whose x-values are equidistant from $x = 2$. That is,

$$f'(2) \approx \frac{f(3) - f(1)}{3 - 1}$$

$$= \frac{9 - 1}{2}$$

$$= \frac{8}{2}$$

$$= 4$$

In this case, our estimate was equal to the tangent-line slope. Let's look at the situation graphically (see Figure 3.15).

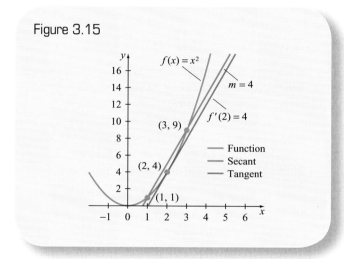

Figure 3.15

When estimating a derivative numerically, we typically select the two closest data points that are horizontally equidistant from our point of interest. Doing so often yields a line that is parallel or near-parallel to the tangent line. Picking two points that are equidistant from the point of interest will tend to give the best estimate of the derivative and can be used as long as the point of interest is not an endpoint. (If the point of interest is an endpoint, we find the slope of the secant line between the endpoint and the next closest point.) If we assume that each output in a table of data represents $f(a)$ for a corresponding input a, then we can symbolically represent the process of numerically estimating a derivative as follows.

If a is the largest domain value in the data set, then

$$f'(a) \approx \frac{f(a) - f(a - h)}{(a) - (a - h)}$$

$$\approx \frac{f(a) - f(a - h)}{h}$$

If a is the smallest domain value in the data set, then

$$f'(a) \approx \frac{f(a + h) - f(a)}{(a + h) - (a)}$$

$$\approx \frac{f(a + h) - f(a)}{h}$$

In both cases, h is the distance between a and the next closest domain value.

Numerical Estimate of the Derivative

The derivative of a function f at a point $(a, f(a))$ may be approximated from a table by

$$f'(a) \approx \frac{f(a + h) - f(a - h)}{(a + h) - (a - h)}$$

$$\approx \frac{f(a + h) - f(a - h)}{2h}$$

where h is the horizontal distance between a and $a + h$.

EXAMPLE 3 Estimating the Derivative from a Table of Data

Use Table 3.8 to estimate how quickly home sales prices in the southern United States were increasing at the end of 2006. (That is, estimate the slope of the tangent line at $(6, 208.2)$.)

Table 3.8

Median Sales Price of a New One-Family House in the Southern United States	
Years (since 2000) (t)	Price (thousands of dollars) (P)
0	148.0
2	163.4
4	181.1
6	208.2
8	203.7

Source: *Statistical Abstract of the United States, 2010*, Table 940

SOLUTION

We'll estimate the slope of the tangent line at $(6, 208.2)$ by calculating the slope of the secant line between $(4, 181.1)$ and $(8, 203.7)$.

$$f'(6) \approx \frac{f(8) - f(4)}{8 - 4} \frac{\text{thousand dollars}}{\text{years}}$$

$$\approx \frac{203.7 - 181.1}{4}$$

$$\approx 5.65 \text{ thousand dollars per year}$$

We estimate that home prices in the southern United States were increasing at a rate of about 5.7 thousand dollars per year at the end of 2006. (Since the original data were accurate to one decimal place, we rounded our final result to one decimal place.)

....

3.3 Exercises

In Exercises 1–10, determine the equation of the tangent line of the function at the given point. Then graph the tangent line and the function together.

1. $f(x) = x^2 - 4x$; $(1, -3)$

2. $f(x) = -x^2 + 6$; $(2, 2)$

3. $g(x) = x^2 + 2x + 1$; $(0, 1)$

4. $g(x) = x^2 - 4$; $(3, 5)$

5. $g(x) = x^2 - 4x - 5$; $(4, -5)$

6. $h(x) = x^3$; $(2, 8)$

7. $h(x) = x^3$; $(0, 0)$

8. $h(x) = x^3$; $(1, 1)$

9. $f(x) = (x - 3)^2$; $(3, 0)$

10. $f(x) = (x + 2)^2$; $(-1, 1)$

In Exercises 11–15, answer the questions by calculating the slope of the tangent line and the tangent-line equation, as appropriate.

11. *Median Price of a New Home* Based on data from 2003 to 2008, the median price of a home in the western region of the United States may be modeled by

$$W(t) = -9.2t^2 + 55t + 253 \text{ thousand dollars}$$

where *t* is the number of years since 2003. (**Source:** Modeled from *Statistical Abstract of the United States, 2010,* Table 940)

How quickly was the median sales price changing in 2009? What was the estimated median sales price in 2010? (Use a tangent-line approximation.)

12. *New One-Family Home Size* Based on data from 2000 to 2008, the average number of square feet in new one-family homes may be modeled by

$$H(t) = -0.0023t^2 + 3447t + 2265 \text{ square feet}$$

where *t* is the number of years after 2000. (**Source:** Modeled from *Statistical Abstract of the United States, 2010,* Table 936)

At what rate was home size changing in 2008? What was the estimated size of a new home in 2009? (Use a tangent-line approximation.)

13. *Student-to-Teacher Ratio* Based on data from 1995 to 2007, the student-to-teacher ratio at private elementary and secondary schools may be modeled by

$$R(t) = 0.00639t^2 - 0.284t + 15.7 \text{ students per teacher}$$

where *t* is the number of years since 1995. (**Source:** Modeled from *Statistical Abstract of the United States, 2010,* Table 245)

According to the model, how quickly was the student-to-teacher ratio changing in 2008? What was the estimated student-to-teacher ratio for 2009? (Use a tangent-line approximation.)

14. *Cassette Tape Shipment Value* Based on data from 1990–2004, the value of music cassette tapes shipped may be modeled by

$$V(s) = -0.00393s^2 + 9.74s - 63.0 \text{ million dollars}$$

where *s* represents the number of music cassette tapes shipped (in millions). (**Source:** Modeled from *Statistical Abstract of the United States, 2006,* Table 1131)

According to the model, at what rate is the value of the cassette tapes shipped changing when the number of cassette tapes shipped is 250 million? What was the estimated shipment value when 251 million cassette tapes were shipped? (Use a tangent-line approximation.)

15. *College Attendance* Based on data from 2000–2002 and Census Bureau projections for 2003–2013, *private* college enrollment may be modeled by

$$P(x) = 0.340x - 457 \text{ thousand students}$$

where *x* is the number of students (in thousands) enrolled in *public* colleges. (**Source:** Modeled from *Statistical Abstract of the United States, 2006,* Table 204)

Calculate the rate of change in *private* college enrollment when 12,752 thousand *public* college students are enrolled. Use a tangent-line approximation to estimate the number of students enrolled in *private* colleges when there are 12,753 thousand *public* college students.

In Exercises 16–20, estimate the specified derivative by using the data in the table. Then interpret the result.

16. *Worker Wages*

Average Annual Salary per Worker: Motion Picture and Sound Recording Industries

Years (since 2000) t	Annual Salary (dollars) $W(t)$
0	55,355
1	54,776
2	54,877
3	55,991
4	60,424
5	62,051
6	65,764
7	67,055

Source: *Statistical Abstract of the United States, 2010,* Table 628

Estimate $W'(6)$.

17. *Ice Cream Prices*

Ice Cream Retail Price

Years (since 2004) t	Price of 1/2 Gallon of Ice Cream $I(t)$
0	3.85
1	3.69
2	3.90
3	4.08
4	4.28

Source: *Statistical Abstract of the United States, 2010,* Table 717

Estimate $I'(3)$.

18. *Undergraduate Tuition at the University of Pittsburgh*

Full-Time Resident Tuition and Fees

Years (since 2004–2005) t	Annual Tuition and Fees (dollars) $f(t)$
0	10,130
1	10,736
2	11,368
3	12,106
4	12,832
5	13,344

Source: University of Pittsburgh

Estimate $f'(4)$ and $f'(5)$. What can you conclude about the rate of increase in tuition and fees at the University of Pittsburgh?

19. *Steak Prices*

Boneless Sirloin Steak Prices

Years (since 2004) t	Price per Pound (dollars) $S(t)$
0	6.09
1	5.93
2	5.79
3	5.91
4	6.07

Source: *Statistical Abstract of the United States, 2010,* Table 717

Estimate $S'(3)$.

20. *Bread Prices*

Whole Wheat Bread Prices

Years (since 2004) t	Price per Loaf (dollars) $B(t)$
0	1.30
1	1.29
2	1.62
3	1.81
4	1.95

Source: *Statistical Abstract of the United States, 2010,* Table 717

Estimate $B'(2)$.

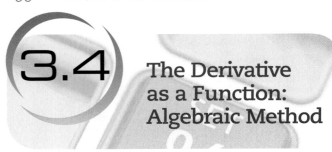

3.4 The Derivative as a Function: Algebraic Method

Based on data from 2000 to 2006, the cumulative number of homicides resulting from a romantic triangle may be modeled by $R(t) = -1.57t^2 + 119t + 125$ homicides between the start of 2000 and the end of year t, where t is the number of years since 2000. (**Source:** Modeled from *Crime in the United States 2006*, Uniform Crime Report, FBI) At what rate was the cumulative number of homicides increasing at the end of 2004, 2005, and 2006? Although we could calculate the derivative at $t = 4$, $t = 5$, and $t = 6$, we can save time by finding the derivative function and then substituting in the various values of t.

In this section, we will introduce the derivative function and show how to find it algebraically. Your skill in finding the derivative at a point will prove to be especially useful in this section. We'll begin with a simple example before returning to the romantic triangle problem.

Shutterstock

EXAMPLE 1 Calculating the Derivative of a Function at Multiple Points

→ Determine the instantaneous rate of change of $f(x) = 3x^2$ at $(1, 3)$, $(3, 27)$, and $(10, 300)$.

SOLUTION

Although we could calculate $f'(1), f'(3)$, and $f'(10)$ individually, it will be more efficient to find the derivative function itself and then substitute in the different values of x.

We begin with the derivative formula; however, instead of substituting a specific value for a, we replace a with the variable x.

$$
\begin{aligned}
f'(x) &= \lim_{h \to 0} \frac{f(x + h) - f(x)}{h} \\
&= \lim_{h \to 0} \frac{(3(x + h)^2) - (3x^2)}{h} \quad \text{Since } f(x) = 3x^2 \\
&= \lim_{h \to 0} \frac{(3(x^2 + 2hx + h^2)) - (3x^2)}{h} \\
&= \lim_{h \to 0} \frac{(3x^2 + 6hx + 3h^2) - (3x^2)}{h} \\
&= \lim_{h \to 0} \frac{6hx + 3h^2}{h} \\
&= \lim_{h \to 0} \frac{h(6x + 3h)}{h} \\
&= \lim_{h \to 0} (6x + 3h) \\
&= 6x + 3(0) \\
&= 6x
\end{aligned}
$$

The result $f'(x) = 6x$ is the derivative function for $f(x) = 3x^2$. It can be used to calculate the instantaneous rate of change of f at any point $(a, f(a))$.

$$
\begin{aligned}
f'(1) &= 6(1) \\
&= 6
\end{aligned}
$$

The instantaneous rate of change of f at $(1, 3)$ is 6.

$$
\begin{aligned}
f'(3) &= 6(3) \\
&= 18
\end{aligned}
$$

The instantaneous rate of change of f at $(3, 27)$ is 18.

$$
\begin{aligned}
f'(10) &= 6(10) \\
&= 60
\end{aligned}
$$

The instantaneous rate of change of f at $(10, 300)$ is 60.

As demonstrated in Example 1, the techniques used to find the derivative function are virtually identical to the procedures used to find the derivative at a point. However, knowing the derivative function allows us to calculate the derivative at a number of different points more quickly than calculating the derivative at each point separately. The derivative function for a function f is called the **derivative of f**.

The Derivative Function

The **derivative function** of a function f is given by

$$f'(x) = \lim_{h \to 0} \frac{f(x + h) - f(x)}{h}$$

if the limit exists.

EXAMPLE 2 Calculating the Instantaneous Rate of Change of a Function at Multiple Points

Based on data from 2000 to 2006, the cumulative number of homicides resulting from a romantic triangle between the start of 2000 and the end of year t may be modeled by

$$R(t) = -1.57t^2 + 119t + 125 \text{ homicides}$$

where t is the number of years since 2000. (**Source:** Modeled from *Crime in the United States 2006*, Uniform Crime Report, FBI) At what rate was the cumulative number of homicides increasing at the end of 2004, 2005, and 2006?

SOLUTION

We will begin by finding the derivative $R'(t)$.

$$R'(t) = \lim_{h \to 0} \frac{R(t + h) - R(t)}{h}$$

Because of the complex nature of $R(t)$, we will first calculate $R(t + h)$ and then substitute the result into the derivative formula.

$$R(t + h) = -1.57(t + h)^2 + 119(t + h) + 125$$

$$= -1.57t^2 (t^2 + 2ht + h^2) + 119t + 119h + 125$$

$$= -1.57t^2 - 3.14ht - 1.57h^2 + 119t + 119h + 125$$

We already know that $R(t) = -1.57t^2 + 119t + 125$.

Substituting both of these quantities into the derivative formula yields

$$R'(t) = \lim_{h \to 0} \frac{R(t + h) - R(t)}{h}$$

$$= \lim_{h \to 0} \frac{-3.14ht - 1.57h^2 + 119h}{h}$$

$$= \lim_{h \to 0} \frac{h(-3.14t - 1.57h + 119)}{h}$$

$$= \lim_{h \to 0} (-3.14t - 1.57h + 119) \quad \text{Since } \frac{h}{h} = 1 \text{ for } h \neq 0$$

$$= -3.14t - 1.57(0) + 119$$

$$= -3.14t + 119$$

So $R'(t) = -3.14t + 119$ homicides per year. We can now compute the instantaneous rate of change in the cumulative number of homicides in 2004, 2005, and 2006.

$$R'(4) = -3.14(4) + 119$$
$$= 106.4$$
$$\approx 106 \text{ homicides per year}$$

$$R'(5) = -3.14(5) + 119$$
$$= 103.3$$
$$\approx 103 \text{ homicides per year}$$

$$R'(6) = -3.14(6) + 119$$
$$= 100.2$$
$$\approx 100 \text{ homicides per year}$$

The cumulative number of homicides resulting from a romantic triangle was increasing at a rate of 106 homicides per year in 2004, 103 homicides per year in 2005, and 100 homicides per year in 2006. According to the model, although the cumulative number of homicides continued to increase, the rate at which these homicides were increasing slowed between the end of 2004 and the end of 2006.

EXAMPLE 3 Finding the Derivative of a Function

Find the derivative of $g(t) = 2t^3 - 4t + 3$.

SOLUTION

We must find $g'(t) = \lim_{h \to 0} \frac{g(t + h) - g(t)}{h}$. We'll first find $g(t + h)$ and then substitute the result into the derivative formula.

$$g(t + h) = 2(t + h)^3 - 4(t + h) + 3$$
$$= 2(t^3 + 3t^2h + 3th^2 + h^3) - 4t - 4h + 3$$
$$= 2t^3 + 6t^2h + 6th^2 + 2h^3 - 4t - 4h + 3$$
$$= g(t) + 6t^2h + 6th^2 + 2h^3 - 4h$$

$$g'(t) = \lim_{h \to 0} \frac{g(t + h) - g(t)}{h}$$
$$= \lim_{h \to 0} \frac{(6t^2h + 6th^2 + 2h^3 - 4h)}{h}$$
$$= \lim_{h \to 0} \frac{h(6t^2 + 6th + 2h^2 - 4)}{h}$$
$$= \lim_{h \to 0} (6t^2 + 6th + 2h^2 - 4)$$
$$= 6t^2 + 6t(0) + 2(0)^2 - 4$$
$$= 6t^2 - 4$$

The derivative of $g(t) = 2t^3 - 4t + 3$ is $g'(t) = 6t^2 - 4$.

Estimating Derivatives

For polynomial functions, all terms in the numerator of the derivative formula without an h will cancel out. This allows us always to eliminate the h in the denominator. However, with some other types of functions, the h in the denominator cannot be eliminated algebraically. In this case, we can estimate the derivative function by substituting in a small positive value (e.g., 0.001) for h. The closer the value of h is to zero, the more accurate the estimate of the derivative will be.

EXAMPLE 4 Calculating the Instantaneous Rate of Change of a Function at Multiple Points

The per capita consumption of bottled water in the United States may be modeled by

$$W = 2.593(1.106)^t \text{ gallons}$$

where t is the number of years since the end of 1980. (**Source:** Modeled from *Statistical Abstract of the United States, 2001,* Table 204)

Determine how quickly bottled water consumption was increasing at the end of 2002, 2004, and 2006.

SOLUTION

We must evaluate $W'(t)$ at $t = 22, 24,$ and 26.

$$W'(t) = \lim_{h \to 0} \frac{W(t + h) - W(t)}{h}$$
$$= \lim_{h \to 0} \frac{(2.593(1.106)^{t+h}) - (2.593(1.106)^t)}{h}$$
$$= \lim_{h \to 0} \frac{(2.593(1.106)^t(1.106)^h) - (2.593(1.106)^t)}{h} \quad \text{Since } (1.106)^{t+h} = (1.106)^t(1.106)^h$$
$$= \lim_{h \to 0} \frac{2.593(1.106)^t((1.106)^h - 1)}{h} \quad \text{Factor out } 2.593(1.106)^t$$
$$= 2.593(1.106)^t \cdot \lim_{h \to 0} \frac{((1.106)^h - 1)}{h}$$

We can move the expression $2.593(1.106)^t$ to the other side of the limit because it does not contain an h. Since

$$\lim_{h \to 0} \frac{((1.106)^h - 1)}{h} \approx \frac{((1.106)^{0.001} - 1)}{0.001}$$
$$\approx 0.1008$$

we have

$$W'(t) = 2.593(1.106)^t \cdot \lim_{h \to 0} \frac{((1.106)^h - 1)}{h}$$
$$\approx 2.593(1.106)^t \cdot (0.1008)$$
$$\approx 0.2614(1.106)^t$$

We will now evaluate the derivative function at $t = 22, 24,$ and 26.

$$W'(22) \approx 0.2614(1.106)^{22}$$
$$\approx 2.398$$

According to the model, bottled water consumption was increasing by 2.398 gallons per year at the end of 2002.

$$W'(24) \approx 0.2614(1.106)^{24}$$
$$\approx 2.934$$

Bottled water consumption was increasing by 2.934 gallons per year at the end of 2004.

$$W'(26) \approx 0.2614(1.106)^{26}$$
$$\approx 3.589$$

Bottled water consumption was increasing at a rate of 3.589 gallons per year at the end of 2006.

Shutterstock

3.4 Exercises

In Exercises 1–5, find the derivative of the function.

1. $f(x) = x^2 - 4x$

2. $g(x) = x^2 + 2x + 1$

3. $g(x) = x^2 - 4x - 5$

4. $j(x) = x^3 + 2$

5. $f(t) = (t - 3)^2$

In Exercises 6–10, find the slope of the tangent line of the function at $x = 1$, $x = 3$, and $x = 5$.

6. $g(x) = 2x^2 + x - 1$

7. $f(x) = x^2 - 2x$

8. $j(x) = -5$

9. $W(x) = -4x + 9$

10. $S(x) = 3x^2 - 2x + 1$

In Exercises 11–13, estimate the derivative of the function. When you are unable to eliminate the h in the denominator of the derivative formula algebraically, use $h = 0.001$.

11. $P(x) = 3^x$

12. $C(x) = -3 \cdot 4^x$

13. $R(x) = 5.042 \cdot (0.98)^x$

In Exercises 14–18, use the derivative function to answer the questions.

14. *Median Price of a New Home* Based on data from 2003 to 2008, the median price of a home in the western region of the United States may be modeled by

$$W(t) = -9.2t^2 + 55t + 253 \text{ thousand dollars}$$

where t is the number of years since 2003. (**Source:** Modeled from *Statistical Abstract of the United States, 2010,* Table 940)

According to the model, was the median sales price changing more quickly at the end of 2006 or the end of 2008?

15. *Median Price of a New Home* Based on data from 2003 to 2008, the median price of a home in the United States may be modeled by

$$U(t) = -5.1t^2 + 33t + 194 \text{ thousand dollars}$$

where t is the number of years since 2003. (**Source:** Modeled from *Statistical Abstract of the United States, 2010,* Table 940)

According to the model, was the median sales price changing more quickly at the end of 2007 or the end of 2009?

16. *Walmart Net Sales* Based on data from 1996–2006, the net sales of Walmart may be modeled by

$$s(t) = 0.8636t^2 + 14.39t + 84.72 \text{ billion dollars}$$

where t is the number of years since 1996. (**Source:** Modeled from Walmart Annual Report, 2006, pp.18–19)

According to the model, how much more rapidly were net sales increasing in 2006 compared to 2005?

17. *Prescription Drug Spending* Based on data from 1990–2003, per capita prescription drug spending may be modeled by

$$P(t) = 2.889t^2 - 2.613t + 158.7 \text{ dollars}$$

where t is the number of years since 1990. (**Source:** *Statistical Abstract of the United States, 2006,* Table 121)

In what year was prescription drug spending increasing twice as fast as it was increasing in 2000?

18. *Oil Production* Based on data from 1985–2004, the difference between U.S. oil field production and net oil imports may be modeled by

$$b(t) = 4.29t^2 - 278t + 2250 \text{ million barrels}$$

where t is the number of years since 1985. (**Source:** Modeled from *Statistical Abstract of the United States, 2007,* Table 881)

At what rate was the difference in U.S. oil field production and net oil imports changing in 2000 and 2005, according to the model?

Exercises 19 and 20 are intended to challenge your understanding.

19. Given $f'(x) = 3x$ and $g(x) = x^2 + 3 + f(x)$, find $g'(x)$.

20. Given $f(x) = x^3 - 3x$, determine where $f'(x) = 0$.

3.5 Interpreting the Derivative

Many of us feel inundated by the advertisements we are sent through the mail. Don't expect this to let up anytime soon: Spending on direct-mail advertising has risen every year since 1990. The amount of money spent on direct-mail advertising may be modeled by

$$A(t) = 70.54t^2 + 1488t + 22{,}828$$
million dollars

where t is the number of years since 1990. (**Source:** Modeled from *Statistical Abstract of the United States, 2001*, Table 1272) According to the model, $A(9) = 41{,}934$ and $A'(9) = 2{,}758$. But what does this mean? In this section, we will discuss how to interpret the meaning of a derivative in the context of a real-life problem.

Recall that the units of the derivative are the units of the output divided by the units of the input.

In this case, the units of A' are $\dfrac{\text{millions of dollars}}{\text{year}}$ or millions of dollars per year. Note that $t = 9$ corresponds to the year 1999. We conclude that in 1999, 41,934 million dollars were spent on direct-mail advertising, and spending was increasing by 2,758 million dollars per year. In other words, according to the model, 41,934 million dollars were spent on direct-mail advertising in 1999 and spending increased by *about*

2,758 million dollars between 1999 and 2000. The terms *about*, *approximately*, or *roughly* must be used when using the second interpretation of the derivative, since we are using a tangent-line approximation to estimate the increase over the next year. (For a graphical discussion of tangent-line approximations, refer to Section 3.3.)

Interpreting the Derivative

Let $f(x)$ be a function. The meaning of $f'(a) = c$ may be written in either of the following two ways:

- When $x = a$, the value of the function f is increasing (decreasing) at a rate of c units of output per unit of input.

- The value of the function f will increase (decrease) by *about* c units of output between a units of input and $a + 1$ units of input.

EXAMPLE 1 Interpreting the Meaning of the Derivative

Based on data from 1940 to 2000, the average monthly Social Security benefit for men may be modeled by

$$P(t) = 0.3486t^2 - 5.505t + 35.13 \text{ dollars}$$

where t is the number of years since 1940. (**Source:** Modeled from Social Security Administration data)

Interpret the meaning of $P(60) = 959.79$ and $P'(60) = 36.33$.

SOLUTION

Since t is the number of years since 1940, $t = 60$ corresponds to 2000. $P(60) = 959.79$ means that (according to the model) the average Social Security benefit for men in 2000 was $959.79.

The units of the derivative are $\dfrac{\text{dollars}}{\text{year}}$. $P'(60) = 36.33$ means that in 2000, the average Social Security benefit for men was increasing by $36.33 per year. In other words, the average Social Security benefit for men was expected to increase by *about* $36.33 between 2000 and 2001.

Shutterstock

EXAMPLE 2 Using a Tangent-Line Approximation to Estimate a Function Value

The average weight of a boy between 2 and 13 years of age may be modeled by

$$W(a) = 0.215a^2 + 2.993a + 23.78 \text{ pounds}$$

where a is the age of the boy in years. (**Source:** Modeled from www.babybag.com data)

Interpret the meaning of $W(10) = 75.21$ and $W'(10) = 7.29$. Then use a tangent-line approximation to estimate $W(11)$.

SOLUTION

Since a is the age of the boy, $a = 10$ corresponds to a 10-year-old boy. $W(10) = 75.21$ means that the average weight of a 10-year-old boy is 75.21 pounds.

The units of the derivative are $\dfrac{\text{pounds}}{\text{year of his age}}$. $W'(10) = 7.29$ means that the average weight of a 10-year-old boy is increasing by 7.29 pounds per year of his age. In other words, the average weight of a boy will increase by *about* 7.29 pounds between his 10th and 11th years.

We use a tangent-line approximation to estimate $W(11)$.

$$W(11) \approx W(10) + W'(10)$$
$$\approx 75.21 + 7.29$$
$$\approx 82.5$$

We estimate that the average weight of an 11-year-old boy is 82.5 pounds.

EXAMPLE 3 Interpreting the Meaning of the Derivative

Based on data from 1992–2005, the number of television sets in U.S. homes may be modeled by

$$T(x) = 0.06680x^3 - 1.276x^2 + 12.46x + 190.7 \text{ million sets}$$

where x is the number of years since 1992. (**Source:** *Statistical Abstract of the United States, 2007,* Table 1111)

Interpret the meaning of $T(12) = 271.9$ and $T'(12) = 10.7$.

SOLUTION

Since t is the number of years, $t = 12$ corresponds with 2004. $T(12) = 271.9$ means that in 2004, there were 271.9 million television sets in U.S. homes.

The units of the derivative are $\dfrac{\text{television sets}}{\text{year}}$.

$T'(12) = 10.7$ means that in 2004, the number of television sets in U.S. homes was increasing at a rate of 10.7 million sets per year. In other words, the number of television sets in homes was expected to increase by about 10.7 million sets between 2004 and 2005.

3.5 Exercises

In Exercises 1–18, interpret the real-life meaning of the indicated values. Answer additional questions as appropriate.

1. *Body Weight* The weight of a girl between 2 and 13 years of age may be modeled by

$$W(a) = 0.289a^2 + 2.464a + 23.10 \text{ pounds}$$

where a is the age of the girl. (**Source:** Modeled from www.babybag.com data)

Interpret the meaning of $W(10) = 76.64$ and $W'(10) = 8.24$. Then estimate $W(11)$.

2. *Body Weight* Compare the results of Example 2 and Exercise 1. Were boys or girls expected to gain more weight between their 10th and 11th years? Explain.

3. *Carbon Monoxide Pollution* Based on data from 1990–2003, carbon monoxide pollution may be modeled by

$$P(t) = -0.248t + 5.99 \text{ parts per million}$$

where t is the number of years since 1990. (**Source:** *Statistical Abstract of the United States, 2006,* Table 359)

Interpret the meaning of $P(15) = 2.27$ and $P'(15) = -0.248$.

4. *Kazakhstan Population* Based on data from 1995–2005, the population of Kazakhstan may be modeled by

$$K(t) = 14{,}825(0.993)^t \text{ thousand people}$$

where t is the number of years since 2005. (**Source:** *World Health Statistics 2006,* World Health Organization)

Interpret the meaning of $K(10) = 13{,}819$ and $P'(10) = -97.08$.

5. *India Population* Based on data from 1995–2005, the population of the India may be modeled by

$$I(t) = 1{,}103{,}371(1.015)^t \text{ thousand people}$$

where t is the number of years since 2005. (**Source:** *World Health Statistics 2006*, World Health Organization)

Interpret the meaning of $I(5) = 1,188,644$ and $I'(5) = 17,697$.

6. *Highway Accidents* Based on data from 2000–2004, the percentage of highway accidents in the United States annually resulting in injuries may be modeled by

$$p(t) = 5.853(0.6521)^t + 44 \text{ percent}$$

where t is the number of years since 2000. (**Source:** Modeled from *Statistical Abstract of the United States, 2007*, Table 1047)

Interpret the meaning of $p(5) = 44.69$ and $p'(5) = -0.295$.

7. *College Enrollment* Based on data from 2001–2003 and projections for 2004–2013, private college enrollment may be modeled by

$$P(x) = 0.340x - 457 \text{ thousand students}$$

where x is the number of public college students enrolled (in thousands). (**Source:** *Statistical Abstract of the United States, 2006*, Table 204)

Interpret the meaning of $P(14,000) = 4303$ and $P'(14,000) = 0.340$.

8. *Heart Disease Death Rate* Based on data from 1974–2003, the death rate due to heart disease (in deaths per 100,000 people) may be modeled by

$$D(p) = 14.08p - 53.87$$

where p is the percentage of people who smoke. (**Source:** Modeled from Centers for Disease Control and Census Bureau data)

Interpret the meaning of $D(20) = 228$ and $D'(20) = 14.08$.

9. *Spending on Food* Based on data from 1995–2004, the amount of money spent by civilian consumers on farm foods away from home may be modeled by

$$a(h) = 0.7905h - 38.36 \text{ billion dollars}$$

where h is the amount of money spent on farm foods at home (in billions of dollars). (**Source:** Modeled from *Statistical Abstract of the United States, 2007*, Table 818)

Interpret the meaning of $a(500) = 356.9$ and $a'(500) = 0.7905$.

10. *Cassette Tape Value* Based on data from 1990–2004, the value of all cassette tapes shipped may be modeled by

$$V(s) = -0.00393s^2 + 9.74s - 63.0 \text{ million dollars}$$

where s is the number of cassette tapes shipped (in millions). (**Source:** *Statistical Abstract of the United States, 2006*, Table 1131)

Interpret the meaning of $V(30) = 226$ and $V'(30) = 9.5$. Then use a tangent line approximation to estimate $V(31)$.

11. *Medicare Enrollees* Based on data from 1980–2004, the number of Medicare enrollees (in millions) may be modeled by

$$M(t) = -0.00472t^2 + 0.663t + 28.4 \text{ million enrollees}$$

where t is the number of years since 1980. (**Source:** Modeled from *Statistical Abstract of the United States, 2006*, Table 132)

Interpret the meaning of $M(25) = 42.0$ and $M'(25) = 0.427$. Then use a tangent line approximation to estimate $M(26)$.

12. *Prescriptions* Based on data from 1995–2004, the number of prescriptions (in millions) may be modeled by

$$p(m) = -43.74m^2 + 3754m - 77,180$$

where m is the number of Medicare enrollees (in millions). (**Source:** Modeled from *Statistical Abstract of the United States, 2006*, Tables 126, 132)

Interpret the meaning of $p(42) = 3331$ and $p'(42) = 79.84$. Then use a tangent line approximation to estimate $p(43)$.

13. *United States Population* Based on data from 1990–2004, the population of the United States may be modeled by

$$P(t) = -19.56t^2 + 3407t + 250,100 \text{ thousand people}$$

where t is the number of years since 1990. (**Source:** *Statistical Abstract of the United States, 2006*, Table 2)

Interpret the meaning of $P(17) = 302,366$ and $P'(17) = 2742$. Then use a tangent line approximation to estimate $P(18)$.

14. *Walmart Net Sales* Based on data from 1996–2006, the net sales of Walmart may be modeled by

$$s(t) = 0.8636t^2 + 14.39t + 84.72 \text{ billion dollars}$$

where *t* is the number of years since 1996. (**Source:** Modeled from Walmart Annual Report, 2006, pp. 18–19)

Interpret the meaning of $s(7) = 227.8$ and $s'(7) = 26.48$. Then use a tangent line approximation to estimate $s(8)$.

15. *Health Care Spending* Based on data from 1960–2003 and projections for 2004–2010, the amount of public money spent on health care may be modeled by

$$G(p) = 0.000172p^2 + 0.712p - 7.01 \text{ billion dollars}$$

where *p* is the amount of private money (out-of-pocket and private insurance) spent on health care (in billions of dollars). (**Source:** *Statistical Abstract of the United States, 2006*, Table 118)

Interpret the meaning of $G(1500) = 1448$ and $G'(1500) = 1.228$. Then use a tangent line approximation to estimate $G(1501)$.

16. *Yogurt Production* Based on data from 1997–2005, the amount of yogurt produced in the United States annually may be modeled by

$$y(x) = 14.99x^2 + 62.14x + 1555 \text{ million pounds}$$

where *x* is the number of years since 1997. (**Source:** Modeled from *Statistical Abstract of the United States, 2007*, Table 846)

Interpret the meaning of $y(13) = 4896$ and $y'(13) = 451.9$. Then use a tangent line approximation to estimate $y(14)$.

17. *Theme Park Tickets* Based on 2007 ticket prices, the cost of an adult Disney Park Hopper Bonus Ticket may be modeled by

$$T(d) = -5.357d^2 + 59.04d + 28.20 \text{ dollars}$$

where *d* is the number of days that the ticket authorizes entrance into Disneyland and Disney California Adventure. (**Source:** www.disneyland.com)

Interpret the meaning of $T(3) = 157.1$ and $T'(3) = 26.90$. Then use a tangent line approximation to estimate $T(4)$.

18. *Life Expectancy* Based on statistical data, the life expectancy of an American female may be modeled by

$$f(m) = 0.045645m^3 - 9.9292m^2 + 720.34m - 17{,}350 \text{ years}$$

where *m* is the life expectancy of an American male (in years). (**Source:** Modeled from *Statistical Abstract of the United States, 2006*, Table 96)

Interpret the meaning of $f(75) = 80.2$ and $f'(75) = 1.22$. Then use a tangent line approximation to estimate $f(76)$.

Exercises 19 and 20 are intended to challenge your understanding of the meaning of derivatives.

19. *AIDS Deaths in the United States* Based on data from 1981 to 2001, the number of adult and adolescent AIDS deaths in the United States may be modeled by a piecewise function *f*, whose graph is shown in the figure along with a scatter plot of the number of deaths as a function of years since the end of 1981.

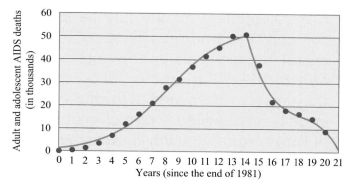

The highest number of AIDS deaths (50,876) occurred in 1995. (**Source:** Centers for Disease Control and Prevention)

Estimate from the graph the year in which f' was the greatest and the year in which f' was the most negative.

20. Let *f* be any continuous, smooth function (no breaks or sharp points) that is defined on the interval $(-\infty, \infty)$. If $f(a) \geq f(x)$ for all *x*, what is the value of $f'(a)$?

Chapter 4

Differentiation
Techniques

A mathematical model may be used to forecast a company's revenue at a given point in time. However, business executives and investors aren't interested only in the dollar amount of a company's revenue; they are also interested in the direction in which revenue is heading. As a rate of change, the derivative of the revenue function shows whether revenues are increasing or decreasing.

4.1 Basic Derivative Rules

4.2 The Product and Quotient Rules

4.3 The Chain Rule

4.4 Exponential and Logarithmic Rules

4.5 Implicit Differentiation

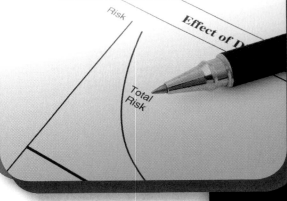

4.1 Basic Derivative Rules

As confidence in the Internet has increased, e-commerce has skyrocketed. Based on data from 2003 to 2007, e-commerce sales may be modeled by

$$S(t) = 0.710t^2 + 9.41t + 40.1 \text{ billion dollars}$$

where t is the number of years since 2003.
(**Source:** Modeled from *Statistical Abstract of the United States, 2010,* Table 1022) How quickly was e-commerce increasing in 2007? How quickly was it expected to increase in 2010?

In the previous chapter, you learned how to find the derivative function by using the limit definition of the derivative. This process is cumbersome and prone to error. In this section, we will introduce some alternative forms of derivative notation and then demonstrate several shortcuts that will greatly enhance our efficiency in calculating derivatives.

Derivative Notation

Up to this point, we have used $f'(x)$ to represent the derivative of a function $f(x)$. In part because calculus was first developed by two people working independently (Newton and

Leibniz), there are multiple ways to represent the same concept.

Note that $\dfrac{d}{dx}$ means "find the derivative with respect to x," whereas $\dfrac{dy}{dx}$ is "the derivative of y with respect to x." The form $\dfrac{dy}{dx}$ is referred to as **Leibniz notation** for the derivative. Throughout this section, we will use various forms of notation for the derivative. Additionally, we will use the term **differentiate** to mean "find the derivative of" and the term **differentiation** to refer to the process of finding the derivative. A function whose derivative exists for all values of x in its domain is said to be **differentiable**.

Derivative Notation

The derivative of a function $y = f(x)$ may be represented by any of the following:

- $f'(x)$, read "f prime of x"

- y', read "y prime"

- $\dfrac{dy}{dx}$, read "dee why dee ex" or "the derivative of y with respect to x"

The Constant Rule

Recall that the derivative represents the slope of the tangent line of the function. Consider the horizontal line $f(x) = c$. The slope of the tangent line of f at (a, c) is

$$f'(a) = \lim_{h \to 0} \frac{f(a + h) - f(a)}{h}$$

$$= \lim_{h \to 0} \frac{c - c}{h}$$

$$= \lim_{h \to 0} \frac{0}{h}$$

$$= 0 \qquad \text{Since dividing 0 by any nonzero number yields 0}$$

Constant Rule

The derivative of a constant function $y = c$ is

$$y' = 0$$

EXAMPLE 1 Finding the Derivative of a Constant Function

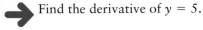 Find the derivative of $y = 5$.

SOLUTION

$$y' = 0$$

The Power Rule

Recall that a power function is a function of the form $f(x) = x^n$, where x is a variable and n is a constant. In Exercises 3.4, you found that the power function $y = x^3$ had derivative $y' = 3x^2$. Similarly, the derivative of $y = x^4$ is $y' = 4x^3$, and the derivative of $y = x^5$ is $y' = 5x^4$. Although the derivative methods previously introduced allow us to find the derivative of any power function, the Power Rule provides a remarkably quick and easy way to calculate the derivative of a power function.

Power Rule

The derivative of a function $y = x^n$, where x is a variable and n is a nonzero constant, is

$$y' = nx^{n-1}$$

EXAMPLE 2 Finding the Derivative of a Power Function

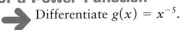 Differentiate $g(x) = x^{-5}$.

SOLUTION

$$g'(x) = -5x^{-5-1}$$
$$= -5x^{-6}$$

EXAMPLE 3 Finding the Derivative of a Power Function

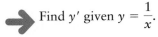 Find y' given $y = \dfrac{1}{x}$.

SOLUTION

Initially, this function doesn't look like a power function. However, recall that $y = \dfrac{1}{x}$ is equivalent to $y = x^{-1}$, which is of the form $y = x^n$.

$$y' = -1x^{-1-1}$$
$$= -x^{-2}$$
$$= -\frac{1}{x^2}$$

EXAMPLE 4 Finding the Derivative of a Power Function with a Rational Exponent

Find $f'(x)$ given $f(x) = \sqrt[3]{x}$.

SOLUTION

Again, $f(x) = \sqrt[3]{x}$ doesn't appear to be of the form $y = x^n$. However, any function of the form $f(x) = \sqrt[n]{x}$ may be rewritten as $f(x) = x^{1/n}$. Therefore, $f(x) = \sqrt[3]{x}$ is equivalent to $f(x) = x^{1/3}$, which is of the form $y = x^n$.

$$f'(x) = \frac{1}{3} x^{1/3 - 1}$$

$$= \frac{1}{3} x^{-2/3}$$

This result may be written in a variety of alternative forms, such as the following:

$$f'(x) = \frac{1}{3x^{2/3}}$$

$$f'(x) = \frac{1}{3\sqrt[3]{x^2}}$$

When working with functions with rational exponents, it is helpful to recall that

$$
\begin{aligned}
y &= \sqrt[n]{x^m} \\
&= x^{m/n} \\
&= (x^m)^{1/n} \\
&= (x^{1/n})^m
\end{aligned}
$$

Each of these forms is a correct way to represent the function $y = \sqrt[n]{x^m}$.

Constant Multiple Rule

The Constant Multiple Rule is used when a constant is multiplied by a function. This rule allows us to factor out the constant, take the derivative of the function, and then multiply the result by the factored-out constant.

Constant Multiple Rule

The derivative of a differentiable function $g(x) = kf(x)$ with constant k is given by

$$g'(x) = kf'(x)$$

EXAMPLE 5 Finding the Derivative of a Power Function

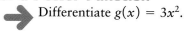 Differentiate $g(x) = 3x^2$.

SOLUTION

$$
\begin{aligned}
g'(x) &= \frac{d}{dx}(3x^2) \\
&= 3 \cdot \frac{d}{dx}(x^2) \qquad \text{Constant Multiple Rule} \\
&= 3(2x) \qquad \text{Power Rule} \\
&= 6x
\end{aligned}
$$

Sum and Difference Rule

The Sum and Difference Rule allows us to calculate the derivative of a function with multiple terms by adding together the derivatives of each of the terms.

Sum and Difference Rule

The derivative of a differentiable function $h(x) = f(x) \pm g(x)$ is given by

$$h'(x) = f'(x) \pm g'(x)$$

EXAMPLE 6 Finding the Derivative of a Sum of Functions

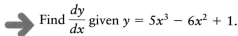 Find $\frac{dy}{dx}$ given $y = 5x^3 - 6x^2 + 1$.

SOLUTION

$$
\begin{aligned}
\frac{dy}{dx} &= \frac{d}{dx}(5x^3 - 6x^2 + 1) \\
&= \frac{d}{dx}(5x^3) - \frac{d}{dx}(6x^2) + \frac{d}{dx}(1) \qquad \text{By the Sum and Difference Rule} \\
&= 15x^2 - 12x + 0 \qquad \text{By the Constant Multiple, Constant, and Power Rules} \\
&= 15x^2 - 12x
\end{aligned}
$$

We will now return to the e-commerce sales function introduced at the beginning of this section.

EXAMPLE 7 Using Derivative Rules to Quickly Calculate an Instantaneous Rate of Change

Based on data from 2003 to 2007, e-commerce sales may be modeled by

$$S(t) = 0.710t^2 + 9.41t + 40.1 \text{ billion dollars}$$

where t is the number of years since 2003. (**Source:** Modeled from *Statistical Abstract of the United States, 2010*, Table 1022) How quickly was e-commerce increasing in 2007? How quickly was it expected to increase in 2010?

SOLUTION

We are asked to evaluate the derivative of the function at $t = 4$ and $t = 7$.

$$S'(t) = \frac{d}{dt}(0.710t^2 + 9.41t + 40.1)$$

$$= \frac{d}{dt}(0.710t^2) + \frac{d}{dt}(9.41t) + \frac{d}{dt}(40.1) \quad \text{Sum and Difference Rule}$$

$$= 2(0.710)t + 1(9.41)t^0 + 0 \quad \text{Power, Constant Multiple, and Constant Rules}$$

$$= 1.420t + 9.41$$

$$S'(4) = 1.420(4) + 9.41$$

$$= 5.68 + 9.41$$

$$\approx 15.1$$

$$S'(7) = 1.420(7) + 9.41$$

$$= 9.94 + 9.41$$

$$\approx 19.4$$

In 2007, e-commerce sales were increasing at a rate of \$15.1 billion per year. In 2010, e-commerce sales were expected to be increasing at a rate of \$19.4 billion per year.

4.1 Exercises

In Exercises 1–5, use the Constant Rule, the Power Rule, the Constant Multiple Rule, and the Sum and Difference Rule (as appropriate) to find the derivative of the function.

1. $f(x) = 5$

2. $v(t) = 5t^3 - 10t^{-2}$

3. $g(x) = -2x + 4x^3$

4. $f(t) = 4t^{-1}$

5. $g(t) = t^3 + 3t^2 + 6t + 6$

In Exercises 6–10, find the derivative of the function at the indicated domain value.

6. $y = \sqrt{x}$; $x = 1$

7. $y = -2.6x^5 + 3x^{-2}$; $x = -1$

8. $y = 1235.3t^3 + 551.23t - 1203.9$; $t = 0$

9. $f(n) = 9n^{1.3} + 5n^{2.1} + 92n^{-0.2}$; $n = 1$

10. $g(x) = x^{-3} + 2.22x^3 - 12.3$; $x = 1$

In Exercises 11–15, use the derivative to find the answer to the question.

11. *Food Spending* Based on data from 1990 to 2004, consumer expenditure for farm foods may be modeled by

$$E(t) = 1.008t^2 + 9.951t + 450.5 \text{ billion dollars}$$

where t is the number of years since 1990. (**Source:** Modeled from *Statistical Abstract of the United States, 2007*, Table 818)

According to the model, at what rate was consumer spending on farm foods changing in 2007?

12. *Average Height of a Girl* The average height of a girl between 2 and 13 years of age may be modeled by

$$H(a) = -0.0392a^2 + 2.987a + 29.69 \text{ inches}$$

where a is the age of the girl. (**Source:** Modeled from www.babybag.com data)

Does the average height of a 4-year-old girl increase faster than the average height of an 11-year-old girl?

13. *United States Population* Based on data from 1990–2004, the population of the United States may be modeled by

$$P(t) = -19.56t^2 + 3407t + 250,100 \text{ thousand people}$$

where t is the number of years since 1990. (**Source:** *Statistical Abstract of the United States, 2006*, Table 2)

According to the model, at what rate was the United States population changing in 2010?

14. *Average Weight of a Boy* The average weight of a boy between 2 and 13 years of age may be modeled by

$$W(a) = 0.215a^2 + 2.993a + 23.78 \text{ pounds}$$

where a is the age of the boy. (**Source:** Modeled from www.babybag.com data)

Does the average weight of a 4-year-old boy increase faster than the average weight of an 11-year-old boy?

15. *Walmart Net Sales* Based on data from 1996–2006, the net sales of Walmart may be modeled by

$$s(t) = 0.8636t^2 + 14.39t + 84.72 \text{ billion dollars}$$

where t is the number of years since 1996. (**Source:** Modeled from Walmart Annual Report, 2006, pp. 18–19)

According to the model, at what rate were net sales increasing in 2008 and 2009?

4.2 The Product and Quotient Rules

Based on data from 2000 to 2008, per capita spending on retail and food services may be modeled by

$$C(t) = -18.90t^3 + 225.7t^2 - 244.1t + 11,760 \text{ dollars}$$

where t is the number of years since 2000. (**Source:** Modeled from *Statistical Abstract of the United States, 2010*, Table 1019)

Based on Census Bureau population projections through 2020, the population of the United States may be modeled by

$$P(t) = 2.71t + 281.7 \text{ million people}$$

where t is the number of years since 2000. (**Source:** Modeled from U.S. Census Bureau data)

At what rate was overall retail and food services spending in the United States changing in 2008? We

Shutterstock

will use the Product Rule to answer this question in Example 4 of this section.

In this section, we will demonstrate how to find the derivative of a product by using the Product Rule and the derivative of a quotient by using the Quotient Rule.

Product Rule

The derivative of a function $h(x) = f(x) \cdot g(x)$ is given by

$$h'(x) = f'(x) \cdot g(x) + g'(x) \cdot f(x)$$

A common error among beginning calculus students is to assume that $\dfrac{d}{dx}(f(x) \cdot g(x)) = f'(x) \cdot g'(x)$. We can easily convince ourselves that this is not true by calculating the derivative of $f(x) = x^2$ using the Power Rule and then by using the erroneous rule.

Using the Power Rule,

$$\frac{d}{dx}(x^2) = 2x$$

Using the erroneous rule,

$$\frac{d}{dx}(x^2) = \frac{d}{dx}(x \cdot x)$$
$$= \frac{d}{dx}(x) \cdot \frac{d}{dx}(x)$$
$$= 1 \cdot 1$$
$$= 1$$

Since $2x \neq 1$, we know that the erroneous rule is not valid. Let's calculate the derivative using the Product Rule.

Using the Product Rule,

$$\frac{d}{dx}(x^2) = \frac{d}{dx}(x \cdot x)$$
$$= \frac{d}{dx}(x) \cdot x + \frac{d}{dx}(x) \cdot x$$
$$= 1 \cdot x + 1 \cdot x$$
$$= 2x$$

The Product Rule yielded the same result as the Power Rule.

EXAMPLE 1 Finding the Derivative of a Product of Functions

Find the derivative of $h(x) = (2x + 4)(5x^3 - 3x)$.

SOLUTION

$$h'(x) = \frac{d}{dx}((2x + 4)(5x^3 - 3x))$$

$$= \left(\frac{d}{dx}(2x + 4)\right) \cdot (5x^3 - 3x) \quad \text{Product Rule}$$

$$+ \left(\frac{d}{dx}(5x^3 - 3x)\right) \cdot (2x + 4)$$

$$= (2) \cdot (5x^3 - 3x) + (15x^2 - 3) \cdot (2x + 4)$$

Power, Constant Multiple, and Constant Rules

$$= (10x^3 - 6x) + (30x^3 + 60x^2 - 6x - 12)$$

$$= 40x^3 + 60x^2 - 12x - 12$$

Often there are multiple ways to calculate the derivative of a function. In general, we will use the simplest techniques possible to find the derivative. However, in order to demonstrate the Product Rule, in this section we may use the Product Rule in places where other methods are typically preferred.

EXAMPLE 2 Finding the Derivative of a Product of Functions

Differentiate $y = (2x^3 - 6x + 5)(x^4 - 2x^2 + 1)$.

SOLUTION

$$\frac{dy}{dx} = \frac{d}{dx}(y)$$

$$= \frac{d}{dx}((2x^3 - 6x + 5)(x^4 - 2x^2 + 1))$$

$$= \left(\frac{d}{dx}(2x^3 - 6x + 5)\right)(x^4 - 2x^2 + 1)$$

$$+ \left(\frac{d}{dx}(x^4 - 2x^2 + 1)\right)(2x^3 - 6x + 5)$$

$$= (6x^2 - 6)(x^4 - 2x^2 + 1)$$

$$+ (4x^3 - 4x)(2x^3 - 6x + 5)$$

$$= (6x^6 - 12x^4 + 6x^2 - 6x^4 + 12x^2 - 6)$$

$$+ (8x^6 - 24x^4 + 20x^3 - 8x^4 + 24x^2 - 20x)$$

$$= 14x^6 - 50x^4 + 20x^3 + 42x^2 - 20x - 6$$

EXAMPLE 3 Finding the Derivative of a Product of Three Functions

Differentiate $R(t) = (5t - 1)(4t + 3)(2t + 6)$.

SOLUTION

This function is different because it is the product of three factors instead of two. Nevertheless, the Product Rule may be generalized to work here. If we have a function of the form $w(x) = f(x) \cdot g(x) \cdot h(x)$, then $w'(x) = f'(x)g(x)h(x) + g'(x)f(x)h(x) + h'(x)f(x)g(x)$. To find each term of the derivative, we calculate the derivative of each factor and then multiply it by the remaining factors. The derivative of $w(x)$ is the sum of the individual terms.

$$R'(t) = 5(4t + 3)(2t + 6) + 4(5t - 1)(2t + 6)$$

$$+ 2(5t - 1)(4t + 3)$$

$$= 5(8t^2 + 24t + 6t + 18) + 4(10t^2 + 30t - 2t - 6)$$

$$+ 2(20t^2 + 15t - 4t - 3)$$

$$= 5(8t^2 + 30t + 18) + 4(10t^2 + 28t - 6)$$

$$+ 2(20t^2 + 11t - 3)$$

$$= (40t^2 + 150t + 90) + (40t^2 + 112t - 24)$$

$$+ (40t^2 + 22t - 6)$$

$$= 120t^2 + 284t + 60$$

EXAMPLE 4 Applying the Product Rule in a Real-World Context

Based on data from 2000 to 2008, per capita spending on retail and food services may be modeled by

$$C(t) = -18.90t^3 + 225.7t^2 - 244.1t + 11{,}760 \text{ dollars}$$

where t is the number of years since 2000. (**Source:** Modeled from *Statistical Abstract of the United States, 2010*, Table 1019)

Based on Census Bureau population projections through 2020, the population of the United States may be modeled by

$$P(t) = 2.71t + 281.7 \text{ million people}$$

where t is the number of years since 2000. (**Source:** Modeled from U.S. Census Bureau data)

At what rate was national retail and food services spending in the United States changing in 2008?

SOLUTION

Per capita means *per person*. The function $C(t) = -18.90t^3 + 225.7t^2 - 244.1t + 11{,}760$ gives the per person spending in dollars. In other words, the units of C are *dollars per person*. The function $P(t) = 2.71t + 281.7$ gives the U.S. population in millions of people. In both functions, t represents the number of years since 2000. We are asked to find the national spending in the United States. To do this, we multiply the spending per person by the total number of people in the country.

We define a new function $S(t) = C(t)P(t)$. What are the units of $S(t)$? We have

$$\left(C \frac{\text{dollars}}{\text{person}}\right)(P \text{ million persons})$$

$$C\frac{\text{dollars}}{\text{person}} \cdot P\frac{\text{million persons}}{1} = CP \text{ million dollars}$$

Therefore, the units of $S(t)$ are millions of dollars.

$$S(t) = C(t) \cdot P(t)$$

$$= (-18.90t^3 + 225.7t^2 - 244.1t + 11{,}760)$$
$$(2.71t + 281.7)$$

We are asked to calculate the instantaneous rate of change of $S(t)$ in 2008. That is, we are to find $S'(8)$. We know from the Product Rule that

$$S'(t) = C'(t) \cdot P(t) + P'(t) \cdot C(t) \frac{\text{million dollars}}{\text{year}}$$

Although we could calculate $S'(8)$ algebraically, the complexity of $S(t)$ would make the process extremely tedious. In cases such as this, it is appropriate to use the power of a graphing calculator, as demonstrated on the Chapter 4 Tech Card. Using the Tech Card, we determine that

$$S'(8) \approx -39{,}896 \frac{\text{million dollars}}{\text{year}}$$

In 2008, national retail and food services spending was decreasing at a rate of nearly \$40 billion per year. The change in spending was due in part to the global economic downturn that took place in the latter part of the first decade of the century.

Although any quotient of two functions may be rewritten as a product, using the Quotient Rule to find the derivative of a quotient is often easier than using the Product Rule.

Quotient Rule

The derivative of a function $h(x) = \dfrac{f(x)}{g(x)}$ is given by

$$h'(x) = \frac{f'(x) \cdot g(x) - g'(x) \cdot f(x)}{(g(x))^2}$$

EXAMPLE 5 Finding the Derivative of a Quotient of Functions

Differentiate $y = \dfrac{6x + 1}{3x - 1}$.

SOLUTION

$$y' = \frac{(6)(3x - 1) - (3)(6x + 1)}{(3x - 1)^2}$$

$$= \frac{18x - 6 - 18x - 3}{(3x - 1)^2}$$

$$= \frac{-9}{(3x - 1)^2}$$

4.2 Exercises

In Exercises 1–10, use the Product or the Quotient Rule to find the derivative of the function. Don't simplify the result.

1. $f(x) = 2x(3x + 4)$

2. $g(x) = 5x(9x - 2)$

3. $f(t) = (2t - 6)(10t + 5)$

4. $s(t) = 2t^3(3t^2 + 7t)$

5. $w(n) = (3n^2 + 8)(n^3 - n)$

6. $D(p) = (p^2 - 7p)(1 - p^3)$

7. $Q(t) = (t^2 + 2t + 1)(t^2 + 4t + 4)$

8. $P(t) = \dfrac{t^2 + 5t + 6}{t^2 + 5t + 4}$

9. $f(x) = \dfrac{x^3 - 1}{x^4 - x^3}$

10. $g(x) = \dfrac{x^3 + 5x^2}{x^2 - 4x + 2}$

In Exercises 11–15, determine the slope of the graph at the indicated domain value.

11. $h(t) = \dfrac{t^2 + 5t + 6}{t}; t = -2$

12. $s(t) = \dfrac{t^4 - 6t^2 + 8}{2t}; t = -1$

13. $f(x) = 2x(3x + 4)(5x + 7); x = 0$

14. $r(x) = (2x + 11)(x - 4)(7x + 3); x = 4$

15. $f(x) = (x + 1)(x - 1)(x^2 + 1); x = -1$

In Exercises 16–18, use the Chapter 4 Tech Card to answer the questions from the real-life scenarios.

16. *Employer Labor Costs* Based on data from 1995 to 1999, the average annual earnings in the lumber and wood products manufacturing industry may be modeled by

$$S(t) = -63.57t^2 + 1253t + 25{,}066$$
$$\text{dollars per employee}$$

and the average number of employees in the lumber and wood products manufacturing industry may be modeled by

$$E(t) = 3.143t^2 + 5.029t + 772.5$$
$$\text{thousand employees}$$

where t is the number of years since 1995. (**Source:** Modeled from *Statistical Abstract of the United States, 2001*, Table 979)

In 1999, at what rate was employer spending on lumber and wood products manufacturing industry employee earnings increasing?

17. *Employer Labor Costs* Based on data from 1995 to 1999, the average annual earnings in the paper and allied products manufacturing industry may be modeled by

$$E(t) = 1335t + 39,408 \text{ dollars per employee}$$

and the average number of employees in the same industry may be modeled by

$$N(t) = t^3 + 5.571t^2 - 12.29t + 684.9$$
$$\text{thousand employees}$$

where t is the number of years since 1995. (**Source:** Modeled from *Statistical Abstract of the United States, 2001*, Table 979)

In 1999, at what rate was employer spending on paper and allied products manufacturing industry employee earnings increasing?

18. *Employer Labor Costs* Based on data from 1995 to 1999, the average annual earnings in the printing and publishing industry may be modeled by

$$E(t) = 88.36t^2 + 1291t + 34,528$$
$$\text{dollars per employee}$$

and the average number of employees in the printing and publishing industry may be modeled by

$$N(t) = -4.417t^3 + 24.93t^2 - 25.30t + 1450$$
$$\text{thousand employees}$$

where t is the number of years since 1995. (**Source:** Modeled from *Statistical Abstract of the United States, 2001*, Table 979)

In 1999, at what rate was employer spending on printing and publishing industry employee earnings increasing?

Exercises 19 and 20 are intended to challenge your understanding of the Product Rule. Solve the problems without using the Tech Card.

19. *Apple Farming* Suppose that a farmer has an apple orchard with 30 trees per acre. The orchard yields

12 bushels per tree. The farmer estimates that for each additional tree planted per acre, the average yield per tree is reduced by 0.1 bushel.

If $y = f(x)$ is the total number of bushels of apples produced per acre when an additional x trees per acre are planted, calculate and interpret the meaning of $f'(40)$, $f'(45)$, and $f'(50)$.

20. *Apple Supplier Prices* A fruit farmer sells apples to a grocery store chain. The amount of apples the store buys depends linearly upon the price per pound that the farmer charges. The farmer estimates that for every $0.02 per pound increase in the price, the store will reduce its order by 44 pounds. The store currently orders 440 pounds per week and pays $0.18 per pound. What price should the farmer charge in order to maximize her revenue from apple sales?

4.3   The Chain Rule

Based on data from 1995 to 1999, the average annual earnings of an employee in the lumber and wood products industry may be modeled by

$$S(n) = -0.7685n^2 + 1295n - 516,565 \text{ dollars}$$

where n is the number of employees in thousands. (**Source:** Modeled from *Statistical Abstract of the United States, 2001*, Table 979)

Based on data from 1995 to 1999, the number of employees in the lumber and wood products industry may be modeled by

$$n(t) = 3.143t^2 + 5.029t + 772.5 \text{ thousand employees}$$

where t is the number of years since 1995. (**Source:** Modeled from *Statistical Abstract of the United States, 2001,* Table 979)

How quickly were the average annual earnings of an employee increasing in 1998? We will answer this question using the Chain Rule.

In this section, we will demonstrate how to find the composition of two functions. We will then demonstrate how to use the Chain Rule to find the derivative of a composition of functions.

Composition of Functions

A composition of two functions occurs when the outputs of one function are the inputs of a second function. Let's consider a common example. Let f represent the freezer function and b represent the blender function.

$b(f(w))$ means that first w is plugged into the function f, and then the result is substituted into b (see Figure 4.1). In our example, water was placed in the freezer, and the result, ice, was put in the blender.

EXAMPLE 1 Finding a Composition of Functions

Let $f(t) = 3t + 2$ and $g(t) = t^2 + 1$. Find $f(g(t))$.

SOLUTION

$$f(g(t)) = f(t^2 + 1)$$
$$= 3(t^2 + 1) + 2$$
$$= 3t^2 + 3 + 2$$
$$= 3t^2 + 5$$

The output of the function g is the input for f
The variable in f is replaced with the output of g

Is $f(g(t))$ the same as $g(f(t))$? Let's return to the freezer and blender example and see (see Figure 4.2).

When we reversed the order of the functions, we got a different result. *Ice* is not the same as *chopped ice*. There are certain functions for which the two different compositions are equal, but in general, $f(g(t)) \neq g(f(t))$.

EXAMPLE 2 Finding a Composition of Functions

Let $f(t) = 3t + 2$ and $g(t) = t^2 + 1$. Find $g(f(t))$.

SOLUTION

$$g(f(t)) = g(3t + 2)$$
$$= (3t + 2)^2 + 1$$
$$= 9t^2 + 12t + 4 + 1$$
$$= 9t^2 + 12t + 5$$

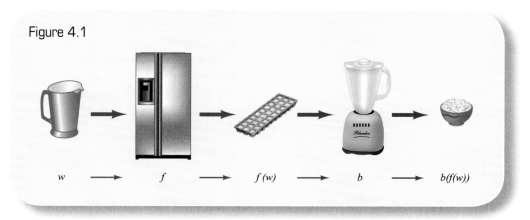

Figure 4.1

$$w \longrightarrow f \longrightarrow f(w) \longrightarrow b \longrightarrow b(f(w))$$

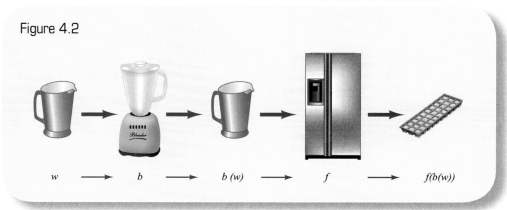

Figure 4.2

$$w \longrightarrow b \longrightarrow b(w) \longrightarrow f \longrightarrow f(b(w))$$

Note:

Although Examples 1 and 2 used the same functions f and g,

f(g(t)) ≠ g(f(t))

Note that although Examples 1 and 2 used the same functions f and g, $f(g(t)) \neq g(f(t))$.

The Chain Rule

The Chain Rule is used whenever we have to find the derivative of a composition of functions.

Chain Rule

The derivative of a function $h(x) = f(g(x))$ is given by

$$h'(x) = f'(g(x)) \cdot g'(x)$$

The Chain Rule tells us to find $f'(x)$, substitute $g(x)$ in for the variable x, and then multiply the result by $g'(x)$.

EXAMPLE 3 Finding the Derivative by Using the Chain Rule

➡ Given $h(x) = (2x + 1)^2 - 4$, find $h'(x)$.

SOLUTION

Because there is a nonconstant function $(2x + 1)$ inside the parentheses, we know that we will need the Chain Rule. The function $h(x) = (2x + 1)^2 - 4$ is a function of the form $h(x) = f(g(x))$. We must determine the functions f and g. Often, selecting the function inside the parentheses as g gives the desired result. In this case, we pick $g(x) = 2x + 1$. Then

$$f(g(x)) = (2x + 1)^2 - 4$$
$$f(g(x)) = (g(x))^2 - 4$$

To find the function f, we simply replace the $g(x)$ in the equation with the variable x. Thus we have

$$f(x) = x^2 - 4$$

We'll first find $f'(x)$ and then use it to calculate $f'(g(x))$.

$$f'(x) = \frac{d}{dx}(x^2 - 4)$$
$$= 2x$$

$$f'(g(x)) = 2 \cdot g(x) \qquad \text{By substitution}$$
$$= 2(2x + 1) \qquad \text{Since } g(x) = 2x + 1$$

Next we'll find $g'(x)$.

$$g'(x) = \frac{d}{dx}(2x + 1)$$

$$= 2$$

The derivative of $f(g(x))$ is

$$f'(g(x)) \cdot g'(x) = (2(2x + 1)) \cdot (2)$$

$$= 4(2x + 1)$$

$$= 8x + 4$$

We'll rework the problem using a slightly different but somewhat easier approach.

$$h(x) = (2x + 1)^2 - 4$$

$$\frac{dh}{dx} = \frac{d}{dx}((2x + 1)^2) - \frac{d}{dx}(4) \quad \text{Sum and Difference Rule}$$

To use the Chain Rule to differentiate $(2x + 1)^2$, we treat the expression $2x + 1$ inside the parentheses as if it were a single variable and use the Power Rule. Then we multiply the result by the derivative of $2x + 1$, the expression inside the parentheses.

$$\frac{dh}{dx} = 2(2x + 1)^{2-1} \cdot \frac{d}{dx}(2x + 1) - 0 \quad \substack{\text{Chain, Power, and} \\ \text{Constant Rules}}$$

$$= 2(2x + 1) \cdot (2)$$

$$= 4(2x + 1)$$

$$= 8x + 4$$

The second technique shown in Example 3 is a demonstration of the Generalized Power Rule.

Generalized Power Rule

Let $u = g(x)$. The derivative of a function $f(x) = u^n$ is given by

$$f'(x) = nu^{n-1}u'$$

The Generalized Power Rule is a special application of the Chain Rule. We'll use this rule in the next two examples.

EXAMPLE 4 Finding the Derivative by Using the Generalized Power Rule

Differentiate $f(x) = 4(2x^2 + x)^3$.

SOLUTION

We pick $u = 2x^2 + x$. Then $f(x) = 4(2x^2 + x)^3$ may be rewritten as $f(x) = 4u^3$. From the Generalized Power Rule, we know that

$$\frac{d}{dx}(f(x)) = \frac{d}{dx}(4u^3)$$

$$= 4\frac{d}{dx}(u^3) \quad \text{Constant Multiple Rule}$$

$$f'(x) = 12u^2u' \quad \text{Generalized Power Rule}$$

We need to write the result in terms of x, not u. Since $u = 2x^2 + x$ and $u' = 4x + 1$, the derivative may be rewritten as

$$f'(x) = 12u^2u'$$

$$= 12(2x^2 + x)^2(4x + 1)$$

$$= 12(x(2x + 1))^2(4x + 1)$$

$$= 12(x^2)(2x + 1)^2(4x + 1)$$

$$= 12x^2(2x + 1)^2(4x + 1)$$

We will often write derivatives in factored form. In real-life optimization problems, we are often interested in the values of x that make the derivative equal to zero. These values are easiest to find if the function is written in factored form.

As demonstrated in the next example, it is necessary to apply the Chain Rule repeatedly for some functions.

It is necessary to apply the Chain Rule repeatedly for some functions.

EXAMPLE 5 Finding the Derivative by Using the Chain Rule Repeatedly

Find the derivative of $S(t) = (3(2t + 1)^2 + 5)^3$.

SOLUTION

We let $u = 3(2t + 1)^2 + 5$. Notice that this function also contains an expression inside parentheses. If we let $v = 2t + 1$, then $u = 3v^2 + 5$.

$S(t) = u^3$

$\dfrac{dS}{dt} = 3u^2 u'$ Generalized Power Rule

$= 3u^2 \dfrac{d}{dt}(u)$ Since $u' = \dfrac{d}{dt}(u)$

$= 3(3v^2 + 5)^2 \cdot \dfrac{d}{dt}(3v^2 + 5)$ Since $u = 3v^2 + 5$

$= 3(3v^2 + 5)^2 \cdot 6vv'$ Generalized Power Rule

$= 3(3v^2 + 5)^2 \cdot 6v \cdot \dfrac{d}{dt}(v)$ Since $v' = \dfrac{d}{dt}(v)$

$= 3(3(2t + 1)^2 + 5)^2 \cdot 6(2t + 1) \cdot \dfrac{d}{dt}(2t + 1)$
    Since $v = 2t + 1$

$= 3(3(2t + 1)^2 + 5)^2 \cdot 6(2t + 1) \cdot (2)$

At this point, we have calculated the derivative. If we want to rewrite the derivative in its simplest factored form, we continue as follows:

$\dfrac{dS}{dt} = 3(3(2t + 1)^2 + 5)^2 \cdot 6(2t + 1) \cdot (2)$

$= 36(3(2t + 1)^2 + 5)^2(2t + 1)$

$= 36(3(4t^2 + 4t + 1) + 5)^2(2t + 1)$

$= 36(12t^2 + 12t + 3 + 5)^2(2t + 1)$

$= 36(12t^2 + 12t + 8)^2(2t + 1)$

$= 36(4(3t^2 + 3t + 2))^2(2t + 1)$

$= 36(4)^2(3t^2 + 3t + 2)^2(2t + 1)$

$= 576(3t^2 + 3t + 2)^2(2t + 1)$

Writing the derivative in factored form required a significant number of additional steps. Although there are some instances that require us to write derivatives in factored form, we will consider the additional steps optional at this point.

Chain Rule: Alternative Form

The Chain Rule may also be expressed using Leibniz notation.

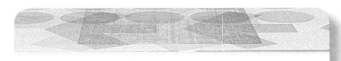

Chain Rule: Alternative Form

If $y = f(x)$ and $x = g(t)$, then $y = f(g(t))$ and

$$\frac{dy}{dt} = \frac{dy}{dx} \cdot \frac{dx}{dt}$$

The units of $\dfrac{dy}{dt}$ are obtained by multiplying the units of $\dfrac{dy}{dx}$ by the units of $\dfrac{dx}{dt}$.

EXAMPLE 6 Finding the Derivative by Using the Alternative Form of the Chain Rule

Based on data from 2000 to 2009, limited service restaurant sales in the United States may be modeled by

$$L(s) = 0.302s - 9.13 \text{ billion dollars}$$

where s is the total sales of all food and drink establishments.

The total sales of all food and drink establishments may be modeled by

$$s(t) = 21.76t + 374.9 \text{ billion dollars}$$

where t is the number of years since 2000. (**Source:** Modeled from *Statistical Abstract of the United States, 2010*, Table 1247)

At what rate were limited service restaurant sales changing in 2009?

SOLUTION

We are asked to find $\dfrac{dL}{dt}$. Recall that the alternative form of the Chain Rule states that $\dfrac{dL}{dt} = \dfrac{dL}{ds} \cdot \dfrac{ds}{dt}$. We will first find $\dfrac{dL}{ds}$ and $\dfrac{ds}{dt}$.

$L(s) = 0.302s - 9.13$

$\dfrac{dL}{ds} = 0.302$ billion dollars in limited service sales per billion dollars in total food and drink establishment sales

This means that for every \$1 billion increase in total food and drink establishment sales, limited service restaurant sales increase by \$0.302 billion.

$$s(t) = 21.76t + 374.9$$

$$\frac{ds}{dt} = 21.76 \frac{\text{billion dollars}}{\text{year}}$$

This means that total food and drink establishment sales increase by \$21.76 billion for each additional year that goes by.

$$\frac{dL}{dt} = \frac{dL}{ds} \cdot \frac{ds}{dt}$$

$$= \left(0.302 \, \frac{\text{billion dollars of limited service sales}}{\text{billion dollars of food establishment sales}} \right)$$

$$\cdot \left(21.76 \, \frac{\text{billion dollars of food establishment sales}}{\text{year}} \right)$$

$$= \left(0.302 \, \frac{\text{billion dollars of limited service sales}}{\text{billion dollars of food establishment sales}} \right)$$

$$\cdot \left(21.76 \, \frac{\text{billion dollars of food establishment sales}}{\text{year}} \right)$$

$$\approx 6.57 \text{ billion dollars per year}$$

We need to determine the rate of change in limited service restaurant sales for 2009 ($t = 9$). Since the derivative is a constant value, the rate of change is constant for all years. Thus in 2009, limited service restaurant sales are predicted to be increasing at a rate of \$6.57 billion per year.

We'll now return to the problem introduced at the beginning of this section.

EXAMPLE 7 Finding the Derivative by Using the Alternative Form of the Chain Rule

Based on data from 1995 to 1999, the average annual earnings of an employee in the lumber and wood products industry may be modeled by

$$S(n) = -0.7685n^2 + 1295n - 516{,}596 \text{ dollars}$$

where n is the number of employees in thousands. (**Source:** Modeled from *Statistical Abstract of the United States, 2001*, Table 979)

Based on data from 1995 to 1999, the number of employees in the lumber and wood products industry may be modeled by

$$n(t) = 3.143t^2 + 5.029t + 772.5 \text{ thousand employees}$$

where t is the number of years since 1995. (**Source:** Modeled from *Statistical Abstract of the United States, 2001*, Table 979)

How quickly were the average annual earnings of an employee increasing in 1998?

SOLUTION

We are asked to find $\dfrac{dS}{dt}$. By the Chain Rule, $\dfrac{dS}{dt} = \dfrac{dS}{dn} \cdot \dfrac{dn}{dt}$.
We will first find $\dfrac{dS}{dn}$ and $\dfrac{dn}{dt}$.

$$S(n) = 0.7685n^2 + 1295n - 516{,}596$$

$$\frac{dS}{dn} = 1.537n + 1295 \frac{\text{dollars}}{\text{thousand employees}}$$

$$n(t) = 3.143t^2 + 5.029t + 772.5$$

$$\frac{dn}{dt} = 6.286t + 5.029 \frac{\text{thousand employees}}{\text{year}}$$

$$\frac{dS}{dt} = \frac{dS}{dn} \cdot \frac{dn}{dt}$$

$$= \left(1.537n + 1295 \frac{\text{dollars}}{\text{thousand employees}} \right)$$

$$\cdot \left(6.286t + 5.029 \frac{\text{thousand employees}}{\text{year}} \right)$$

$$= (-1.537n + 1295) \cdot (6.286t + 5.029)$$

$$\frac{\text{dollars}}{\text{thousand employees}} \quad \frac{\text{thousand employees}}{\text{year}}$$

$$= (-1.537(3.143t^2 + 5.029t + 772.5) + 1295) \cdot (6.286t + 5.029) \text{ dollars per year}$$

$$\approx (-4.8308t^2 - 7.7296t - 1187.3325 + 1295) \cdot (6.286t + 5.029)$$

$$= (-4.8308t^2 - 7.7296t + 107.6675) \cdot (6.286t + 5.029) \text{ dollars per year}$$

We must find $\left. \dfrac{dS}{dt} \right|_{t=3}$, since $t = 3$ in the year 1998. (The notation $\left. \dfrac{dS}{dt} \right|_{t=3}$ means the value of $\dfrac{dS}{dt}$ when $t = 3$.)

$$\left. \frac{dS}{dt} \right|_{t=3} = (-4.8308(3)^2 - 7.7296(3) + 107.6675) \cdot (6.286(3) + 5.029)$$

$$= (-43.4772 - 23.1888 + 107.6675) \cdot (23.887)$$

$$= (41.0015) \cdot (23.887)$$

$$\approx 979.40 \text{ dollars per year}$$

According to the model, the average annual earnings of a lumber and wood products industry employee were increasing at a rate of roughly $979 per year in 1998.

....

4.3 Exercises

In Exercises 1–3, find the function $f(g(x))$. Then simplify the result.

1. $f(x) = 2x + 3, g(x) = x^2 + 4$

2. $f(x) = 7x - 4, g(x) = x^2 - 2x + 1$

3. $f(x) = 3x^2 - 11x, g(x) = x^3$

In Exercises 4 and 5, the function is of the form $h(x) = f(g(x))$. Determine $f(x)$ and $g(x)$.

4. $h(x) = (x + 1)^2 - 6(x + 1) + 9$

5. $h(x) = 2 \ln(x^2 + 4)$

In Exercises 6–15, use the Chain Rule in finding the derivative of the function.

6. $f(x) = 5(3x + 1)^2$

7. $g(t) = 4(6t - 7)^3 + 2$

8. $h(t) = 3(t^3 - t^2 + 1)^4$

9. $S(n) = (n^3 - 3n^2)^2$

10. $f(n) = 5(n^2 + 2n + 1)^2 + 6$

11. $g(x) = (4(3x + 5)^2 + x)^2$

12. $f(x) = (-5(4x - 1)^2 + 2x)^2$

13. $s(t) = (2t + 1)^2(6t)$

14. $f(t) = (5t^2 + t)(3t + 2)^2$

15. $h(x) = (4x + 1)^2(-x + 2)^2$

In Exercises 16–18, determine the slope of the graph at the indicated domain value.

16. $g(x) = (x^2 + 3)(x^2 - x)^{-1}; x = 2$

17. $R(x) = (3x + 1)(2x + 5)^{-1}; x = 2$

18. $f(x) = (5(4x + 1)^2 + 2)^2; x = -1$

In Exercises 19 and 20, find the equation of the tangent line at the indicated domain value.

19. $s(t) = (3t + 1)^2(-8t + 4)^{-2}; t = 0$

20. $p(x) = ((x + 1)^2 + 2)^3 + 3; x = -1$

In Exercise 21, use the Chain Rule to answer the question.

21. *Average Height of a Girl* The average height of a girl between 2 and 13 years of age may be modeled by

$$F(m) = 1.071m - 3.554 \text{ inches}$$

where m is the average height of a boy of the same age.

The average height of a boy between 2 and 13 years of age may be modeled by

$$m(a) = -0.0507a^2 + 2.997a + 30.44 \text{ inches}$$

where a is the age of the boy in years. (**Source:** Modeled from www.babybag.com data)

According to the models, how quickly is the height of a girl increasing when she is 10 years old?

Exercises 22 and 23 are intended to challenge your understanding of the Chain Rule.

22. Find the derivative of $f(x) = ((x^2 + 1)^2 + 1)^2 + 1$.

23. Given that $\dfrac{dy}{dt} = 5$ people per day, $\dfrac{dt}{dx} = 2$ days per year, and $\dfrac{dp}{dy} = 4$ sales per person, calculate $\dfrac{dp}{dx}$, including units.

4.4 Exponential and Logarithmic Rules

The per capita consumption of bottled water has risen dramatically over the past two decades, from around 9 gallons in 1990 to more than 27 gallons in 2006. Based on data from 1990 to 2006, the annual per capita bottled water consumption may be modeled by

$$W(t) = 8.42(1.0755)^t \text{ gallons}$$

where t is the number of years since 1990. (**Source:** Modeled from *Statistical Abstract of the United States, 2010*, Table 207) According to the model, at what rate was consumption increasing in 2010?

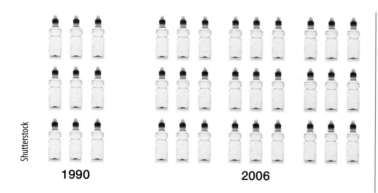

1990   **2006**

Questions such as these may be answered using the Exponential Rule for derivatives.

In this section, we will introduce the Exponential Rule and the Logarithmic Rule for derivatives. These rules will allow us to calculate the exact derivative of these types of functions quickly and accurately instead of relying on a numerical estimate as we have done previously.

Exponential Rule

Recall that an exponential function may be written in the form $y = ab^x$, where a and b are constants and b is positive and not equal to 1. Since the variable of an exponential function is in the exponent, the Power Rule cannot be used to find the derivative of this type of function.

Exponential Rule

The derivative of a function $f(x) = b^x$ is given by

$$f'(x) = (\ln b)b^x$$

This rule is not entirely obvious. To better understand its origin, let's consider the limit definition of the derivative of an exponential function.

$$f'(x) = \lim_{h \to 0} \frac{f(x + h) - f(x)}{h}$$

$$= \lim_{h \to 0} \frac{b^{x+h} - b^x}{h} \qquad \text{Since } f(x) = b^x$$

$$= \lim_{h \to 0} \frac{b^x b^h - b^x}{h} \qquad \text{Algebraic rules of exponents}$$

$$= \lim_{h \to 0} \frac{b^x(b^h - 1)}{h}$$

$$= b^x \lim_{h \to 0} \frac{(b^h - 1)}{h} \qquad \text{Since } b^x \text{ is independent of the value of } h$$

Although we cannot simplify $\lim\limits_{h \to 0} \dfrac{(b^h - 1)}{h}$ algebraically, it turns out that $\lim\limits_{h \to 0} \dfrac{(b^h - 1)}{h} = \ln(b)$. Consequently,

$$f'(x) = b^x \lim_{h \to 0} \frac{(b^h - 1)}{h}$$

$$= (\ln b)b^x$$

EXAMPLE 1 Finding the Derivative of an Exponential Function

Find the derivative of $g(x) = 2^x$.

SOLUTION

$$g'(x) = (\ln 2)2^x$$

Exponential functions frequently use the number e as the base. Recall that $e \approx 2.7182818$.

EXAMPLE 2 Finding the Derivative of an Exponential Function

Differentiate $y = e^x$.

SOLUTION

$$\frac{dy}{dx} = (\ln e)e^x$$

$$= 1 \cdot e^x \qquad \text{Since } \ln e = 1$$

$$= e^x$$

The derivative of e^x is e^x! This quirky result makes the function $y = e^x$ one of the most popular functions in calculus. For the function $f(x) = e^x$, the instantaneous rate of change in the function at a point $(a, f(a))$ is the same as $f(a)$. In other words, the slope of the tangent line to the graph of $f(x) = e^x$ at $x = a$ is e^a.

Exponential Rule: Special Case

The derivative of a function $f(x) = e^x$ is given by

$$f'(x) = e^x$$

EXAMPLE 3 Finding the Derivative of an Exponential Function

Calculate $\dfrac{d}{dx}(3 \cdot 4^x)$.

SOLUTION

$$\frac{d}{dx}(3 \cdot 4^x) = 3 \cdot \frac{d}{dx}(4^x) \qquad \text{Constant Multiple Rule}$$

$$= 3(\ln 4)4^x \qquad \text{Exponential Rule}$$

EXAMPLE 4 Finding the Derivative by Using the Exponential and Chain Rules

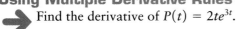 Find y' given $y = 3^{x^2+x}$.

SOLUTION

This function is different because there is a function in the exponent rather than a single variable. Consequently, we'll have to use the Chain Rule in conjunction with the Exponential Rule to find the derivative.

$$y' = (\ln 3)3^{x^2+x} \cdot \frac{d}{dx}(x^2 + x) \qquad \text{Exponential and Chain Rules}$$

$$= (\ln 3)3^{x^2+x} \cdot (2x + 1) \qquad \text{Power Rule}$$

$$= (2x + 1)(\ln 3)3^{x^2+x}$$

In general, a function of the form $y = e^u$, where u is some function of x, will have the derivative $y' = e^u \cdot u'$. This is referred to as the Generalized Exponential Rule; it is a special application of the Chain Rule together with the Exponential Rule.

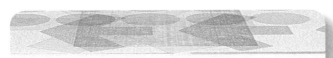

Generalized Exponential Rule

Let $u = g(x)$. The derivative of a function $f(x) = b^u$ is given by

$$f'(x) = (\ln b)b^u \, b'$$

If $b = e$, the rule simplifies to

$$f'(x) = e^u u'$$

EXAMPLE 5 Finding the Derivative by Using Multiple Derivative Rules

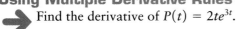 Find the derivative of $P(t) = 2te^{3t}$.

SOLUTION

Although this function looks relatively simple, there is a lot going on. To find the derivative, we must use the Constant Multiple Rule, the Power Rule, the Product Rule, and the Generalized Exponential Rule.

$$\frac{dP}{dt} = \frac{d}{dt}(2t) \cdot e^{3t} + \frac{d}{dt}(e^{3t}) \cdot (2t) \qquad \text{Product Rule}$$

$$= 2e^{3t} + (e^{3t} \cdot 3) \cdot (2t) \qquad \text{Constant Multiple, Generalized Exponential, and Power Rules}$$

$$= 2e^{3t} + 6te^{3t}$$

$$= e^{3t}(2 + 6t)$$

$$= 2e^{3t}(3t + 1)$$

EXAMPLE 6 Finding the Derivative by Using the Generalized Exponential Rule

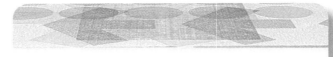 The annual per capita bottled water consumption in the U.S. may be modeled by

$$W(t) = 8.42(1.0755)^t \text{ gallons}$$

where t is the number of years since 1990. According to the model, at what rate was consumption changing in 2010?

SOLUTION

$$\frac{dW}{dt} = 8.42\ln(1.0755)(1.0755)^t$$

$$\approx 0.6129(1.0755)^t \text{ gallons per year} \qquad \text{Evaluate ln(1.0755) and multiply by 8.42}$$

In 2010, $t = 20$, so

$$\left.\frac{dW}{dt}\right|_{t=20} = 0.6129(1.0755)^{20} \text{ gallons per year}$$

$$\approx 2.6$$

In 2010, the per capita bottled water consumption was increasing by 2.6 gallons per year.

Logarithmic Rule

What is the derivative of a function of the form $y = \log_b x$? The answer is not obvious. We will state the rule here and show how the formula was determined when we discuss implicit differentiation in a subsequent section.

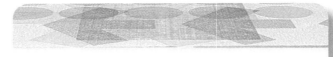

Logarithmic Rule

The derivative of a function $f(x) = \log_b x$ is given by

$$f'(x) = \frac{1}{\ln b} \cdot \frac{1}{x}$$

EXAMPLE 7 Finding the Derivative of a Logarithmic Function

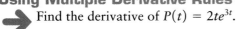 Differentiate $y = \ln x$.

SOLUTION

Recall that $\ln x = \log_e x$.

$$y' = \frac{1}{\ln e} \cdot \frac{1}{x}$$

$$= \frac{1}{1} \cdot \frac{1}{x} \qquad \text{Since ln } e = 1$$

$$= \frac{1}{x}$$

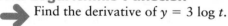

Logarithmic Rule: Special Case

The derivative of a function $f(x) = \ln x$ is given by

$$f'(x) = \frac{1}{x}$$

EXAMPLE 8 Finding the Derivative of a Logarithmic Function

Find the derivative of $y = 3 \log t$.

SOLUTION

Recall that $\log t$ means $\log_{10} t$.

$$\frac{dy}{dt} = \frac{d}{dt}(3 \log t)$$

$$= 3 \cdot \frac{d}{dt}(\log_{10} t) \qquad \text{Constant Multiple Rule}$$

$$= 3 \cdot \frac{1}{\ln 10} \cdot \frac{1}{t} \qquad \text{Logarithmic Rule}$$

$$= \frac{3}{\ln 10} \cdot \frac{1}{t}$$

Generalized Logarithmic Rule

The derivative of a function $f(x) = \ln u$ where u is a function of x is given by

$$f'(x) = \frac{1}{u}u'.$$

The Generalized Logarithmic Rule is simply a combination of the Logarithmic Rule and the Chain Rule.

EXAMPLE 9 Finding the Derivative by Using the Generalized Logarithmic Rule

Find $\dfrac{dy}{dx}$ for $y = \log(x^2)$.

SOLUTION

Since y is the composition of $f(x) = \log x$ and $g(x) = x^2$, we must use the Chain Rule in conjunction with the Logarithmic Rule.

$$\frac{dy}{dx} = \left(\frac{1}{\ln 10} \cdot \frac{1}{x^2}\right) \cdot \frac{d}{dx}(x^2)$$
Generalized Logarithmic Rule

$$= \left(\frac{1}{\ln 10} \cdot \frac{1}{x^2}\right) \cdot (2x) \qquad \text{Power Rule}$$

$$= \frac{1}{\ln 10} \cdot \frac{2x}{x^2}$$

$$= \frac{2}{(\ln 10)x}$$

EXAMPLE 10 Applying the Logarithmic Rule in a Real-World Context

Based on data from the National Science Foundation, the percentage of Americans who use the Internet may be modeled by

$$p(t) = 2.092 + 30.44\ln t \text{ percent}$$

where t is years since 1995. (**Source:** Modeled from Science and Engineering Indicators 2006, National Science Foundation, Table 7-8)

Calculate and interpret the meaning of $p'(16)$.

SOLUTION

$$p'(t) = 30.44\left(\frac{1}{t}\right) \qquad \begin{array}{l}\text{Constant and}\\\text{Logarithmic Rules}\end{array}$$

$$p'(16) = \frac{30.44}{16} \frac{\text{percentage points}}{\text{year}}$$

$$= 1.90 \text{ percentage points per year}$$

In 2011 ($t = 16$), the percentage of Americans who use the Internet is predicted to be increasing at a rate of 1.90 percentage points per year.

4.4 Exercises

In Exercises 1–10, use the Exponential Rule or the Logarithmic Rule, as appropriate, to find the derivative of the function. Additional derivative rules may also be needed.

1. $y = 4^x$

2. $g(t) = 3(5)^t$

3. $s(t) = 4 \ln t$

4. $f(t) = 3 \log t$

5. $y = 5e^x + \ln x$

6. $P(n) = 4n^2(2^n)$

7. $f(x) = 5e^x \ln x$

8. $g(x) = e^{2x-1}$

9. $y = 2^{x^2 + 5x}$

10. $h(x) = \ln(x^3 + 3x)$

In Exercises 11–15, determine the slope of the function at the indicated domain value.

11. $S(n) = \log(n^2 - 4); n = 3$

12. $w(t) = \log(2^t); t = -1$

13. $H(t) = 3^{\ln t}; t = e$

14. $f(t) = 3 \log(5^t); t = 0$

15. $C(n) = e^{n^2 + 2n - 1}; n = 1$

In Exercises 16–18, determine the equation of the tangent line at the indicated domain value.

16. $f(t) = 3^t; t = 1$

17. $g(x) = \ln(4x + 1); x = 1$

18. $f(x) = \dfrac{\ln x}{x}; x = e$

In Exercises 19 and 20, answer the questions by using the Exponential Rule or the Logarithmic Rule, as appropriate.

19. *College Tuition* Based on data from 1994 to 2001, the quarterly tuition of a resident student at Green River Community College may be modeled by

$$E(t) = 430.6(1.042)^t \text{ dollars}$$

where t is the number of years since 1994. (**Source:** Modeled from Green River Community College data)

How quickly was tuition increasing in 2001?

20. *School Spending* Based on data from the 1990–1991 through the 2002–2003 school year, the total school expenditure, in billions of dollars, may be modeled by

$$T(E) = -2915 + 366.6 \ln E$$

where E is the per pupil expenditure in dollars. (**Source:** Modeled from National Center for Education Statistics, Table 34)

Calculate and interpret the meaning of $T'(9000)$.

4.5 Implicit Differentiation

The intent of this section is to give you the skills necessary to work the related-rate problems in the next chapter. We will show you how to use *implicit differentiation* to find the derivative of functions and nonfunctions. Knowing how to do implicit differentiation is a key skill that is necessary for our discussion of related rates.

The equation of a circle may be written as $x^2 + y^2 = r^2$, where r is the radius of the circle. A circle is not a function because it fails the Vertical Line Test. However, we can draw tangent lines to the graph of a circle. Since the derivative is the slope of the tangent line, we should be able to find the derivative at any point on the circle. However, since most x-values have two different y-values, the derivative will be a function of both x and y.

Consider $x^2 + y^2 = 1$ and its associated graph (Figure 4.3). What is the slope of each of the tangent

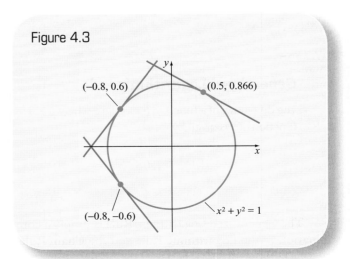

Figure 4.3

lines shown? In other words, what is the derivative at the indicated points?

We will find the derivative by using **implicit differentiation.** We begin by taking the derivative of both sides of the equation with respect to x. Since y is a function of x, $\frac{d}{dx}(y) = \frac{dy}{dx}$. This fact will become important as we move through the problem.

$$x^2 + y^2 = 1$$

$$\frac{d}{dx}(x^2 + y^2) = \frac{d}{dx}(1)$$

$$\frac{d}{dx}(x^2) + \frac{d}{dx}(y^2) = 0 \qquad \text{Sum and Difference and Constant Rules}$$

$$2x + 2y \cdot \frac{d}{dx}(y) = 0 \qquad \text{Power and Chain Rules}$$

$$2x + 2y \cdot \frac{dy}{dx} = 0$$

$$2y\frac{dy}{dx} = -2x$$

$$\frac{dy}{dx} = \frac{-2x}{2y}$$

$$\frac{dy}{dx} = -\frac{x}{y}$$

Note that the derivative is a function of both x and y. We can now calculate the slope of the tangent line at each of the points shown.

At $(-0.8, 0.6)$, $\frac{dy}{dx} = -\frac{-0.8}{0.6} \approx 1.333.$

At $(-0.8, -0.6)$, $\frac{dy}{dx} = -\frac{-0.8}{-0.6} \approx -1.333.$

At $(0.5, 0.866)$, $\frac{dy}{dx} = -\frac{0.5}{0.866} \approx -0.5774.$

Steps of Implicit Differentiation

1. Differentiate both sides of the equation with respect to x. $\left(\text{Recall } \frac{d}{dx}(y) = \frac{dy}{dx}. \right)$

2. Algebraically isolate the $\frac{dy}{dx}$ term.

EXAMPLE 1 Finding $\frac{dy}{dx}$ by Using Implicit Differentiation

Differentiate $xy = 1$ using implicit differentiation.

SOLUTION

$$\frac{d}{dx}(xy) = \frac{d}{dx}(1)$$

$$\frac{d}{dx}(x) \cdot y + \frac{d}{dx}(y) \cdot x = 0 \qquad \text{Product Rule}$$

$$1 \cdot y + \frac{dy}{dx} \cdot x = 0 \qquad \text{Power Rule}$$

$$y + x\frac{dy}{dx} = 0$$

$$x\frac{dy}{dx} = -y$$

$$\frac{dy}{dx} = -\frac{y}{x}$$

If we first solve the equation for y and then differentiate, we get a very different-looking result.

$$xy = 1$$

$$y = \frac{1}{x}$$

$$\frac{d}{dx}(y) = \frac{d}{dx}\left(\frac{1}{x}\right)$$

$$\frac{dy}{dx} = \frac{d}{dx}(x^{-1})$$

$$= -x^{-2} \qquad \text{Power Rule}$$

$$= -\frac{1}{x^2}$$

Is this answer equivalent to our implicit differentiation result? Let's see.

$$\frac{dy}{dx} = -\frac{y}{x}$$

$$= -\frac{1}{x}y$$

$$= -\frac{1}{x} \cdot \frac{1}{x} \qquad \text{Since } y = \frac{1}{x}$$

$$= -\frac{1}{x^2}$$

The two equations are equivalent. When using implicit differentiation, it is common for correct answers to look very different from each other.

EXAMPLE 2 Finding $\dfrac{dy}{dx}$ by Using Implicit Differentiation

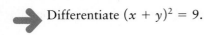 Differentiate $(x + y)^2 = 9$.

SOLUTION

$$\frac{d}{dx}(x + y)^2 = \frac{d}{dx}(9)$$

$$2(x + y) \cdot \frac{d}{dx}(x + y) = 0 \qquad \text{Constant and Chain Rules}$$

$$2(x + y) \cdot \left(1 + \frac{dy}{dx}\right) = 0 \qquad \text{Power Rule}$$

$$\left(1 + \frac{dy}{dx}\right) = \frac{0}{2(x + y)}$$

$$1 + \frac{dy}{dx} = 0$$

$$\frac{dy}{dx} = -1$$

For all values of x and y on the graph of $(x + y)^2 = 9$, the slope of the tangent line is -1.

Let's solve the original equation for y.

$$(x + y)^2 = 9$$

$$\sqrt{(x + y)^2} = \sqrt{9}$$

$$(x + y) = \pm 3$$

$$y = -x \pm 3$$

When we graph $y = -x - 3$ and $y = -x + 3$ simultaneously, we see a pair of parallel lines with slope -1 (Figure 4.4). This is consistent with the implicit differentiation result.

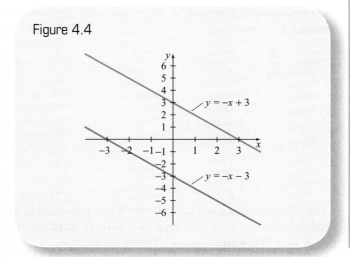

Figure 4.4

EXAMPLE 3 Finding $\dfrac{dy}{dx}$ by Using Implicit Differentiation

Find the derivative of $y^2 + 2y + 1 = x$ with respect to x.

SOLUTION

$$\frac{d}{dx}(y^2 + 2y + 1) = \frac{d}{dx}(x)$$

$$2y \cdot \frac{d}{dx}(y) + 2 \cdot \frac{d}{dx}(y) + 0 = 1 \qquad \text{Power, Chain, and Constant Rules}$$

$$2y \cdot \frac{dy}{dx} + 2 \cdot \frac{dy}{dx} = 1$$

$$\frac{dy}{dx}(2y + 2) = 1$$

$$\frac{dy}{dx} = \frac{1}{(2y + 2)}$$

EXAMPLE 4 Proving $\dfrac{d}{dx}\,(\log_b(x)) = \dfrac{1}{\ln(b)\cdot x}$

Prove that the derivative of $y = \log_b(x)$ is $y' = \dfrac{1}{\ln(b)\cdot x}$.

SOLUTION

We will do some clever manipulations with the function and then use the Exponential Rule to find the derivative.

Recall that according to the definition of logarithms, the following two equations are equivalent:

$$y = \log_b(x) \text{ and}$$

$$b^y = x$$

Differentiating the second equation with respect to x yields

$$\frac{d}{dx}(b^y) = \frac{d}{dx}(x)$$

$$(\ln b)b^y \cdot \frac{d}{dx}(y) = 1 \qquad \text{Chain, Exponential, and Power Rules}$$

$$(\ln b)b^y \cdot \frac{dy}{dx} = 1 \qquad \text{Chain Rule}$$

$$\frac{dy}{dx} = \frac{1}{(\ln b)b^y}$$

$$= \frac{1}{(\ln b)x} \qquad \text{Since } b^y = x$$

EXAMPLE 5 Finding $\dfrac{dy}{dx}$ by Using Implicit Differentiation

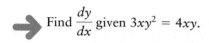 Find $\dfrac{dy}{dx}$ given $3xy^2 = 4xy$.

SOLUTION

$$\frac{d}{dx}(3xy^2) = \frac{d}{dx}(4xy)$$

$$\frac{d}{dx}(3x) \cdot y^2 + \frac{d}{dx}(y^2) \cdot 3x = \frac{d}{dx}(4x) \cdot y + \frac{d}{dx}(y) \cdot 4x$$

Product Rule

$$(3)y^2 + \left(2y\frac{dy}{dx}\right) \cdot 3x = (4)y + \frac{dy}{dx} \cdot 4x$$

Constant Multiple, Power, and Chain Rules

$$3y^2 + 6xy\frac{dy}{dx} = 4y + 4x\frac{dy}{dx}$$

$$6xy\frac{dy}{dx} - 4x\frac{dy}{dx} = 4y - 3y^2$$

$$\frac{dy}{dx}(6xy - 4x) = 4y - 3y^2$$

$$\frac{dy}{dx} = \frac{4y - 3y^2}{6xy - 4x}$$

4.5 Exercises

In Exercises 1–10, use implicit differentiation to find $\dfrac{dy}{dx}$. Then evaluate the derivative function at the designated point.

1. $2x^2 + y^2 = 32; (4, 0)$

2. $x^2 - y^2 = 9; (5, 4)$

3. $x^2y - 3 = y^2; (2, 1)$

4. $4y^2 = x^2; (4, 2)$

5. $xy = 6; (2, 3)$

6. $x^2y - xy^2 = 20; (5, 4)$

7. $9x - y^3 = 0; (3, 3)$

8. $xy^2 + xy + y = -1; (3, -1)$

9. $20y - x^2y^2 = 75; (1, 5)$

10. $xy^3 - x^3y = 30; (-2, -3)$

In Exercises 11–15, use implicit differentiation to find $\dfrac{dy}{dx}$.

11. $y^2 = x$

12. $ye^y = 2$

13. $\ln y = xy$

14. $2xy - y^2 = 9$ $\qquad x(x^2 - y^2) = y$

15. $\ln(xy^2) = y$

In Exercises 16–18, determine at what points, if any, $\dfrac{dy}{dx} = 0$.

16. $x^2 + y^2 = 1$

17. $x^2 + 2y^2 = y$

18. $x^2 + xy = y^2 - 5$

Exercises 19 and 20 are intended to challenge your understanding of implicit differentiation.

19. Given $V = \pi r^2 h$, write $\dfrac{dr}{dt}$ in terms of $\dfrac{dV}{dt}, \dfrac{dh}{dt}, h,$ and r. (*Hint:* Differentiate both sides with respect to t.)

20. Given the equation $ye^x = xe^y$, find the solution to the equation $\dfrac{dy}{dx} = 0$, if it exists.

Chapter 5

Derivative Applications

Businesses survive by being profitable. A savvy business owner effectively analyzes the factors that contribute to the financial success or failure of his business. Using mathematical models, he may forecast what prices, production levels, shipment schedules, and other such elements will result in maximum profits. Although no mathematical model is perfect at forecasting future results, a model can help a business owner make informed decisions.

5.1 Maxima and Minima

5.2 Applications of Maxima and Minima

5.3 Concavity and the Second Derivative

5.4 Related Rates

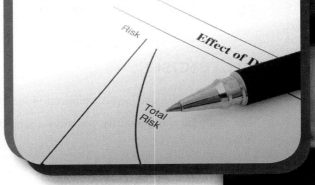

5.1 Maxima and Minima

Based on data from 1995 to 2005, the juvenile arrest rate for curfew violations may be modeled by

$$A(t) = 0.892t^3 - 13.5t^2 + 22.3t + 579$$

where A is the juvenile arrest rate per 100,000 juveniles and t is the number of years since 1995. (**Source:** Modeled from Office of Justice Programs, U.S. Department of Justice statistics)

In what year between 1995 and 2005 did the juvenile arrest rate for curfew violations reach a maximum? Questions such as these may be answered using the concepts of relative and absolute extrema.

In this section, we will informally discuss continuity. In addition, we will discuss how to find *relative extrema* and *absolute extrema*.

Continuity

The notion of continuity is best understood graphically. Loosely speaking, a function is said to be **continuous** if its graph can be drawn by a pencil without lifting the pencil from the page. If there is a break in the graph, the graph is said to be **discontinuous**. This loose definition of continuity will be sufficient for our purposes. (The formal definition of continuity, which relies heavily on the concept of limits, may be obtained from a traditional calculus text.)

Many functions are continuous. For example, linear, polynomial, exponential, and logarithmic functions are all continuous. Frequently, functions that are discontinuous have domain restrictions or are defined piecewise. Consider the piecewise function $f(x) = \begin{cases} 2x + 6 & \text{if } x < 1 \\ 3x + 2 & \text{if } x \geq 1 \end{cases}$ shown in Figure 5.1. Since there is a break in the graph at $x = 1$, the graph is said to be discontinuous.

Figure 5.1

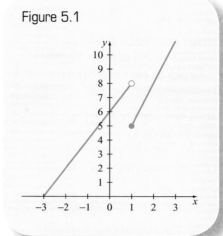

Relative and Absolute Extrema

An **extremum** is a maximum or minimum value of a function. The plural of *extremum* is **extrema**.

Relative Extrema

- A **relative maximum** of a continuous function f occurs at a point $(c, f(c))$ if $f(x) \leq f(c)$ for all x in some interval (a, b) containing c. (That is, $a < c < b$.)

- A **relative minimum** of a continuous function f occurs at a point $(c, f(c))$ if $f(x) \geq f(c)$ for all x in some interval (a, b) containing c. (That is, $a < c < b$.)

Absolute Extrema

- An **absolute maximum** of a function f occurs at a point $(c, f(c))$ if $f(x) \leq f(c)$ for all x in the domain of f.

- An **absolute minimum** of a function f occurs at a point $(c, f(c))$ if $f(x) \geq f(c)$ for all x in the domain of f.

According to the definition, a relative extremum may not occur at an endpoint. We will explore the concept of relative extrema by looking at the graph of $f(x) = x^3 - 3x^2$ on the domain $[-2, 4]$ (Figure 5.2).

Figure 5.2

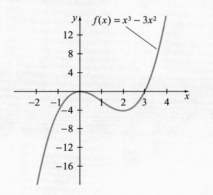

Graphically, we are looking for the peaks and valleys of the graph. In this case, a peak occurs at $(0, 0)$ and a valley occurs at $(2, -4)$. A relative maximum occurs at the point $(0, 0)$, since $y = 0$ is larger than all y-values of the function *nearby* $x = 0$. (When we use the term *nearby,* we mean an arbitrarily small open interval (a, b) surrounding $x = 0$.) A relative minimum occurs at the point $(2, -4)$, since $y = -4$ is smaller than all y-values of the function nearby $x = 2$.

It is important to note that unlike relative extrema, *absolute extrema may occur at endpoints.*

To find the absolute maximum of the function $f(x) = x^3 - 3x^2$ on the domain $[-2, 4]$ we look for the largest y-value. Since f is continuous (there are no breaks in the graph of f), we need only check the relative maxima and the y-values of the endpoints. The

graph of f has a relative maximum at $(0, 0)$ and endpoints at $(-2, -20)$ and $(4, 16)$. Since $(4, 16)$ has the largest y-value, $y = 16$ is the absolute maximum.

To find the absolute minimum, we are looking for the smallest y-value. Since f is continuous, we need only check the relative minima and the endpoints. The graph of f has a relative minimum at $(2, -4)$ and endpoints at $(-2, -20)$ and $(4, 16)$. Since $(-2, -20)$ has the smallest (most negative) y-value, $y = -20$ is the absolute minimum (Figure 5.3).

Figure 5.3

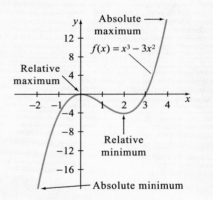

Precisely determining relative extrema is difficult to do graphically, especially when extrema occur at irrational points such as $(\pi, \sqrt{2})$. Fortunately, by using the derivative, we can quickly find relative extrema.

The derivative of $f(x) = x^3 - 3x^2$ is $f'(x) = 3x^2 - 6x$. Recall that the derivative at a point represents the slope of the tangent line (or slope of the graph) at that point. If the slope is positive, the graph is increasing. If the slope is negative, the graph is decreasing. If the slope is zero, the graph is neither increasing nor decreasing but remains flat. In other words, the graph has a horizontal tangent line if and only if $f'(x) = 0$. For $f(x) = x^3 - 3x^2$, we have the results shown in Table 5.1.

Observe that $f'(x) = 0$ where the relative extrema occurred $((0, 0)$ and $(2, -4))$.

Table 5.1

x	$f(x)$	$f'(x)$	Slope of Tangent Line	Graph of f
−2	−20	24	Positive	Increasing
−1	−4	9	Positive	Increasing
0	0	0	Zero	Flat
1	−2	−3	Negative	Decreasing
2	−4	0	Zero	Flat
3	0	9	Positive	Increasing
4	16	24	Positive	Increasing

The Value of the Derivative at a Relative Extremum

Let f be a continuous function. If a relative extremum of f occurs at $(c, f(c))$, then $f'(c) = 0$ or $f'(c)$ is undefined.

The converse is not true. That is, if $f'(c) = 0$ or $f'(c)$ is undefined, then a relative extremum is not guaranteed to occur at $(c, f(c))$. We will demonstrate this in Example 2.

EXAMPLE 1 Determining from a Graph Where $f'(x) = 0$

The graph of $f(x) = x^3 - 12x$ on the interval $[-3, 3]$ is shown (Figure 5.4). Determine the points on the graph where $f'(x) = 0$.

Figure 5.4

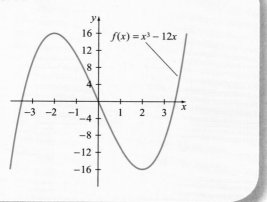

SOLUTION

Since the graph is a smooth curve (no sharp points), the derivative will equal zero at the relative extrema. It appears that a relative maximum occurs when $x = -2$ and a relative minimum occurs when $x = 2$. We will now use the derivative to confirm our observation.

$$f(x) = x^3 - 12x$$
$$f'(x) = 3x^2 - 12$$
$$f'(2) = 3(2)^2 - 12$$
$$= 0$$
$$f'(-2) = 3(-2)^2 - 12$$
$$= 0$$

The results confirm our graphical estimate. At the point $(-2, 16)$ and the point $(2, -16)$, $f'(x) = 0$.

EXAMPLE 2 Determining from a Graph Where $f'(x) = 0$

The graph of $f(x) = x^4 - 4x^3$ on the interval $[-3, 5]$ is shown in Figure 5.5. Determine the points on the graph where $f'(x) = 0$.

Figure 5.5

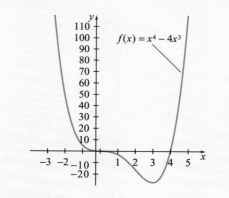

SOLUTION

The graph is a continuous, smooth curve. We are looking for places on the graph where the function has a horizontal tangent line ($f'(x) = 0$). It looks like the graph has a horizontal tangent line at $x = 0$ and $x = 3$. We will confirm this result by evaluating $f'(x)$ at $x = 0$ and $x = 3$.

$$f(x) = x^4 - 4x^3$$
$$f'(x) = 4x^3 - 12x^2$$
$$= 4x^2(x - 3)$$

$$f'(0) = 4(0)^2(0 - 3)$$

$$= 0$$

$$f'(3) = 4(3)^2(3 - 3)$$

$$= 0$$

So $f'(x) = 0$ at $(0, 0)$ and $(3, -27)$.

As discussed previously, the fact that $f'(c) = 0$ does not guarantee that a relative extremum occurs at $(c, f(c))$. In this example, a relative minimum occurred at $(3, -27)$, but a relative extremum did *not* occur at $(0, 0)$.

In Example 3, we will use the absolute value function. Recall that the absolute value function is formally defined as a piecewise function.

$$|x| = \begin{cases} x & \text{if } x \geq 0 \\ -x & \text{if } x < 0 \end{cases}$$

The graph of the absolute value function looks like a V (Figure 5.6).

Figure 5.6

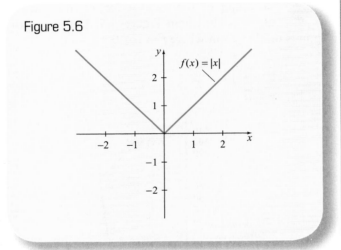

Since the graph is decreasing for $x < 0$, the slope of the tangent line of f is negative for $x < 0$. Since the graph is increasing for $x > 0$, the slope of the tangent line of f is positive for $x > 0$. But what is the slope of the tangent line at $x = 0$? Recall that the derivative of a function f at $x = a$ is formally defined as

$$f'(a) = \lim_{h \to 0} \frac{f(a + h) - f(a)}{h}$$

For $f(x) = |x|$, the derivative of the function at $x = 0$ is

$$f'(0) = \lim_{h \to 0} \frac{|0 + h| - |0|}{h}$$

$$= \lim_{h \to 0} \frac{|h|}{h}$$

As h gets close to 0, what value does $\dfrac{|h|}{h}$ approach? We first observe that $h \neq 0$ because division by 0 is not defined. From the piecewise definition of the absolute value function, we know

$$\frac{|h|}{h} = \begin{cases} \dfrac{h}{h} & \text{if } h > 0 \\ \text{undefined} & \text{if } h = 0 \\ \dfrac{-h}{h} & \text{if } h < 0 \end{cases}$$

Simplifying the equation yields

$$\frac{|h|}{h} = \begin{cases} 1 & \text{if } h > 0 \\ \text{undefined} & \text{if } h = 0 \\ -1 & \text{if } h < 0 \end{cases}$$

For positive values of h, $\dfrac{|h|}{h} = 1$. For negative values of h, $\dfrac{|h|}{h} = -1$. Since $\dfrac{|h|}{h}$ doesn't approach a constant value as h nears 0, $\lim_{h \to 0} \dfrac{|h|}{h}$ does not exist. That is, for $f(x) = |x|$, $f'(0)$ is undefined. In other words, the derivative of the absolute value function does not exist at $x = 0$. In general, the derivative of any function does not exist wherever the graph of a function has a "sharp point."

EXAMPLE 3 Finding the Relative Extrema of a Function

The graph of $f(x) = -|x| + 2$ is shown in Figure 5.7. Find the relative extrema of f on the interval $[-3, 3]$.

Figure 5.7

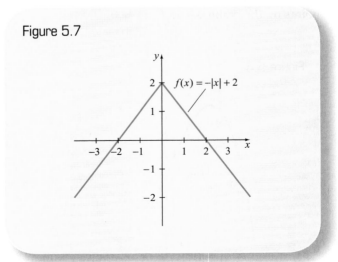

SOLUTION

From the graph, we can see that the function has a relative maximum at $(0, 2)$. Therefore, $f'(x) = 0$ or $f'(x)$ is undefined at that point. At any sharp point on a graph, the derivative is undefined. Therefore, $f'(0)$ is undefined. Does $f'(x) = 0$ anywhere? From the graph, we see that f is increasing for all $x < 0$. Therefore, $f'(x)$ is positive for $x < 0$. Similarly, f is decreasing for all $x > 0$. Therefore, $f'(x)$ is negative for $x > 0$. Hence, $f'(x) \neq 0$ anywhere. Thus the only relative extremum occurs at $(0, 2)$.

Critical Values

In Example 2, we saw that relative extrema may occur where $f'(x) = 0$. In Example 3, we saw that relative extrema may also occur where $f'(x)$ is undefined. These two situations represent two different types of *critical values* of a function.

Critical Values

Let f be a continuous function on the interval (a, b) with $a < c < b$. If

- $f'(c) = 0$ or

- $f'(c)$ is undefined

then c is called a **critical value** of f.

If $f'(c) = 0$, the critical value is called a **stationary value**. If $f'(c)$ is undefined, the critical value is called a **singular value**. The corresponding **critical point** $(c, f(c))$ is called a **stationary point** if $f'(c) = 0$ and a **singular point** if $f'(c)$ is undefined.

Existence of Relative Extrema

Let f be a continuous function on the interval (a, b). All relative extrema of f occur at critical values of f.

EXAMPLE 4 Finding Critical Values of a Function

Find the critical values of $f(x) = x^3 - 3x$ on the interval $[-2, 4]$.

SOLUTION

To find the stationary values, we set the derivative equal to zero and solve for x.

$$f'(x) = 3x^2 - 3$$

$$0 = 3x^2 - 3 \qquad \text{Set } f'(x) = 0$$

$$0 = 3(x^2 - 1)$$

$$0 = 3(x - 1)(x + 1)$$

$$x - 1 = 0 \qquad x + 1 = 0 \qquad \text{Set each factor equal to 0}$$
$$\qquad\qquad\qquad\qquad\qquad\qquad \text{and solve for } x$$

$$x = 1 \qquad\quad x = -1$$

The stationary values are $x = 1$ and $x = -1$. Since the derivative is defined for all values of x, there are no singular values.

EXAMPLE 5 Finding Critical Points of a Function

Find the critical points of $f(x) = x^{1/3}$ on the interval $[-2, 2]$.

SOLUTION

$$f'(x) = \frac{1}{3} x^{-2/3}$$

$$= \frac{1}{3x^{2/3}}$$

$f'(x) \neq 0$ for all values of x, since a fraction can equal zero only if its numerator is zero. Therefore, f does not have any stationary points.

A rational expression $\dfrac{g(x)}{h(x)}$ is undefined when $h(x)$ is equal to zero. What value of x makes the denominator of $f'(x)$ equal zero?

$$0 = 3x^{2/3} \qquad \text{Set the denominator equal to 0}$$

$$0 = x^{2/3} \qquad \text{Divide by 3}$$

$$(0)^3 = (x^{2/3}) \qquad \text{Cube both sides of the equation}$$

$$0 = x^2 \qquad (x^{2/3})^3 = x^{(2/3)3} = x^2$$

$$x = 0 \qquad \text{Take the square root}$$

Therefore, $f'(0)$ is undefined. A singular point may occur at $x = 0$. We need only confirm that the function

f is defined at $x = 0$. Since $f(0) = 0$, $x = 0$ is in the domain of f and is indeed a singular value of f. The point $(0, 0)$ is a singular point of f.

The First Derivative Test

Is there a nongraphical way to tell whether a relative maximum or a relative minimum occurs at a critical value? Fortunately, yes. Recall that when a function is increasing, its derivative is positive. When a function is decreasing, its derivative is negative. If the derivative changes sign at a critical value, a relative extremum occurs at that value. The First Derivative Test will allow us to determine where relative extrema occur for *nonconstant* functions.

If f is a constant function ($f(x) = k$ for some constant k), then $f'(x) = 0$ for all x and all points of the function are critical points. Furthermore, since $k \leq f(x) \leq k$ for all x, both relative maxima and relative minima occur at all points of f. Since most real-life data models are not constant functions, we will focus our discussion on *nonconstant* functions.

First Derivative Test

Let $(c, f(c))$ be a critical point of a nonconstant, continuous function f.

- If f' changes from positive to negative at $x = c$, then a relative maximum occurs at $(c, f(c))$.

- If f' changes from negative to positive at $x = c$, then a relative minimum occurs at $(c, f(c))$.

- If f' doesn't change sign at $x = c$, then a relative extremum does not occur at $(c, f(c))$.

It is often helpful to develop a *sign chart* when determining where relative extrema occur. We will demonstrate this with an example and then detail the specific steps of the process.

EXAMPLE 6 Finding Relative and Absolute Extrema

Find the relative and absolute extrema of $f(x) = x^4 - 4x^3$ on the interval $[-1, 5]$.

SOLUTION

We will first find the relative extrema by using the First Derivative Test.

$$f'(x) = 4x^3 - 12x^2$$
$$= 4x^2(x - 3)$$

$4x^2 = 0$	$x - 3 = 0$	Set each factor equal to 0
$x = 0$	$x = 3$	Solve for x

Setting the derivative equal to zero, we see that $x = 0$ and $x = 3$ are critical values of f. We begin by drawing a number line with the critical values clearly marked.

We will evaluate $f'(x)$ at an x-value in each of the following three intervals: $[-1, 0)$, $(0, 3)$, and $(3, 5]$. (These intervals may be equivalently expressed as $-1 \leq x < 0$, $0 < x < 3$, and $3 < x \leq 5$, respectively.) We are not concerned with the actual value of the derivative. We simply want to know if the derivative is positive or negative.

From the interval $[-1, 0)$, we pick $x = -1$

$$f'(-1) = 4(-1)^2(-1 - 3)$$ The product is equivalent to $(4)(1)(-4)$

$$f'(-1) < 0$$ Since there are an odd number of negative factors

On the interval $[-1, 0)$, the derivative is negative. We update the chart to reflect our finding.

From the interval $(0, 3)$, we pick $x = 1$.

$$f'(1) = 4(1)^2(1 - 3)$$ The product is equivalent to $(4)(1)(-2)$

$$f'(-1) < 0$$ Since there are an odd number of negative factors

On the interval $(0, 3)$, the derivative is negative. We update the chart to reflect our finding.

From the interval $(3, 5]$, we pick $x = 4$.

$$f'(4) = 4(4)^2(4 - 3)$$ The product is equivalent to $(4)(16)(1)$

$$f'(4) > 0$$ Since there are no negative factors

On the interval $(3, 5]$, the derivative is positive. We update the chart to reflect our finding.

At the critical value $x = 0$, the derivative did not change sign, so a relative extremum does not occur there. Graphically speaking, the graph of f is decreasing through the whole interval $[1, 3]$. At the critical value $x = 3$, the derivative changed sign from negative to positive, so a relative minimum occurs there.

$$f' \xleftarrow{\quad \underset{0}{-} \quad\quad \underset{3}{\overset{\text{min}}{-}} \; + \quad}$$

Graphically speaking, the graph of f changed from decreasing to increasing at $x = 3$.

To find absolute extrema, we consider the relative extrema and evaluate the function f at the endpoints (see Table 5.2).

Table 5.2

x	$f(x)$	
-1	5	
3	-27	Relative minimum, absolute minimum
5	125	Absolute maximum

The largest value of $f(x)$ in the table is the absolute maximum, and the smallest value is the absolute minimum. In this case, the relative and absolute minimum occurs at $(3, -27)$ and is equal to -27. The absolute maximum occurs at $(5, 125)$ and is equal to 125. There is no relative maximum. (Recall from our definition of

a relative extremum that a relative extremum may not occur at an endpoint. This definition allows us to use critical values to find all relative extrema.)

EXAMPLE 7 Finding Relative and Absolute Extrema

Find the absolute and relative extrema of $g(x) = x^3 - 2x^2 - 5x + 6$ on the interval $[-3, 4]$.

SOLUTION

$$g'(x) = 3x^2 - 4x - 5$$

We set $g'(x) = 0$.

$$3x^2 - 4x - 5 = 0$$

This is a quadratic equation in the form $ax^2 + bx + c = 0$. Since the derivative doesn't factor, we may use the Quadratic Formula to find the stationary points (or use a calculator or computer).

$$x = \frac{-b \pm \sqrt{b^2 - 4ac}}{2a}$$
$$= \frac{-(-4) \pm \sqrt{(-4)^2 - 4(3)(-5)}}{2(3)}$$
$$= \frac{4 \pm \sqrt{76}}{6}$$
$$= \frac{4 \pm 2\sqrt{19}}{6}$$
$$= \frac{2}{3} \pm \frac{\sqrt{19}}{3}$$

Since $\frac{2}{3} + \frac{\sqrt{19}}{3} \approx 2.120$ and $\frac{2}{3} - \frac{\sqrt{19}}{3} \approx -0.7863$, the stationary values are $x \approx -0.7863$ and $x \approx 2.120$. The derivative is defined for all values of x, so there are no singular values. We now construct a sign chart.

$$g' \xleftarrow{\quad \underset{-0.7863}{} \quad\quad \underset{2.120}{} \quad}$$

To determine the sign of the derivative $g'(x) = 3x^2 - 4x - 5$ in each interval, we evaluate the derivative at $x = -1$, $x = 0$, and $x = 3$.

Constructing a Derivative Sign Chart

1. Label a number line with the critical values of the function f.

2. Write f' next to the number line.

3. Evaluate f' at a value in each of the number line intervals.

4. Record a "+" if the derivative is positive and a "−" if the derivative is negative.

5. Use the First Derivative Test to determine where relative maxima and minima occur and record the results on the chart.

$$g'(-1) = 3(-1)^2 - 4(-1) - 5$$
$$= 3 + 4 - 5$$
$$g'(-1) > 0$$
$$g'(0) = 3(0)^2 - 4(0) - 5$$
$$= 0 - 0 - 5$$
$$g'(0) < 0$$
$$g'(3) = 3(3)^2 - 4(3) - 5$$
$$= 27 - 12 - 5$$
$$g'(3) > 0$$

We record the sign of the derivative in each interval on the sign chart.

$$g' \xleftarrow[]{} \overset{+ \quad \text{max}}{\underset{-0.7863}{|}} \overset{- \quad \quad \text{min} \quad +}{\underset{2.120}{|}} \xrightarrow[]{}$$

A relative maximum occurs at

$$x = \frac{2}{3} - \frac{\sqrt{19}}{3} \approx -0.7863$$

and a relative minimum occurs at

$$x = \frac{2}{3} + \frac{\sqrt{19}}{3} \approx 2.120.$$

To find the absolute extrema, we consider the relative extrema and evaluate the function g at the endpoints, as shown in Table 5.3.

Table 5.3

x	$g(x)$	
–3	–24	Absolute minimum
–0.7863	8.209	Relative maximum
2.120	–4.061	Relative minimum
4	18	Absolute maximum

The absolute minimum occurs at $(-3, -24)$, although there is a relative minimum at $(2.120, -4.061)$. The absolute maximum occurs at $(4, 18)$, although there is a relative maximum at $(-0.7863, 8.209)$.

It is often helpful to use a graphing calculator to calculate values of the derivative, especially when evaluating the derivative at other than whole-number values. The instructions on the Chapter 5 Tech Card detail how to use your calculator to evaluate the derivative at a point quickly.

EXAMPLE 8 Interpreting Extrema in a Real-World Context

Based on data from 1995 to 2005, the juvenile arrest rate for curfew violations may be modeled by

$$A(t) = 0.892t^3 - 13.5t^2 + 22.3t + 579$$

where A is the juvenile arrest rate per 100,000 juveniles and t is the number of years since 1995. (**Source:** Modeled from Office of Justice Programs, U.S. Department of Justice statistics)

Find the relative and absolute extrema of $A(t)$ on the interval [0, 10] and interpret the meaning of the results.

SOLUTION

We must first find the critical values of $A(t)$.

$$A'(t) = 2.676t^2 - 27.0t + 22.3 \qquad \text{Set } A'(t) \text{ equal to 0}$$

$$0 = 2.676t^2 - 27.0t + 22.3$$

Using the graphing calculator, we determine $A'(0.908) = 0$ and $A'(9.18) = 0$.

The critical values are $t \approx 0.91$ and $t \approx 9.18$. Using these values together with the interval test values of $t = 0$, $t = 5$, and $t = 10$, we construct the first derivative sign chart.

$$A' \xleftarrow[]{} \overset{+ \quad \text{max of } A \quad - \quad \text{min of } A \quad +}{\underset{0.91 \qquad \qquad 9.18}{|\qquad \qquad |}} \xrightarrow[]{}$$

A has a relative minimum at $t \approx 9.18$ and a relative maximum at $t \approx 0.91$. Evaluating $A(t) = 0.892t^3 - 13.5t^2 + 22.3t + 579$ at the critical values and the endpoints, we construct a table of values, as shown in Table 5.4.

Table 5.4

t	$A(t)$	
0	579	
0.91	588.8	Relative and absolute maximum
9.18	336.1	Relative and absolute minimum
10	344	

According to the model, the juvenile arrest rate for curfew violations was increasing between the end of 1995 ($t = 0$) and late October 1996 ($t = 0.91$). It reached a relative high of 558.8 arrests per 100,000 juveniles in late October 1996. From late October 1996 to early March 2005, the juvenile arrest rate was decreasing. It reached a relative low of 336.1 arrests per 100,000 juveniles in early March 2005. Between early March 2005 and the end of 2005, the juvenile arrest rate increased. The highest juvenile arrest rate between 1995 and 2005 was 558.8 arrests per 100,000 juveniles in the one-year period ending in late October 1995, and the lowest juvenile arrest rate between 1995 and 2005 was 336.1 arrests per 100,000 juveniles in the one-year period ending in early March 2005.

EXAMPLE 9 Interpreting Extrema in a Real-World Context

Based on data from 1990 to 2006, the annual per capita bottled water consumption may be modeled by

$$W(t) = 8.42(1.0755)^t \text{ gallons}$$

where t is the number of years since 1990. (**Source:** Modeled from *Statistical Abstract of the United States, 2010*, Table 207) In what year between 1990 and 2006 was per capita bottled water consumption the greatest?

SOLUTION

We are asked to find the absolute maximum of $W(t) = 8.42(1.0755)^t$ on the interval $[0, 16]$. We first observe that W is an exponential function. Recall that the derivative of an exponential function $y = ab^x$ is $y' = a \ln(b)b^x$. Since $a \neq 0$ and $\ln(b) \neq 0$, $y' = 0$ only when $b^x = 0$. But no value of x can make $b^x = 0$. Similarly, no value of x can make b^x undefined. Therefore, an exponential function $y = ab^x$ doesn't have any critical values. Thus exponential functions do not have any relative extrema. So to determine when the per capita bottled water consumption was the greatest, we only need to evaluate $W(t) = 8.42(1.0755)^t$ at the endpoints, as shown in Table 5.5.

Table 5.5

t	$W(t)$	
0	8.42	Absolute minimum
16	26.98	Absolute maximum

According to the model, per capita bottled water consumption reached a maximum of 26.98 gallons at the end of 2006.

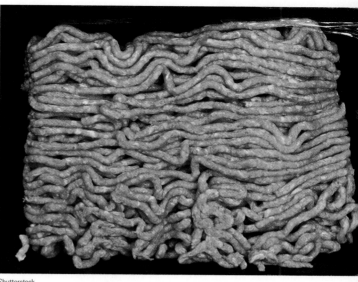
Shutterstock

EXAMPLE 10 Forecasting Maximum Revenue

Based on data from 2002 to 2008, the price of lean ground beef may be modeled by

$$p(t) = 0.020t^3 - 0.170t^2 + 0.436t + 2.63 \text{ dollars per pound}$$

where t is the number of years since 2002. (**Source:** *Statistical Abstract of the United States, 2010*, Table 717)

According to the model, in what year between 2002 and 2008 was the price of lean ground beef at a minimum? Answer the question by using the graph of the derivative.

SOLUTION

The derivative of the price function is

$$p'(t) = 0.060t^2 - 0.340t + 0.436$$

and represents the instantaneous rate of change in the price of lean ground beef in dollars per year.

The graph of $p'(t)$ on the interval $[0, 6]$ is shown in Figure 5.8.

Using instructions from the Chapter 5 Tech Card, we determined that the horizontal intercepts of $p'(t)$ occur at $t = 1.96$ and $t = 3.71$. In other words, $p'(1.96) = 0$ and $p'(3.71) = 0$. That is, the horizontal intercepts of $p'(t)$ are the critical points of $p(t)$.

From the graph of $p'(t)$, we can readily construct the sign chart of $p'(t)$. When the graph of $p'(t)$ is below the horizontal axis, $p'(t) < 0$. When the graph of $p'(t)$ is above the horizontal axis, $p'(t) > 0$.

Figure 5.8

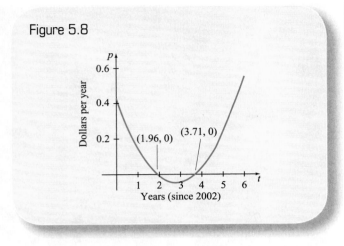

Years (since 2002)

A relative maximum occurs at $t = 1.96$, and a relative minimum occurs at $t = 3.71$. We evaluate the function R at the endpoints and the relative extrema, as shown in Table 5.6.

Table 5.6

t	$p(t)$	
0	2.63	Absolute minimum
1.96	2.98	Relative maximum
3.71	2.93	Relative minimum
6	3.45	Absolute maximum

According to the model, the maximum price of $3.45 per pound occurred at the end of 2008.

5.1 Exercises

In Exercises 1–5, determine the points on the graph where $f'(x) = 0$ or $f'(x)$ is undefined.

1. $f(x) = x^2 - 2x$

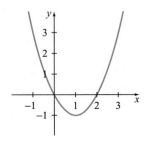

2. $f(x) = x^3 + 1$

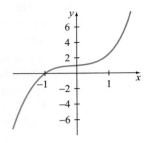

3. $f(x) = -x^2 + 4x - 4$

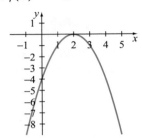

4. $f(x) = -|x - 2| + 4$

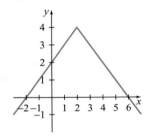

5. $f(x) = x^5 - 5x^4 + 200$

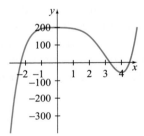

In Exercises 6–10, find the critical values of the function. Then classify the critical values as stationary values or singular values. Recall that for a function f, a is a stationary value if $f'(a) = 0$ and is a singular value if $f(a)$ is defined but $f'(a) = $ undefined.

6. $f(x) = x^2 - 4x$

7. $h(t) = t^3 - 6t^2$

8. $g(x) = 3x^{1/3} - x$

9. $h(x) = \dfrac{x}{x^2 + 1}$

10. $C(p) = 4p^3 - 12p + 7$

In Exercises 11–18, construct a sign chart for the derivative. Then determine the relative and absolute extrema of the function on the specified domain. Use the First Derivative Test as appropriate. (*Note:* Exercises 11–15 use the same functions as Exercises 6–10.)

11. $f(x) = x^2 - 4x; [-1, 5]$

12. $h(t) = t^3 - 6t^2; [-1, 7]$

13. $g(x) = 3x^{1/3} - x; [-3, 3]$

14. $h(x) = \dfrac{x}{x^2 + 1}; [-2, 2]$

15. $C(p) = 4p^3 - 12p + 7; [-3, 3]$

16. $f(x) = x^3 - 4x; [-3, 3]$

17. $h(t) = t \ln(t); [0.1, 3]$

18. $g(x) = e^x - \ln x; [0.1, 3]$

In Exercises 19 and 20, the number of tickets sold, $N(p)$, is given as a function of the ticket price p. Determine what ticket price will maximize the revenue from ticket sales.

19. $N(p) = -60p + 3000$

20. $N(p) = 625e^{-0.025p}$

In Exercises 21–23, use the methods discussed in this section to answer the questions.

21. *Vegetable Exports* Based on data from 1990 to 2005, the U.S. exports of frozen and canned vegetables may be modeled by

$$V(t) = 0.1657t^4 - 5.112t^3 + 44.84t^2 - 44.49t + 529.0 \text{ thousand metric tons}$$

where t is the number of years since 1990. (**Source:** Modeled from *Statistical Abstract of the United States, 2007,* Table 819)

Between 1990 and 2005, when were vegetable exports decreasing?

22. *Farm Land* Based on data from 1974 to 2002, the amount of farm land in the U.S. may be modeled by

$$F(t) = -0.001723t^4 + 0.1266t^3 - 3.041t^2 + 23.28t + 964.9 \text{ million acres}$$

where t is the number of years since 1970. (**Source:** Modeled from *Statistical Abstract of the United States, 2007,* Table 802)

Find the relative and absolute extrema of $F(t)$ on the interval $[0, 28]$. Then interpret the practical meaning of the extrema.

23. *DVD Market Value* Based on data from 1998 to 2001, the total dollar value of DVDs shipped by manufacturers may be modeled by

$$V(x) = -0.2173x^2 + 25.84x - 0.02345 \text{ million dollars}$$

where x is the number of DVDs shipped in millions. (**Source:** Modeled from Recording Industry Association of America data)

According to the model, how many DVDs should be shipped in order to maximize the value of the manufacturers' DVD shipments?

5.2 Applications of Maxima and Minima

The equation that expresses the price of an item as a function of the number of items sold is referred to as a **demand equation**. Based on data from 2007 and 2008, the demand equation for cassette tapes may be modeled by

$$p = -8.33q + 8.17$$

where q is the number of cassette tapes sold (in millions) and p is the cassette tape price. (**Source:** Modeled from Recording Industry Association of America data)

What price will maximize the revenue from cassette tapes? Questions such as these can be answered using the notion of the derivative.

In this section, we will further illustrate applications of maxima and minima. Specifically, we will look at the business concepts of revenue, cost, and profit as well as marginal revenue, marginal cost, and marginal profit. We will also explore area and volume problems.

Revenue, Cost, and Profit

A company's **revenue** is the total amount of money it brings in. The company's **cost** is the total amount of money it spends; this includes both fixed costs and variable costs. **Fixed costs** are costs that the company incurs regardless of production. **Variable costs** are typically expenses that vary based on the level of production. **Profit** is the difference between the company's revenue and its cost. The **break-even point** is the production level that results in revenue equaling cost.

Marginal **revenue** is an approximation of the additional revenue generated if one more unit is produced and sold. **Marginal cost** is the approximate cost incurred by producing one more unit, and **marginal profit** is the approximate profit resulting from the production and sale of one more unit. In order to maximize profits, product manufacturers should continue to increase production as long as the marginal profit is positive. If they continue production when the marginal profit turns negative, they will actually lose money by producing more items.

Marginal profit is determined by taking the derivative of the profit function. This relationship between the derivative and marginal profit is seen by looking at the difference quotient of the profit function $P(x)$. Suppose we wanted to estimate the additional profit earned by selling one more item after 100 items had been sold. We know that $P(101) - P(100)$ is the additional profit earned by selling the 101st item. Note that

$$P(101) - P(100) = \frac{P(101) - P(100)}{101 - 100}$$

$$\approx \lim_{h \to 0} \frac{P(100 + h) - P(100)}{(100 + h) - 100}$$

$$\approx \lim_{h \to 0} \frac{P(100 + h) - P(100)}{h}$$

$$\approx P'(100)$$

It may seem illogical to evaluate the derivative at $x = 100$ when we could calculate the exact amount of the additional profit earned by producing and selling the 101st item by evaluating $P(101) - P(100)$. However, if P is a complicated function, it is often easier to evaluate the derivative at a single value instead of calculating P at two different values.

In a similar manner, we can see that marginal cost is the derivative of the cost function and marginal revenue is the derivative of the revenue function. We'll begin by working a straightforward example to demonstrate the concepts of marginal revenue and marginal profit.

EXAMPLE 1　Forecasting Maximum Profit

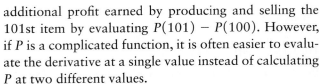 Suppose that the demand equation for a certain brand of squirt gun is given by

$$p = -x + 15$$

where x is the number of squirt guns sold (in thousands) and p is the price per squirt gun (in dollars).

The cost to produce x thousand squirt guns is

$$C(x) = 5x \text{ thousand dollars}$$

(a) If the company is currently producing 7,000 squirt guns per year, should it increase or decrease squirt gun production? Explain using the concepts of marginal revenue and marginal profit.

(b) Determine the production level that will maximize profit.

SOLUTION

(a) The revenue function $R(x)$ is the product of the price p and the number of squirt guns sold, x.

$$R(x) = \left(p \, \frac{\text{dollars}}{\text{squirt gun}} \right) \cdot (x \text{ thousand squirt guns})$$

$$= (-x + 15)x \, \frac{(\text{dollars})(\text{thousand } \cancel{\text{squirt guns}})}{\cancel{\text{squirt gun}}}$$

$$= -x^2 + 15x \text{ thousand dollars}$$

The marginal revenue is the derivative of the revenue function.

$$R'(x) = -2x + 15 \, \frac{\text{thousand dollars}}{\text{thousand squirt guns}}$$

Evaluating the marginal revenue at a production level of 7,000 squirt guns yields

$$R'(7) = -2(7) + 15$$

$$= 1 \frac{\text{thousand dollars}}{\text{thousand squirt guns}}$$

Increasing production from 7,000 squirt guns to 8,000 squirt guns will increase revenue by about $1,000.

Profit is the difference between revenue and cost. The profit function is

$$P(x) = R(x) - C(x)$$

$$= (-x^2 + 15x) - (5x)$$

$$= -x^2 + 10x \text{ thousand dollars}$$

Evaluating marginal profit function is

$$P'(x) = -2x + 10 \frac{\text{thousand dollars}}{\text{thousand squirt guns}}$$

Evaluating the marginal profit at a production level of 7,000 squirt guns yields

$$P'(7) = -2(7) + 10$$

$$= -4$$

thousand dollars per thousand squirt guns

Increasing production from 7,000 squirt guns to 8,000 squirt guns will decrease profit by about $4,000.

Although marginal revenue is positive at a production level of 7,000 squirt guns, marginal profit is negative at that production level. This means that any additional revenue brought in won't be enough to cover the cost of producing the extra squirt guns. The company should reduce production.

(b) The company wants to maximize profit. When profit is maximized, the marginal profit will be equal to 0. Recall that the profit function is

$$P(x) = -x^2 + 10x \text{ thousand dollars}$$

and the marginal profit function is the derivative

$$P'(x) = -2x + 10$$

thousand dollars per thousand squirt guns

Setting the marginal profit to zero yields

$$0 = -2x + 10$$

$$2x = 10$$

$$x = 5 \text{ thousand squirt guns}$$

Does a maximum or a minimum of the profit function occur at the critical value $x = 5$? Observe that the graph of $P(x) = -x^2 + 10x$ is a concave down parabola. A concave down parabola has one relative maximum and no relative minima. The relative maximum is also an absolute maximum. Consequently, an absolute maximum of the profit function $P(x) = -x^2 + 10x$ occurs at the critical value $x = 5$. At a production level of 5,000 squirt guns, the profit will be maximized. The maximum profit is

$$P(5) = -(5)^2 + 10(5)$$

$$= 25 \text{ thousand dollars}$$

We can verify the accuracy of our result by evaluating the profit function at other production levels on either side of $x = 5$.

$$P(4) = -(4)^2 + 10(4)$$

$$= 24 \text{ thousand dollars}$$

$$P(7) = -(7)^2 + 10(7)$$

$$= 21 \text{ thousand dollars}$$

Increasing or decreasing the production level from 5,000 squirt guns will decrease profit.

EXAMPLE 2 Forecasting Maximum Revenue

Based on data from 2007 and 2008, the demand equation for cassette tapes may be modeled by

$$p = -8.33q + 8.17$$

where q is the number of cassette tapes (in millions) and p is the price per cassette. (**Source:** Modeled from Recording Industry Association of America data)

What price will maximize the revenue from cassette tape sales? In order for the price and quantity to be positive, q must be positive and at most $0.98.

SOLUTION

Revenue is the product of the price and the quantity sold. That is, $R = pq$.

$$R = pq$$

$$R = (-8.33q + 8.17)q \quad \text{Since } p = -8.33q + 8.17$$

$$R = -8.33q^2 + 8.17q$$

The marginal revenue is $\dfrac{dR}{dq}$.

$$\frac{dR}{dq} = -16.66q + 8.17 \text{ million}$$
$$\text{dollars per million cassettes}$$

We want to know when the marginal revenue is zero.

$$0 = -16.66q + 8.17$$

$$16.66q = 8.17$$

$$q = 0.49$$

Evaluating the derivative on either side of the critical value $q = 0.49$, we get

$$R'(0) = -16.66(0) + 8.17$$

$$= 8.17$$

$$R'(1) = -16.66(1) + 8.17$$

$$= -8.49$$

$$\begin{array}{ccc} & & \text{max} \\ \dfrac{dR}{dq} & + & \text{of } R \quad - \\ & & \xleftarrow{\hspace{3cm}} \\ & & 0.49 \end{array}$$

The revenue function has a relative maximum when the number of cassettes sold is 0.49 million, as shown in the sign chart and Table 5.7.

Table 5.7

q	R(q)	
0.01	0.08	
0.49	2.00	Relative and absolute maximum
0.98	0.01	

According to the model, the maximum revenue from cassette tape sales will be generated when 0.49 million cassettes are sold. This was predicted to occur when the cassette price was $4.08, according to the demand equation.

Area and Volume

Shipping companies such as United Parcel Service (UPS) and the U.S. Postal Service (USPS) classify packages by size and weight.

USPS advises customers that Parcel Post is the best value when shipping a single package or a small number of packages. A Parcel Post package "can weigh up to 70 pounds and measure up to 130 inches in length and girth combined." (**Source:** U.S. Postal Service) The length of a package is the length of the longest side, and the girth of a package is the distance around its thickest part. Businesses are often interested in fitting the maximum amount of their product into a box whose dimensions meet the postal service guidelines. They can save a substantial amount of money in shipping costs by conforming to the postal service standard.

EXAMPLE 3 Minimizing Use of Resources

Moving companies are hired by businesses and private individuals to pack, load, move, and unload household goods. As part of their service, movers bring cardboard boxes that can be easily constructed and easily broken down. Boxes are purchased by the moving company from a box manufacturer.

In an effort to cut costs, a box manufacturer wants to design a closable box with a square base that has a volume of 8 cubic feet and uses the least amount of cardboard. The box must be at least 24 inches in height and will be constructed according to the design shown in Figure 5.9. What are the dimensions of the box that uses the least material?

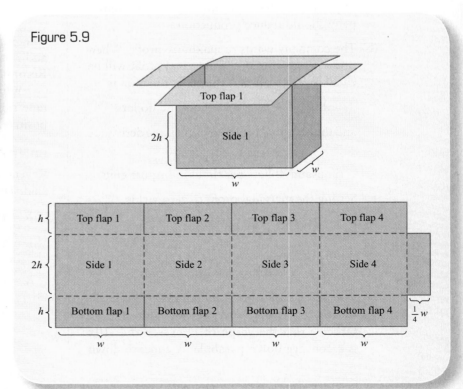

Figure 5.9

SOLUTION

The surface area of the cardboard material is

$$A = (h + 2h + h)(w + w + w + w) + \frac{1}{4}w(2h)$$

$$= (4h)(4w) + \frac{1}{2}hw$$

$$= 16hw + \frac{1}{2}hw$$

$$= 16.5hw \text{ square feet}$$

Since we want to minimize the amount of cardboard used, we must minimize the surface area. We are unable to find A' because it is a function of two variables. We must use a secondary equation to write one variable in terms of the other.

The volume of the box is

$$V = w \cdot w \cdot 2h$$

$$= 2hw^2$$

Since the box has a volume of 8 cubic feet, we have

$$8 = 2hw^2$$

$$h = \frac{8}{2w^2}$$

$$= 4w^{-2}$$

Substituting this result into the surface area equation, we get

$$A = 16.5hw$$

$$= 16.5(4w^{-2})w \quad \text{Since } h = 4w^{-2}$$

$$= 66w^{-1}$$

Do we have any restrictions on w? Clearly, $w > 0$, since the box must have a positive width. (If $w = 0$, the surface area function is undefined.) Additionally, the box is required to have a height of at least 24 inches (2 feet). That is,

$$2h \geq 2$$

$$h \geq 1$$

We can write $h \geq 1$ in terms of w .

$$h \geq 1$$

$$4w^{-2} \geq 1 \quad \text{Since } h = 4w^{-2}$$

$$4 \geq w^2$$

$$4 - w^2 \geq 0$$

Graphically speaking, we want to know when the graph of $f(w) = 4 - w^2$ is on or above the horizontal axis. (This is when $f(w) \geq 0$.) Rewriting $f(w)$ in the standard

form of a quadratic equation ($f(w) = aw^2 + bw + c$) yields $f(w) = -w^2 + 4$. Since $a < 0$, the parabola is concave down. Since $c = 4$, the graph has a vertical intercept of $(0, 4)$. Since $f(-2) = 0$ and $f(2) = 0$, $(-2, 0)$ and $(2, 0)$ are the horizontal intercepts of $f(w) = -w^2 + 4$ (Figure 5.10).

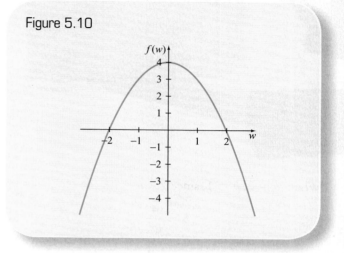

Figure 5.10

From the graph, we see that when $-2 \leq w \leq 2$, $f(w) = -w^2 + 4 \geq 0$. Recall that since w is the width, we also required $w > 0$. Thus the domain of the surface area function is $(0, 2]$.

To minimize the surface area, we will differentiate A.

$$A(w) = 66w^{-1}$$

$$A'(w) = -66w^{-2}$$

$$= -\frac{66}{w^2}$$

$A'(w)$ is negative for all values of w in the domain of A, so there are no critical values. Since the derivative is negative, the surface area function is decreasing over its entire domain. Consequently, the smallest area will occur at the rightmost endpoint, $w = 2$. This may be easily confirmed by looking at a table of values for A (Table 5.8).

Table 5.8

w	A(w)	
0	Undefined	
0.5	132	
1.0	66	
1.5	44	
2.0	33	Absolute minimum

Shutterstock

Figure 5.11

When $w = 2$, $h = 1$, since

$$h = \frac{4}{w^2}$$

$$= \frac{4}{(2)^2}$$

$$= \frac{4}{4}$$

$$= 1$$

The box has dimensions $w \times w \times 2h$, so the optimal dimensions are $2' \times 2' \times 2'$.

EXAMPLE 4 Minimizing Landscaping Costs

A landscape designer offers an economy garden package to her clients. The rectangular garden has flowers along the front of the garden and shrubs along the back and sides (Figure 5.11).

The designer charges $25 per linear foot for the shrubs and $6 per linear foot for the flowers. A client wants to install a 400-square-foot garden. What will be the dimensions of the least expensive garden?

SOLUTION

We begin by labeling the figure. We let l represent the length of the garden and w represent the width of the garden (Figure 5.12). The area of the garden is given by

$$A = lw$$

$$400 = lw$$

Figure 5.12

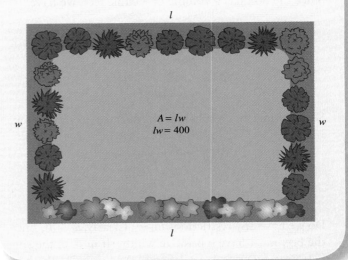

$$A = lw$$
$$lw = 400$$

Assuming that the shrubs extend across the back of the garden and fully down both sides, the cost of the shrubs (in dollars) is

$$S = 25(w + l + w)$$

$$= 25l + 50w$$

Assuming that the flowers extend across the front of the garden, the cost of the flowers (in dollars) is

$$F = 6l$$

Shutterstock

The combined cost for the garden is

$$C = (25l + 50w) + 6l$$

$$= 31l + 50w$$

We want to minimize this cost. We can't differentiate C because it contains two input variables, l and w. However, recall that

$$lw = 400$$

$$l = \frac{400}{w}$$

Therefore, the cost function may be rewritten as

$$C = 31l + 50w$$

$$= 31\left(\frac{400}{w}\right) + 50w$$

$$= 12{,}400w^{-1} + 50w$$

Differentiating the cost function and setting it equal to zero yields

$$C'(w) = -12{,}400w^{-2} + 50$$

$$0 = -12{,}400w^{-2} + 50 \quad \text{Set } C'(w) = 0$$

$$\frac{12{,}400}{w^2} = 50$$

$$12{,}400 = 50w^2$$

$$w^2 = 248$$

$$w \approx 15.75 \quad \text{Since } w > 0, \text{ we omit the solution } w \approx -15.75$$

Applying the First Derivative Test, we determine that a relative minimum occurs at $w = 15.75$. (*Note:* $C'(15) < 0$ and $C'(16) > 0$.) To find the corresponding value of l, we substitute the value of w into the area equation.

$$l = \frac{400}{w}$$

$$= \frac{400}{15.75}$$

$$= 25.40$$

The dimensions that will minimize the cost are 25.40 feet by 15.75 feet. The minimum cost is

$$C = 31l + 50w$$

$$= 31(25.40) + 50(15.75)$$

$$= \$1574.90$$

The shape of the garden (drawn to scale) is shown in Figure 5.13. The client may want to change the dimensions; however, any change in dimensions will result in a higher cost.

Figure 5.13   25.40 feet × 15.75 feet

5.2 Exercises

In Exercises 1–20, use the derivative techniques demonstrated in this section to determine the answer to each question.

1. **Company Profit** In 2001, the Kellogg Company introduced a new breakfast cereal: Special K Red Berries cereal. It quickly achieved an impressive 1 percent market share during its first six months on the market and helped boost corporate sales. Based on data from 1999 to 2001, the net sales of the Kellogg Company may be modeled by

   $$R(t) = 964.1t^2 - 993.6t + 6984 \text{ million dollars}$$

   where t is the number of years since 1999. The cost of goods sold may be modeled by

   $$C(t) = 399.8t^2 - 379.9t + 3325.1 \text{ million dollars}$$

   where t is the number of years since 1999. (**Source:** Modeled from Kellogg Company 2001 Annual Report, pp. 7, 27)
   (a) Find the equation for the gross profit. (This is net sales minus the cost of goods sold.)
   (b) Between 1999 and 2001, in which year was gross profit minimized?

2. *Forecasting Maximum Revenue* Based on data from 2003–2005, a demand function model for the Apple iPod is

$$p = \frac{52.1}{0.00484q + 0.152} \text{ dollars}$$

where q is the number of iPods sold (in millions) in a year. (**Source:** Modeled from Apple Computer Corporation 2005 Annual Report, pp. 31–32) Assume that between 0 and 200 million iPods will be sold in a given year. According to the model:
   (a) What is the revenue function for the iPod?
   (b) What is the marginal revenue function for the iPod?
   (c) What is the maximum revenue the company will earn from the iPod in a single year?
   (d) What will be the price per iPod when the revenue is maximized?

3. *Forecasting Maximum Profit* Based on data from 2003–2005, a revenue function model for the Apple iPod is

$$R(q) = \frac{5.21q}{0.00484q + 0.152}$$

where q is the number of iPods sold (in millions) in a year. (**Source:** Modeled from Apple Computer Corporation 2005 Annual Report, pp. 31–32)
   Suppose that the cost to produce q million iPods is given by

$$C(q) = 35q \text{ million dollars}$$

Assume that between 0 and 200 million iPods will be sold in a given year. According to the model:
   (a) What is the maximum profit?
   (b) What price yields the maximum profit?

4. *Maximizing Package Volume* A rectangular package has a square end (see the figure). The sum of the length and girth of the package is equal to 130 inches, the maximum sum allowed for Parcel Post packages shipped by the U.S. Postal Service. (The *girth* is the distance around the thickest part.) (**Source:** www.usps.com)

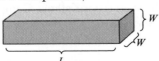

   (a) Find an equation that relates the width of the package to the volume of the package.
   (b) Determine what width maximizes the volume of the package.

5. *Company Revenue* In its 2001 annual report, the Coca-Cola Company reported that "Our worldwide unit case volume increased 4 percent in 2001, on top of a 4 percent increase in 2000. The increase in unit case volume reflects consistent performance across certain key operations despite difficult global economic conditions. Our business system sold 17.8 billion unit cases in 2001." (**Source:** Coca-Cola Company 2001 Annual Report, p. 46) A unit case is equivalent to 24 eight-ounce servings of finished beverage.
   (a) Based on data from 1999 to 2001, the net operating revenue of the Coca-Cola Company may be modeled by

$$R(s) = -0.4029s^2 + 14.44s - 109.2 \text{ billion dollars}$$

      where s is the number of unit cases sold (in billions). Find the marginal revenue function.
   (b) According to the model, at what unit case production level will revenue be maximized?

6. *Company Profit* Based on data from 1999 to 2001, the net operating revenues of the Coca-Cola Company may be modeled by

$$R(t) = -201t^2 + 806t + 19{,}284 \text{ million dollars}$$

   and the cost of goods sold may be modeled by

$$C(t) = -177.5t^2 + 372.5t + 6009 \text{ million dollars}$$

   where t is the number of years since 1999.
   (a) Find the gross profit function.
   (b) According to the model, in what year is the gross profit projected to reach a maximum?
   (c) Does the result of part (b) seem reasonable? Explain.

7. *Company Costs* Frito-Lay North America, a subsidiary of PepsiCo, produces Doritos, Cheetos, Fritos corn chips, and a variety of other salty, sweet, or grain-based snacks. Based on data from 1999 to 2001, the net sales (revenue) of Frito-Lay North America may be modeled by

$$R(t) = -168t^2 + 907t + 8232 \text{ million dollars}$$

   and the operating profit (earnings before interest and taxes) may be modeled by

$$P(t) = -47.5t^2 + 283.5t + 1679 \text{ million dollars}$$

   where t is the number of years since 1999. (**Source:** Modeled from 2001 PepsiCo Annual Report, pp. 23, 44)
   (a) In what year are net sales projected to reach a maximum?

(b) Find the cost function for Frito-Lay North America.

(c) According to the model, in what year are costs expected to reach a maximum?

(d) Compare the results of parts (a) and (c). Do the results seem reasonable?

8. *Maximizing Package Volume* A rectangular package is twice as wide as it is tall (see the figure). The sum of the length and girth of the package is equal to 108 inches, the maximum sum allowed for standard packages shipped by the U.S. Postal Service. (The *girth* is the distance around the thickest part.) (**Source:** www.usps.com)

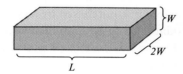

(a) Find an equation that relates the width of the package to the volume of the package.

(b) Determine what width maximizes the volume of the package.

9. *Employee Wages* Based on data from 1991 to 2001, the average wage of a Ford Motor Company employee may be modeled by

$$W(t) = -0.003931t^4 + 0.1005t^3 - 0.8295t^2 + 3.188t + 16.48 \text{ dollars per hour}$$

where t is the number of years since 1990. (**Source:** Modeled from Ford Motor Company 2001 Annual Report, p. 71)

Was the average wage changing more rapidly in 1995 or 2000?

10. *Pool Size Optimization* A pool builder makes two types of economy-priced pools: square and circular. He estimates the price of the job by multiplying the perimeter or circumference of the pool (in linear feet) by a fixed price per linear foot. A homeowner wants her pool to have the maximum amount of water surface area for the lowest possible price.

(a) Should the homeowner have a square or a circular pool built? Explain. (The area of a square with a perimeter p is $S = \frac{1}{16}p^2$.

The area of a circle with circumference c is

$$A = \frac{1}{4\pi}c^2.)$$

(b) Shortly before construction, the homeowner decides to increase the 100-foot distance around the pool by 1 foot. For both shapes of pool, use the derivative to approximate how much the surface area of the pool will increase by increasing the perimeter (circumference) by 1 foot.

(c) Does your result in part (b) confirm your conclusion in part (a)? Explain.

11. *Apple Farming* Historically, many apple farmers spaced trees 40 feet by 40 feet apart (27 per acre). Trees typically took 25 years to reach their maximum production of 500 bushels per acre. (A bushel is about 44 pounds.) In recent years, agriculturists have created dwarf and semidwarf varieties that allow trees to be spaced 10 feet by 10 feet apart. (**Source:** USDA)

Suppose that a farmer has an apple orchard with 40 trees per acre. The orchard yields 10 bushels per tree. The farmer estimates that for each additional tree planted (per acre), the average yield per tree is reduced by 0.1 bushel.

If $y = f(x)$ is the total number of bushels of apples produced per acre when an additional x trees per acre are planted, determine how many additional trees should be planted in order to maximize the number of bushels of apples produced.

12. *Apple Farming* Suppose that a farmer has an apple orchard with 30 trees per acre. The orchard yields 12 bushels per tree. The farmer estimates that for each additional tree planted (per acre), the average yield per tree is reduced by 0.1 bushel.

If $y = f(x)$ is the total number of bushels of apples produced per acre when an additional x trees per acre are planted, determine how many additional trees should be planted in order to maximize the number of bushels of apples produced.

13. *Apple Supplier Prices* A fruit farmer sells apples to a grocery store chain. The amount of apples the store buys depends linearly on the price per pound the farmer charges. The farmer estimates that for every $0.02 per pound increase in the price, the store will reduce its order by 50 pounds. The store presently orders 500 pounds per week and pays $0.18 per pound. What price should the farmer charge in order to maximize her revenue from apple sales?

14. **Apple Retailer Prices** A grocery store has priced apples at $0.70 per pound and sells 1,000 pounds per week. The amount of apples the store sells depends linearly on the price per pound the store charges. The store manager estimates that for every $0.04 per pound increase in the price, the store will reduce its sales by 100 pounds. What price should the store charge for its apples in order to maximize revenue from apple sales?

15. **Company Sales** Based on data from 1993 to 2002, the annual sales of Starbucks Corporation may be modeled by

$$S(t) = 29.23t^2 + 79.33t + 177.4 \text{ million dollars}$$

where t is the number of years since 1993. (**Source:** Modeled from data at www.starbucks.com)

According to the model, how much more rapidly were sales increasing in 2001 than in 1999?

16. **Company Sales** Based on data from 1991 to 2001, the franchised sales of McDonald's Corporation may be modeled by

$$S(t) = -41.293t^2 + 1729.0t + 11,139 \text{ million dollars}$$

where t is the number of years since 1990. (**Source:** Modeled from data at www.mcdonalds.com)

Calculate the instantaneous rate of change of sales in 1999 and 2001. Explain the financial significance of the result.

17. **Fleet Vehicle Sales** A business that owns or leases 10 or more vehicles may qualify for auto manufacturer fleet purchase incentives. Incentives vary from $300 to $7,000 on new vehicles. In August 2003, the manufacturer's fleet incentive for a 2004 Chevrolet Cavalier was $1,400. (**Source:** www.fleet-central.com)

An auto dealer offers an additional discount to fleet buyers who purchase one or more new Cavaliers. To encourage sales, the dealer reduces the after-incentive price of each car by x%, where x is the total number of cars purchased. Assuming that the pre-incentive price of a 2004 Chevrolet Cavalier is $14,400, how many vehicles would the dealer need to sell in order to maximize revenue?

18. **Fleet Vehicle Sales** In August 2003, the manufacturer's fleet incentive for a 2004 Buick Century was $1,800. (**Source:** www.fleet-central.com)

An auto dealer offers an additional discount to fleet buyers who purchase one or more new Buick Century vehicles. To encourage sales, the dealer reduces the after-incentive price of each car by $200x$ dollars, where x is the total number of cars purchased. Assuming that the pre-incentive price of a 2004 Buick Century is $19,000, how many vehicles would the dealer need to sell in order to maximize revenue?

19. **Pricing Analysis** Based on the results of Exercises 17 and 18, what additional restrictions (if any) should the dealer place on his advertised discount in order to ensure that selling additional vehicles won't reduce his revenue?

20. **Profit Lost Due to Waste** A fruit vendor purchases x pounds of fruit from his supplier. He estimates that $0.2x$% of each pound of produce he buys spoils before it is purchased by a consumer. For each pound of fruit he sells, he makes a profit of $0.50. How many pounds of fruit should he buy in order to maximize profit?

5.3 Concavity and the Second Derivative

Based on data from the 1992–93 through 2003–04 school years, the number of school-related homicides of students between the ages of 5 to 18 years may be modeled by

$$H(t) = -0.6776t^4 + 18.57t^3 - 155.2t^2 + 249.3t + 2705 \text{ homicides}$$

where t is the number of school years since 1992–93. (**Source:** Modeled from National Center for Educational Statistics, Indicators of School Crime and Safety: 2006, Table 1.1)

In what year(s) between 1992–93 and 2003–04 were homicides decreasing at the fastest rate, and in what year were they decreasing at the slowest rate? Questions such as these may be answered using the *second derivative*.

In this section, we will introduce the second derivative and discuss the graphical concepts of concavity and inflection points. We will use these concepts to investigate the relationship between position, velocity, and acceleration. We will also introduce the Second Derivative Test as an alternative means of finding relative extrema. We will conclude with curve-sketching techniques.

Concavity

The term **concavity** refers to the curvature of a graph. A graph is said to be **concave up** if it curves upward and **concave down** if it curves downward. A simple rhyme is helpful in remembering the meaning of the terms.

Concave up is like a cup.
Concave down is like a frown.

We will explore the concept of concavity by looking at the graph of the function $f(x) = x^3 - 12x$ on the interval $[-3, 3]$ (Figure 5.14).

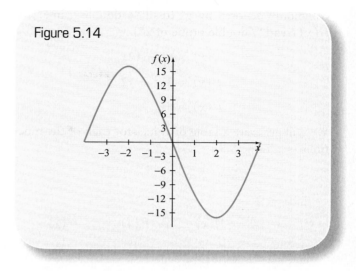

Figure 5.14

From $x = -3$ to $x = 0$, the graph is curved downward (looks like a frown). At $x = 0$, the concavity changes. From $x = 0$ to $x = 3$, the graph is curved upward (looks like a cup). Is there a way to determine the concavity of a graph algebraically? Let's see. We'll begin by finding the derivative of the function and then the derivative of the derivative. The derivative of the derivative is called the **second derivative** and is

Concave Up

Concave Down

Positive

Negative

Shutterstock

commonly denoted by y'' (read "y double prime") or $f''(x)$ (read "f double prime of x").

$$f(x) = x^3 - 12x$$

$$f'(x) = 3x^2 - 12$$

$$f''(x) = 6x$$

We will generate a table of values for each of the functions (see Table 5.9).

Table 5.9

x	$f(x)$	$f'(x)$	$f''(x)$
-3	9	15	-18
-2	16	0	-12
-1	11	-9	-6
0	0	-12	0
1	-11	-9	6
2	-16	0	12
3	-9	15	18

We can learn much about the graph of the function from this table.

The graph of the function has x-intercepts when $f(x) = 0$. The x-intercept of the function on the domain $[-3, 3]$ occurs at $x = 0$, since $f(0) = 0$.

The graph of the function has relative extrema when $f'(x) = 0$ and the derivative changes sign. A relative maximum occurs at $x = -2$, since $f'(-2) = 0$ and the derivative changes from positive to negative at $x = -2$. A relative minimum occurs at $x = 2$, since $f'(2) = 0$ and the derivative changes from negative to positive at $x = 2$.

We observe that on the interval $[-3, 0)$, $f''(x) < 0$, and on the interval $(0, 3]$, $f''(x) > 0$. At $x = 0$, $f''(x) = 0$. Observe that the graph of f is concave down when $f''(x) < 0$ and is concave up when $f''(x) > 0$. The graph changes concavity from down to up at $x = 0$. The point on the graph where a function changes concavity is called an **inflection point**.

By referring back to the earlier rhyme, we can come up with a clever strategy to remember how to determine the concavity of a graph.

When we're positive, we are up (like the cup).
When we're negative, we're down and we frown.

Admittedly, the rhyme is a bit silly; however, its sheer wackiness will make the concept easier to remember.

Concavity and Inflection Points of a Graph

1. A continuous function f is *concave up* at a point $(c, f(c))$ if $f''(c) > 0$.

2. A continuous function f is *concave down* at a point $(c, f(c))$ if $f''(c) < 0$.

3. A continuous function f has an *inflection point* at $(c, f(c))$ if
   - $f''(c) = 0$ or $f''(c)$ is undefined and
   - $f''(c)$ changes sign at $x = c$

A common error made by learners in their search for inflection points is to assume that if $f''(c) = 0$, then $(c, f(c))$ is an inflection point. Although $f''(c) = 0$ is a necessary condition for an inflection point, it is not a sufficient condition. Consider the function $f(x) = x^4$ with its associated second derivative $f''(x) = 12x^2$. Although $f''(0) = 0$, the point $(0, 0)$ is not an inflection point of f since the graph of f is concave up on both sides of the point (Figure 5.15).

Figure 5.15

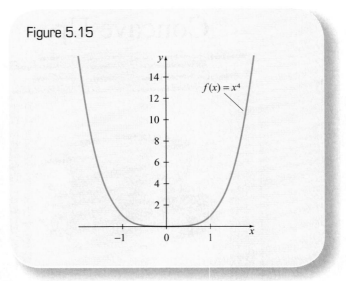

EXAMPLE 1 Determining the Concavity of a Graph

The graph of $f(x) = x^4 - 4x^3$ on the interval $[-3, 5]$ is shown in Figure 5.16. Determine where the graph is concave up and where the graph is concave down. Identify all inflection points.

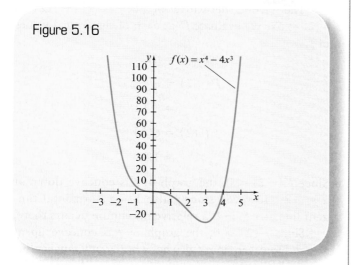

Figure 5.16

$f(x) = x^4 - 4x^3$

SOLUTION

The graph appears to be concave up from $[-3, 0)$, concave down from $(0, 2)$, and concave up again from $(2, 5]$. It is difficult to determine the exact location of inflection points visually, so we will proceed algebraically to verify our visual conclusions.

$$f(x) = x^4 - 4x^3$$

$$f'(x) = 4x^3 - 12x^2$$

$$f''(x) = 12x^2 - 24x$$

In order to identify potential inflection points, we must set $f''(x)$ equal to zero.

$$0 = 12x^2 - 24x$$

$$0 = 12x(x - 2)$$

$$12x = 0 \qquad\qquad x - 2 = 0$$
$$\text{or}$$
$$x = 0 \qquad\qquad x = 2$$

Inflection points may occur at $x = 0$ or $x = 2$; however, we must verify that $f''(x)$ changes sign at each of these points. We will do this using a sign chart for f''. We evaluate $f''(x)$ using points on either side of the potential inflection points.

$$f''(x) = 12x\,(x - 2)$$

$$f''(-1) = 12(-1)(-1 - 2) > 0$$

$$f''(1) = 12(1)(1 - 2) < 0$$

$$f''(3) = 12(3)(3 - 2) > 0$$

We update the sign chart for f'' with the results.

f'' + | − | +
0   2

Since $f''(x)$ changes sign at $x = 0$ and $x = 2$, inflection points occur at each of these points. Since $f''(x) > 0$ on $[-3, 0)$ and on $(2, 5]$, it is concave up on those intervals. Since $f''(x) < 0$ on $(0, 2)$, it is concave down on that interval. Our algebraic analysis confirms our graphical estimation.

EXAMPLE 2 Using the Second Derivative to Find Extrema of the First Derivative

Based on data from the 1992–93 through 2003–04 school years, the number of school-related homicides of students between the ages of 5 to 18 years may be modeled by

$$H(t) = -0.6776t^4 + 18.57t^3 - 155.2t^2$$
$$+ 249.3t + 2705 \text{ homicides}$$

where t is the number of school years since 1992–93. (**Source:** Modeled from National Center for Educational Statistics, Indicators of School Crime and Safety: 2006, Table 1.1)

In what year(s) between 1992–93 and 2003–04 were homicides decreasing at the fastest rate? At what rate was the number of homicides decreasing?

SOLUTION

The rate of change of the homicide function is $H'(t)$.

$$H'(t) = -2.7104t^3 + 55.71t^2$$
$$- 310.4t + 249.3 \frac{\text{homicides}}{\text{year}}$$

This function tells us the rate of change in the number of homicides. We want to determine the extrema of $H'(t)$. To do this, we differentiate $H'(t)$.

$$H''(t) = -8.1312t^2 + 111.42t$$
$$- 310.4 \frac{\text{homicides per year}}{\text{year}}$$

We set $H''(t)$ equal to zero and solve using technology.

$$t = 3.89; 9.81$$

We must evaluate $H''(t)$ on either side of the critical points of $H'(t)$ and construct a sign chart for $H''(t)$.

H'' − | + | −
min of H' max of H'
3.89   9.81

The relative extrema of $H'(t)$ are in fact the inflection points of $H(t)$. Inflection points tell us when the rate of change of $H(t)$ is accelerating or decelerating at the greatest rate. Let's generate a table of values for H, H', and H''. Tables such as Table 5.10 may be quickly generated using the instructions found on your Chapter 5 Tech Card. The end points and inflection points of $H(t)$ are highlighted in the table.

Table 5.10

Years (since 1992–93) (t)	H(t) (homicides)	H'(t) (homicides per year)	H''(t) (homicides per year)
0	2705	249.3	−310.4
2	2721	−170.3	−120.1
3.89	2264	−274.7	0.0
4	2234	−274.4	5.2
6	1747	−193.0	65.4
8	1499	−56.2	60.6
9.81	1471	6.8	0.1 ≈ 0.0
10	1472	5.9	−9.3
11	1464	−31.7	−68.7

When $H'(t) > 0$, the number of homicides is increasing. When $H'(t) < 0$, the number of homicides is decreasing. Notice that the most negative value of $H'(t)$ occurs when $t = 3.89$. That is, 3.89 years after the 1992–93 school year the number of homicides was dropping at a rate of 274.7 homicides per year. This is the most dramatic annual decline in the number of homicides between 1992–93 and 2003–04.

The Second Derivative Test

As we've seen, the second derivative may be used to determine the concavity of the graph and find the location of inflection points. The second derivative may also be used to find relative extrema. Consider the function $f(x) = x^3 - 12x$. Differentiating the function yields

$$f'(x) = 3x^2 - 12$$

$$0 = 3(x^2 - 4) \quad \text{Set } f'(x) = 0 \text{ to find critical values}$$

$$0 = 3(x - 2)(x + 2)$$

$$x - 2 = 0 \qquad x + 2 = 0$$
$$\text{or}$$
$$x = 2 \qquad x = -2$$

The function f has critical values $x = -2$ and $x = 2$, since $f'(-2) = 0$ and $f'(2) = 0$. Graphically speaking, the graph of $f(x) = x^3 - 12x$ is flat (has a horizontal tangent line) at $x = -2$ and $x = 2$.

The second derivative of $f(x) = x^3 - 12x$ is $f'(x) = 6x$. We evaluate f'' at each of the critical values of f.

$$f''(x) = 6x$$

$$f''(-2) = 6(-2)$$

$$= -12$$

$$f''(2) = 6(2)$$

$$= 12$$

Since $f''(-2) < 0$, the graph of f is concave down at $x = -2$. However, since f also has a horizontal tangent line at $x = -2$, a relative maximum occurs there.

Since $f''(2) > 0$, the graph of f is concave up at $x = 2$. However, since f also has a horizontal tangent line at $x = 2$, a relative minimum occurs there (Figure 5.17).

Figure 5.17

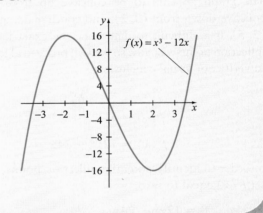

These results are summarized in the Second Derivative Test.

The Second Derivative Test
Let f be a continuous function with $f'(c) = 0$. If

- $f''(c) > 0$, then a relative minimum of f occurs at $x = c$.

- $f''(c) < 0$, then a relative maximum of f occurs at $x = c$.

- $f''(c) = 0$, then the test is inconclusive.

EXAMPLE 3 Finding Relative Extrema with the Second Derivative Test

→ Use the Second Derivative Test to find the relative extrema of the function $f(x) = x^5 - 5x^4$.

SOLUTION

$$f'(x) = 5x^4 - 20x^3$$

$$0 = 5x^3(x - 4) \qquad \text{Set } f'(x) = 0 \text{ and find critical values}$$

$$5x^3 = 0 \qquad\qquad x - 4 = 0$$
$$\text{or}$$
$$x = 0 \qquad\qquad\quad x = 4$$

The critical values are $x = 0$ and $x = 4$. To apply the Second Derivative Test, we calculate $f''(x)$ and evaluate it at the critical values, $x = 0$ and $x = 4$. Writing the second derivative in factored form will make it easier to evaluate.

$$f''(x) = 20x^3 - 60x^2$$
$$= 20x^2(x - 3)$$
$$f''(0) = 20(0)^2(0 - 3)$$
$$= 0$$
$$f''(4) = 20(4)^2(4 - 3)$$
$$= 320$$

Since $f''(4) > 0$, a relative minimum occurs at $x = 4$. The relative minimum is $(4, -256)$.

At $x = 0$, the Second Derivative Test is inconclusive. What is happening at $x = 0$? Let's evaluate $f'(x)$ at points on either side of the critical value $x = 0$ to see if the slope of f changes sign at $x = 0$. (This is the First Derivative Test.)

$$f'(-1) = 5(-1)^3(-1 - 4)$$
$$= 25$$

Since $f'(-1) > 0$, the graph of f is increasing at $x = -1$.

$$f'(1) = 5(1)^3(1 - 4)$$
$$= -15$$

Since $f'(1) < 0$, the graph of f is decreasing at $x = 1$. Since f' changes from positive to negative at $x = 0$, a relative maximum occurs at $x = 0$. The relative maximum is $(0, 0)$. The graph of the function (Figure 5.18) confirms our conclusion.

Although calculations are often simpler when we use the Second Derivative Test, we may have to revert back to the First Derivative Test if the Second Derivative Test yields an inconclusive result.

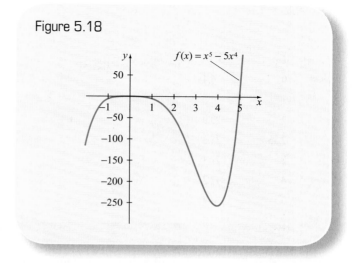

Figure 5.18

Point of Diminishing Returns

In business, we are often interested in knowing when the rate of change in sales reaches a maximum value. The point at which this occurs is called the **point of diminishing returns**. In terms of calculus, we want to know when the derivative $S'(x)$ of a function $S(x)$ attains a maximum.

EXAMPLE 4 Finding the Point of Diminishing Returns

→ Based on data from 2005 to 2008, cumulative sales of the Nintendo DS game console may be modeled by

$$S(t) = \frac{103.1}{1 + 17.68e^{-1.216t}} \text{ million units}$$

where t is the number of years since 2005. (**Source:** Modeled from Nintendo Annual Report data)

Find the point of diminishing returns for cumulative Nintendo DS game console sales and determine the cumulative sales and the rate at which cumulative sales are changing at that point.

SOLUTION

The rate of change in cumulative Nintendo DS sales is $S'(t)$.

$$S(t) = \frac{103.1}{1 + 17.68e^{-1.216t}}$$

$$S'(t) = \frac{-(17.68e^{-1.216t}(-1.216))(103.1)}{(1 + 17.68e^{-1.216t})^2}$$

$$= \frac{2216.53e^{-1.216t}}{(1 + 17.68e^{-1.216t})^2}$$

$$= \frac{2216.53}{e^{1.216t}(1 + 17.68e^{-1.216t})^2} \text{ million units per year}$$

Table 5.11

t	$S'(t)$
0	6.4
1	16.9
2	29.9
2.36	31.3
3	27.1
4	13.2
5	4.7

Relative and absolute maximum of $S'(t)$

Observe that $S'(t) > 0$ for all values of t. This means that $S(t)$ is increasing everywhere.

Due to the complexity of the derivative $S'(t)$, we will use technology to determine when $S'(t)$ attains a maximum. Graphing $S'(t)$ on the graphing calculator and using the calculator *maximum* command, we determine that $S'(t)$ has a maximum at $t = 2.36$. At this value of t, $S''(t) = 0$.

Thus the point of diminishing returns occurs at $t = 2.36$. The point of diminishing returns is $(2.36, S(2.36)) = (2.36, 51.5)$. According to the model, cumulative Nintendo DS sales were 51.5 million units in early 2008 $(t = 2.36)$. Although cumulative sales are continually increasing (since $S'(t) > 0$), the rate at which cumulative sales are increasing begins to slow after $(t = 2.36)$. Since $S'(2.36) = 31.3$ million units per year, the maximum rate at which cumulative sales are increasing is 31.3 million units per year.

We verify that the maximum rate of change occurs at $(t = 2.36)$ by evaluating $S'(t)$ at surrounding points (see Table 5.11) and by graphing the result.

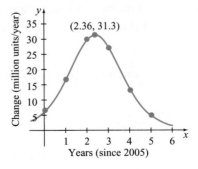

At no other point are cumulative Nintendo DS sales increasing more rapidly than at $t = 2.36$ when cumulative sales were increasing at a rate of 31.3 million units per year.

Position, Velocity, and Acceleration

The velocity of an object is the rate of change in its position over time. The acceleration of an object is the rate of change in its velocity over time. These relationships between position, velocity, and acceleration are nicely captured using the derivative concept.

Relationship between Position, Velocity, and Acceleration

Let $s(t)$ be the function that describes the position of an object at time t. Then

- $v(t) = s'(t)$, where $v(t)$ is the velocity of the object at time t.

- $a(t) = v'(t) = s''(t)$, where $a(t)$ is the acceleration of the object at time t.

EXAMPLE 5 Using a Position Function to Determine Velocity and Acceleration

The author tracked his mileage as he drove through a residential area and into a cemetery. Every 15 seconds he recorded the mileage (accurate to 0.05 mile). Based on 1.5 minutes of data, his distance from a stoplight at the bottom of a hill is given by the position function

$$s(t) = -0.5333t^4 + 1.333t^3 \\ - 0.7667t^2 + 0.3167t \text{ miles}$$

where t is in minutes.

Determine the velocity and acceleration functions. Then calculate the author's velocity and acceleration at 1 minute into his timed trip and at 1.25 minutes into his trip.

SOLUTION

$$v(t) = s'(t)$$

$$v(t) = \frac{d}{dt}(-0.5333t^4 + 1.333t^3 - 0.7667t^2 + 0.3167t)$$

$$v(t) = -2.1332t^3 + 3.999t^2 - 1.5334t + 0.3167 \frac{\text{miles}}{\text{minute}}$$

$$a(t) = v'(t)$$

$$a(t) = \frac{d}{dt}(-2.1332t^3 + 3.999t^2 - 1.5334t + 0.3167)$$

$$a(t) = -6.3996t^2 + 7.998t - 1.5334 \frac{\text{miles per minute}}{\text{minute}}$$

To calculate the velocity and acceleration one minute into the trip, we evaluate $v(1)$ and $a(1)$.

$$v(1) = -2.1332(1)^3 + 3.999(1)^2$$
$$- 1.5334(1) + 0.3167 \frac{\text{miles}}{\text{minute}}$$

$$= 0.6491 \frac{\text{miles}}{\text{minute}}$$

$$= 0.6491 \frac{\text{miles}}{\text{minute}} \cdot \frac{60 \text{ minutes}}{1 \text{ hour}}$$

$$= 38.946 \text{ miles per hour}$$

$$\approx 39 \text{ miles per hour}$$

$$a(1) = -6.3996(1)^2 + 7.998(1)$$
$$- 1.5334 \frac{\text{miles per minute}}{\text{minute}}$$

$$= 0.065 \frac{\text{miles per minute}}{\text{minute}}$$

$$= 0.065 \frac{\text{miles}}{\text{minute}} \cdot \frac{1}{\text{minute}}$$

$$= 0.065 \frac{\text{miles}}{\text{minute}} \cdot \frac{1}{\text{minute}} \cdot \frac{60 \text{ minutes}}{1 \text{ hour}}$$

$$= 3.9 \frac{\text{miles}}{\text{hour}} \cdot \frac{1}{\text{minute}}$$

$$\approx 4 \text{ miles per hour per minute}$$

At 1 minute into his trip, the author was traveling at approximately 39 miles per hour and was accelerating at a rate of 4 miles per hour per minute. That is, if he maintained his current rate of acceleration for the next minute, his speed would increase by about 4 miles per hour.

To calculate the velocity and acceleration 1.25 minutes into the trip, we evaluate $v(1.25)$ and $a(1.25)$.

$$v(1.25) = -2.1332(1.25)^3 + 3.999(1.25)^2$$
$$- 1.5334(1.25) + 0.3167 \frac{\text{miles}}{\text{minute}}$$

$$= 0.4820 \frac{\text{miles}}{\text{minute}}$$

$$= 0.4820 \frac{\text{miles}}{\text{minute}} \cdot \frac{60 \text{ minutes}}{1 \text{ hour}}$$

$$\approx 29 \text{ miles per hour}$$

$$a(1.25) = -6.3996(1.25)^2 + 7.998(1.25)$$
$$- 1.5334 \frac{\text{miles per minute}}{\text{minute}}$$

$$= -1.535 \frac{\text{miles per minute}}{\text{minute}}$$

$$= -1.535 \frac{\text{miles}}{\text{minute}} \cdot \frac{1}{\text{minute}} \cdot \frac{60 \text{ minutes}}{1 \text{ hour}}$$

$$= -92.10 \frac{\text{miles}}{\text{hour}} \cdot \frac{1}{\text{minute}}$$

$$\approx -92 \text{ miles per hour per minute}$$

At 1 minute and 15 seconds into his trip, he was traveling at approximately 29 miles per hour and was decelerating at a rate of 92 miles per hour per minute. That is, if he maintained his current rate of deceleration for the next minute, his speed would decrease by about 92 miles per hour. The rapid rate of deceleration indicates that he is braking, probably in preparation for turning into the cemetery. (Since he was traveling at 29 mph, we know that he won't be able to maintain the high rate of deceleration for an entire minute. Otherwise, his velocity would turn negative, indicating that he was moving away from the cemetery and back toward the stoplight.)

Curve Sketching

Although nowadays it is customary to graph functions using a graphing calculator, knowing how to graph functions by hand will greatly increase your understanding of calculus. If you know the x- and y-intercepts, relative and absolute extrema, and the concavity and inflection points of a function, it is relatively easy to come up with a good sketch of the curve. We demonstrate the curve-sketching process in the following.

EXAMPLE 6 Sketching a Function

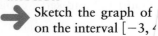 Sketch the graph of on the interval $[-3, 4]$.

SOLUTION

We will set $f(x) = 0$ to find critical values, and f'' inflection points may occur

X- AND Y-INTERCEPTS

$$f(x) = x^3 - 2x^2 - 4x + 8$$

$0 = (x^3 - 2x^2) - (4x - 8)$ Set $f(x) = 0$ and group terms

$0 = x^2(x - 2) - 4(x - 2)$ Factor each group

$0 = (x - 2)(x^2 - 4)$ Factor $(x - 2)$ out of each term

$0 = (x - 2)(x - 2)(x + 2)$ Factor $(x^2 - 4)$

$0 = (x - 2)^2(x + 2)$ Group like terms

$x - 2 = 0$ or $x + 2 = 0$

$x = 2$ $x = -2$

The graph of $f(x) = x^3 - 2x^2 - 4x + 8$ has x-intercepts at $(-2, 0)$ and $(2, 0)$. To find the y-intercept, we evaluate $f(x) = x^3 - 2x^2 - 4x + 8$ at $x = 0$.

$$f(0) = (0)^3 - 2(0)^2 - 4(0) + 8$$
$$= 8$$

y-intercept is $(0, 8)$. We plot these points (Figure 5.19).

TIVE AND ABSOLUTE EXTREMA

critical values, we find $f'(x)$ and set it equal

$x^3 - 2x^2 - 4x + 8$

$3x^2 - 4x - 4$ Differentiate $f(x)$

$x^2 - 4x - 4$ Set $f'(x) = 0$

$+ 2)(x - 2)$ Factor

Figure 5.19

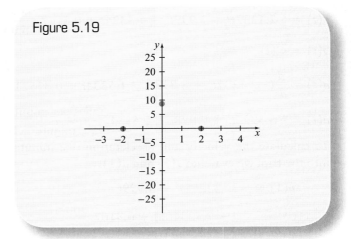

$3x + 2 = 0$ or $x - 2 = 0$

$3x = -2$ $x = 2$

$x = -\dfrac{2}{3}$

$f(x)$ has critical values at $x = -\dfrac{2}{3}$ and $x = 2$. We construct a sign chart for $f'(x)$ to determine where the relative extrema occur.

We evaluate $f(x)$ at the critical values and the endpoints (see Table 5.12).

Table 5.12

x	$f(x) = x^3 - 2x^2 - 4x + 8$	
-3	-25	Absolute minimum
$-\dfrac{2}{3}$	$9\dfrac{13}{27} \approx 9.5$	Relative maximum
2	0	Relative minimum
4	24	Absolute maximum

We plot the corresponding points on the graph of f (Figure 5.20).

From the sign chart of $f'(x)$, we can also determine the intervals on which f is increasing or decreasing. Since $f'(x) > 0$ on $\left[-3, -\dfrac{2}{3}\right)$ and $(2, 4]$, f is increasing on those intervals. Similarly, since $f'(x) < 0$ on the interval $\left(-\dfrac{2}{3}, 2\right)$, f is decreasing on that interval. We won't update the graph of f with this information yet, since we still don't know the concavity of the graph on these intervals.

Figure 5.20

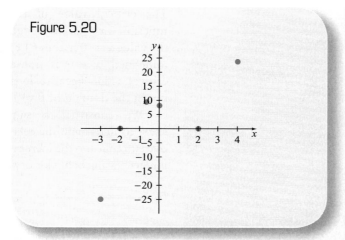

Figure 5.21

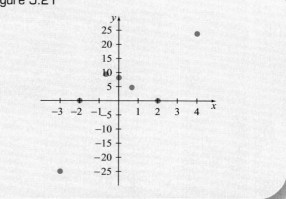

INFLECTION POINTS

To find the inflection points of f, we set $f''(x) = 0$.

$$f'(x) = 3x^2 - 4x - 4$$

$$f''(x) = 6x - 4 \qquad \text{Differentiate } f'(x)$$

$$0 = 6x - 4 \qquad \text{Set } f''(x) = 0$$

$$4 = 6x$$

$$x = \frac{2}{3} \qquad \text{Simplify } \frac{4}{6} \text{ to } \frac{2}{3}$$

An inflection point *may* occur at $x = \frac{2}{3}$. We'll construct a sign chart for $f''(x)$ and see if the concavity of f changes at $x = \frac{2}{3}$.

$$f''(x) \quad \xleftarrow{\qquad - \quad \overset{\text{inflection}}{\underset{\frac{2}{3}}{\text{point of } f}} \quad + \qquad} $$

Since $f''(x) < 0$ on the interval $\left[-3, \frac{2}{3}\right)$, the graph of f is concave down on that interval. Since $f''(x) > 0$ on the interval $\left(\frac{2}{3}, 4\right]$, the graph of f is concave up on that interval. An inflection point occurs at $x = \frac{2}{3}$. We give the coordinate of the inflection point in Table 5.13.

Table 5.13

x	$f(x) = x^3 - 2x^2 - 4x + 8$	
$\dfrac{2}{3}$	$4\dfrac{20}{27} \approx 4.7$	Inflection point

We update the graph with the inflection point (Figure 5.21).

Referring to our earlier observations regarding the increasing/decreasing behavior and concavity of f, we finish the graph (Figure 5.22).

Figure 5.22

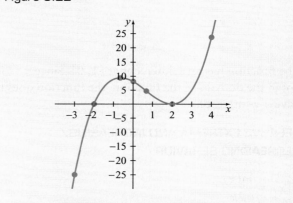

EXAMPLE 7 Sketching a Function

Using the curve-sketching techniques of calculus, graph the function $f(x) = \dfrac{\ln(x)}{2x}$ on the domain $[1, 8]$.

SOLUTION

X- AND Y-INTERCEPTS

$$f(x) = \frac{\ln(x)}{2x}$$

$$0 = \frac{\ln(x)}{2x} \qquad \text{Set } f(x) = 0$$

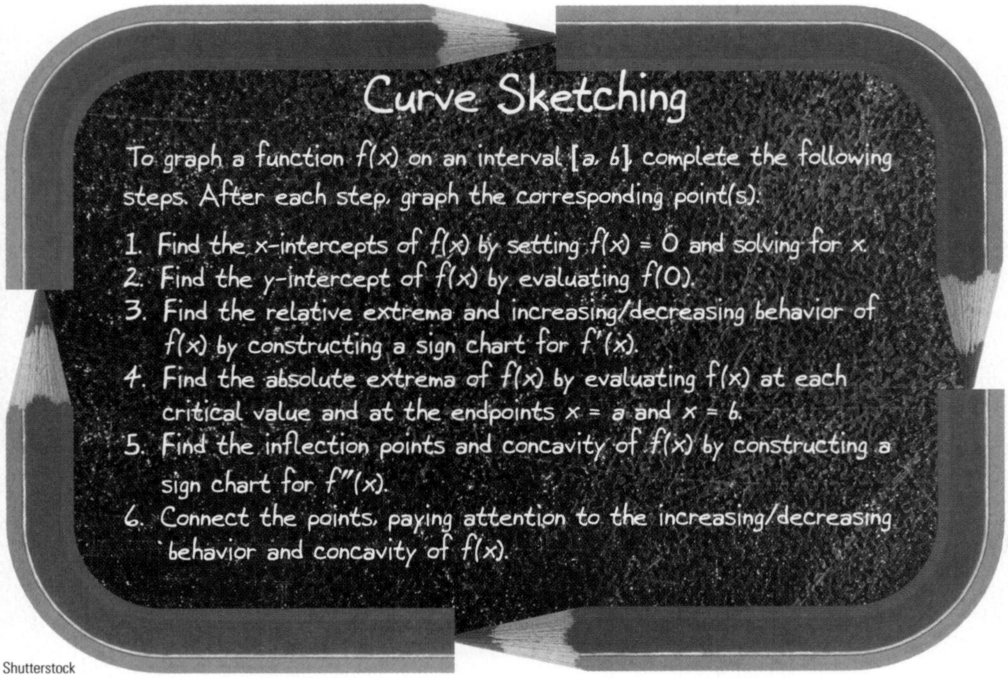

Curve Sketching

To graph a function $f(x)$ on an interval $[a, b]$, complete the following steps. After each step, graph the corresponding point(s).

1. Find the x-intercepts of $f(x)$ by setting $f(x) = 0$ and solving for x.
2. Find the y-intercept of $f(x)$ by evaluating $f(0)$.
3. Find the relative extrema and increasing/decreasing behavior of $f(x)$ by constructing a sign chart for $f'(x)$.
4. Find the absolute extrema of $f(x)$ by evaluating $f(x)$ at each critical value and at the endpoints $x = a$ and $x = b$.
5. Find the inflection points and concavity of $f(x)$ by constructing a sign chart for $f''(x)$.
6. Connect the points, paying attention to the increasing/decreasing behavior and concavity of $f(x)$.

Shutterstock

The critical value of the function is $x = e$. (Although the value $x = 0$ makes $f'(x)$ undefined, $x = 0$ is not a critical value because it is not in the domain of $f(x)$.)

We construct a sign chart to determine the relative extrema and the increasing/decreasing behavior of f.

$$f'(x) \xleftarrow{\qquad \overset{+}{\qquad} \overset{\text{max of } f}{\underset{e}{|}} \overset{-}{\qquad} \qquad} $$

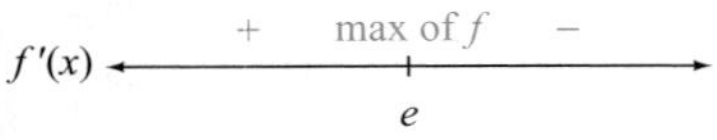

The graph of f is increasing on the interval $[1, e)$ and decreasing on the interval $(e, 8]$. A relative maximum occurs at $x = e$. We evaluate f at the critical value and endpoints (see Table 5.14).

$$0 = \ln(x) \qquad \text{Multiply both sides by } 2x$$

$$e^0 = x \qquad \text{Rewrite as an exponential function}$$

$$x = 1 \qquad \text{Since } e^0 = 1$$

The function has an x-intercept at $(1, 0)$. Since $x = 0$ is not in the domain of the function, the function does not have a y-intercept.

RELATIVE EXTREMA AND INCREASING/ DECREASING BEHAVIOR

$$f(x) = \frac{\ln(x)}{2x}$$

$$f'(x) = \frac{\left(\frac{1}{x}\right)(2x) - (2)(\ln(x))}{(2x)^2} \qquad \text{Apply the Quotient Rule}$$

$$= \frac{2 - 2\ln(x)}{4x^2}$$

$$0 = \frac{2 - 2\ln(x)}{4x^2} \qquad \text{Set } f'(x) = 0$$

$$0 = 2 - 2\ln(x) \qquad \text{Set numerator equal to 0}$$

$$2 = 2\ln(x)$$

$$1 = \ln(x)$$

$$e^1 = e^{\ln(x)} \qquad \text{Exponentiate both sides}$$

$$e = x \qquad \text{Since } e^{\ln(x)} = x$$

Table 5.14

x	$f(x) = \dfrac{\ln(x)}{2x}$	
1	0	Absolute minimum
e	$\dfrac{1}{2e} \approx 0.18$	Relative maximum Absolute maximum
8	$\dfrac{\ln(8)}{16} \approx 0.13$	

We create a graph of the points (Figure 5.23).

Figure 5.23

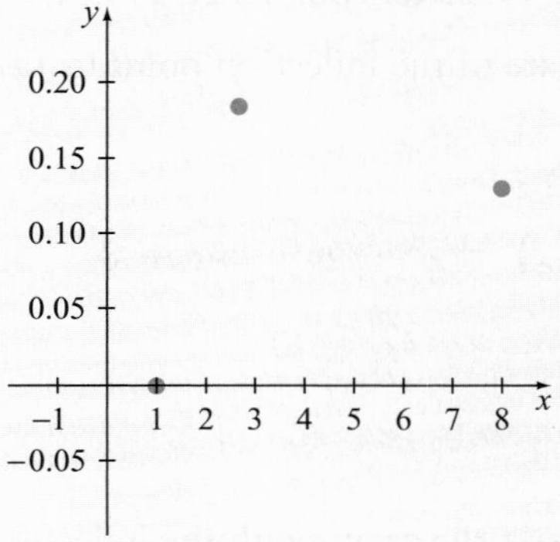

INFLECTION POINTS AND CONCAVITY

$$f'(x) = \frac{2 - 2\ln(x)}{4x^2}$$

$$f''(x) = \frac{\left(-\frac{2}{x}\right)(4x^2) - 8x(2 - 2\ln(x))}{(4x^2)^2}$$ Apply the Quotient Rule

$$= \frac{-8x - 16x + 16x\ln(x)}{16x^4}$$

$$= \frac{-24x + 16x\ln(x)}{16x^4}$$

$$= \frac{-8x(3 - 2\ln(x))}{16x^4}$$ Factor the numerator

$$0 = \frac{-8x(3 - 2\ln(x))}{16x^4}$$ Set $f'(x) = 0$

$$0 = -8x(3 - 2\ln(x))$$ Multiply both sides by $16x^4$

$$-8x = 0 \qquad \text{or} \qquad 3 - 2\ln(x) = 0$$

$$x = 0 \qquad\qquad\qquad 3 = 2\ln(x)$$

$$\frac{3}{2} = \ln(x)$$

$$x = e^{3/2}$$

Since $x = 0$ is not in the domain of the function, $x = 0$ is not a point of inflection. We construct a sign chart for $f''(x)$.

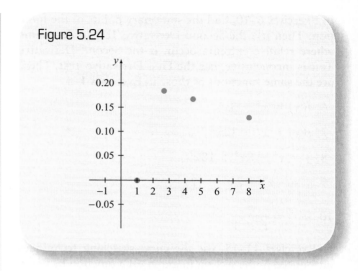

The graph of f is concave down on $[1, e^{3/2})$ and concave up on $(e^{3/2}, 8]$. An inflection point occurs at $x = e^{3/2} \approx 4.5$ (see Table 5.15).

Table 5.15

x	$f(x) = \dfrac{\ln(x)}{2x}$	
$e^{3/2}$	$\dfrac{3}{4e^{3/2}} \approx 0.17$	Inflection point

We update the graph with the inflection point (Figure 5.24).

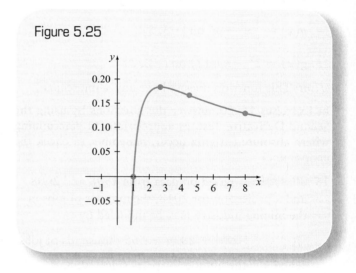

Figure 5.24

Paying attention to the increasing/decreasing behavior and concavity of f, we finish the graph (Figure 5.25).

Figure 5.25

5.3 Exercises

In Exercises 1–5, use the second derivative to find the inflection points of the function algebraically.

1. $f(x) = x^3 - 3x^2$

2. $f(x) = x^4 - 12x^2$

3. $y = x^4 - 8x^3 + 18x^2$

4. $y = -x^4 - 2x^3 + 12x^2$

5. $h(t) = \dfrac{t^2}{t^2 + 1}$

In Exercises 6–10, find the stationary points of the function. Then use the Second Derivative Test to determine where relative extrema occur. If the Second Derivative Test is inconclusive, use the First Derivative Test. These are the same functions as those in Exercises 1–5.

6. $f(x) = x^3 - 3x^2$

7. $f(x) = x^4 - 12x^2$

8. $y = x^4 - 8x^3 + 18x^2$

9. $y = -x^4 - 2x^3 + 12x^2$

10. $h(t) = \dfrac{t^2}{t^2 + 1}$

In Exercises 11–15, use the curve-sketching techniques introduced in the section to graph each function by hand.

11. $f(x) = \dfrac{1}{3}x^3 - x^2$ on $[-2, 4]$

12. $f(x) = -x^3 + 3x + 1$ on $[-2, 2]$

13. $g(x) = -x^3 + 12x$ on $[-4, 4]$

14. $g(x) = \dfrac{1}{5}x^5 - x^4$ on $[-3, 5]$

15. $g(x) = 2^x - x\ln(2)$ on $[-2, 2]$

(*Hint:* This function doesn't have any x-intercepts.)

In Exercises 16–20, answer the questions by using the Second Derivative Test, as appropriate. In determining where absolute extrema occur, remember to check the endpoints.

16. *Mining Jobs* Based on data from 1998 and 2008 and a projection for 2018, the number of jobs in the mining industry may be modeled by

$$j(t) = -1.28t^2 + 28.0t + 565 \text{ thousands of jobs}$$

where t is the number of years since 1998. (**Source:** Modeled from U.S. Bureau of Labor Statistics data)

What is the maximum and what is the minimum number of mining jobs projected to be available between 1998 and 2018?

17. *Ship and Boat Building Jobs* Based on data from 1998 and 2008 and a projection for 2018, the number of jobs in the ship and boat building industry may be modeled by

$$j(t) = -0.0990t^2 + 1.28t + 154 \text{ thousands of jobs}$$

where t is the number of years since 1998. (**Source:** Modeled from U.S. Bureau of Labor Statistics data)

What is the maximum and what is the minimum number of ship and boat building jobs projected to be available between 1998 and 2018?

18. *Dry Cleaning and Laundry Services* Based on data from 1998 and 2008 and a projection for 2018, the number of jobs in the dry cleaning and laundry services industry may be modeled by

$$j(t) = 0.307t^2 - 7.90t + 383 \text{ thousands of jobs}$$

where t is the number of years since 1998. (**Source:** Modeled from U.S. Bureau of Labor Statistics data)

What is the maximum and what is the minimum number of dry cleaning and laundry services jobs projected to be available between 1998 and 2018?

19. *Fishing, Hunting, and Trapping Jobs* Based on data from 1998 and 2008 and a projection for 2018, the number of jobs in the fishing, hunting, and trapping industry may be modeled by

$$j(t) = 0.0545t^2 - 1.63t + 57.8 \text{ thousands of jobs}$$

where t is the number of years since 1998. (**Source:** Modeled from U.S. Bureau of Labor Statistics data)

What is the maximum and what is the minimum number of fishing, hunting, and trapping jobs projected to be available between 1998 and 2018?

20. *Love Triangles* Based on data from 2000 to 2006, the cumulative number of homicides resulting from a romantic triangle may be modeled by

$$R(t) = -1.57t^2 + 119t + 125 \text{ homicides}$$

between the start of 2000 and the end of year t, where t is the number of years since 2000. (**Source:** Modeled from *Crime in the United States 2006*, Uniform Crime Report, FBI)

In what year between 2000 and 2006 did the cumulative number of such homicides increase most rapidly? How many homicides due to a romantic triangle occurred that year?

5.4 Related Rates

On April 20, 2010, an oil well blowout in the Gulf of Mexico killed 11 people and spawned the largest offshore oil spill in American history. The spill continued for weeks as industry experts struggled to cap the oil well 5,000 feet below the water's surface. Although early estimates were sketchy, the U.S. Geological Survey Director issued an official spill flow rate estimate of 12,000 to 19,000 barrels per day on May 27, 2010. The environmental and economic impact of the spill was massive. Early estimates of the economic impact exceeded $10 billion.

When an oil spill occurs, many questions immediately arise. How quickly is the oil spilling into the water? How rapidly is the area covered by the oil expanding? If the oil continues to spill at its current rate, how much area will be covered an hour from now? These questions may be addressed using the concept of *related rates*.

In this section, we will demonstrate how related rates can be used to measure how the rate of change in one variable affects the rate of change in another variable. We will use the implicit differentiation techniques covered in Section 4.5 extensively in this section.

In reality, most oil spills aren't circular due to the motion of ocean waves. However, to keep things simple, we will explore the oil spill context assuming the spill is circular.

The area of a circular oil spill with radius r feet is given by the formula $A = \pi r^2$. If the area of the circle is increasing at a rate of 100 square feet per minute, how quickly is the radius changing? Since the radius rate of change function is the derivative of the radius function, we are looking for $\dfrac{dr}{dt}$ with units $\dfrac{\text{feet}}{\text{minute}}$. The rate of change in the area is given by $\dfrac{dA}{dt}$. The units of $\dfrac{dA}{dt}$ are

$\dfrac{\text{square feet}}{\text{minute}}$. Notice that both A and r are differentiated with respect to t. Using implicit differentiation, we have

$$A = \pi r^2$$

$$\frac{d}{dt}(A) = \frac{d}{dt}(\pi r^2)$$

$$\frac{dA}{dt} = \pi \frac{d}{dt}(r^2)$$

$$\frac{dA}{dt} = \pi \left(2r \frac{dr}{dt} \right)$$

$$\frac{dA}{dt} = 2\pi r \frac{dr}{dt}$$

$$\frac{dr}{dt} = \frac{1}{2\pi r} \frac{dA}{dt}$$

Thus the rate of change in the radius is the rate of change in the area divided by the product of 2π and the current radius. For this problem, we indicated that the area was increasing at a rate of 100 square feet/minute. That is, $\dfrac{dA}{dt} = 100$.

Therefore, the rate of change in the radius is

$$\frac{dr}{dt} = \frac{1}{2\pi r} \frac{dA}{dt}$$

$$= \frac{1}{2\pi r}(100)$$

$$= \frac{50}{\pi r}$$

When the radius is 1 foot, $\dfrac{dr}{dt} = \dfrac{50}{\pi(1)} \approx 15.9$ feet per minute. When the radius is 2 feet, $\dfrac{dr}{dt} = \dfrac{50}{\pi(2)} \approx 7.96$ feet per minute. When the radius is 1,000 feet, $\dfrac{dr}{dt} = \dfrac{50}{\pi(1000)} \approx 0.01592$ feet per minute. Assuming that the shape of the spill remains circular, we can calculate the rate of change in the radius for any value of the radius.

Related rates are most often used when comparing the changes in two related quantities over time. In the next several examples, we will illustrate how related rates are used.

EXAMPLE 1 Interpreting the Meaning of $\frac{dV}{dt}$, $\frac{dr}{dt}$, and $\frac{dh}{dt}$

The volume of space occupied by a liquid in a cylindrical can is given by $V = \pi r^2 h$, where r is the radius of the can and h is the depth of the liquid in the can. Differentiate V with respect to t and interpret the meaning of $\frac{dV}{dt}$, $\frac{dr}{dt}$, and $\frac{dh}{dt}$. Assume that the radius and height are measured in centimeters and that time is measured in seconds. Assuming that the radius of the can is fixed, simplify the result for $\frac{dV}{dt}$.

SOLUTION

$$\frac{d}{dt}(V) = \frac{d}{dt}(\pi r^2 h)$$

$$\frac{dV}{dt} = \pi \frac{d}{dt}(r^2 h) \qquad \text{Constant Multiple Rule}$$

$$= \pi\left(2r\frac{dr}{dt} \cdot h + \frac{dh}{dt} \cdot r^2\right) \qquad \text{Product and Chain Rules}$$

$$= \pi r\left(2h\frac{dr}{dt} + r\frac{dh}{dt}\right)$$

$\frac{dr}{dt}$ is the rate of change in the radius in centimeters per second.

$\frac{dh}{dt}$ is the rate of change in depth of the liquid in centimeters per second.

$\frac{dV}{dt}$ is the rate of change in the volume of the liquid in the can in cubic centimeters per second. It is a function of the radius, the depth, and the rates of change in the radius and the depth. Notice that if the can has a fixed radius, then $\frac{dr}{dt} = 0$ and

$$\frac{dV}{dt} = \pi r\left(2h(0) + r\frac{dh}{dt}\right)$$

$$= \pi r^2 \frac{dh}{dt}$$

EXAMPLE 2 Using Related Rates in a Real-World Context

The volume of a cylindrical can is given by $V = \pi r^2 h$, where r is the radius and h is the height of the interior of the can from lid to lid. A #10 can has a radius of 3.625 inches (9.208 centimeters). Suppose that the can is being filled with rice and that the rice is being poured into the can at a constant rate of 20 cups per minute. At what rate is the height of the rice in the can changing? Will it take more or less than a minute to fill the can?

SOLUTION

You may immediately ask, "Why do we care?" Admittedly, we may not care if we have to fill only one can. However, if we worked in a cannery and wanted to program a machine to fill thousands of cans, the problem would immediately become meaningful. Failure to program the machine with the correct fill rate would result in overfilled cans or insufficiently filled cans. Both of these would hamper the cannery's effort to create a uniform product with minimum waste.

Since we are filling #10 cans, the volume equation becomes

$$V = \pi(9.208)^2 h$$

$$= 84.79\pi h \text{ cubic centimeters}$$

and the rate of change in the volume is given by

$$\frac{dV}{dt} = 84.79\pi \frac{dh}{dt} \frac{\text{cubic centimeters}}{\text{minute}}$$

One of the challenges of the standard measurement system (inches, feet, miles, cups, gallons, etc.) is that conversion between quantities (e.g., cups to cubic inches) is difficult. For this reason, we will convert standard measurements to their metric equivalents. One cup is equivalent to 236.6 cubic centimeters. Since the volume is changing at a rate of 20 cups per minute, we have

$$\frac{dV}{dt} = 20 \frac{\text{cups}}{\text{minute}} \cdot \frac{236.6 \text{ cubic centimeters}}{1 \text{ cup}}$$

$$= 4732 \frac{\text{cubic centimeters}}{\text{minute}}$$

We must find $\frac{dh}{dt}$.

$$\frac{dV}{dt} = 84.79\pi \frac{dh}{dt}$$

$$4732 = 266.4 \frac{dh}{dt}$$

$$\frac{dh}{dt} = \frac{4732}{266.4}$$

$$= 17.76 \text{ centimeters per minute}$$

The height of the rice in the can is increasing by about 17.76 centimeters per minute. Since the interior height of a #10 can is 16.83 centimeters (6.625 inches), it will take less than a minute to fill the can. (To be precise, it will take 57 seconds.)

EXAMPLE 3 Using Related Rates in a Real-World Context

A circular above ground swimming pool has a 12-foot radius and may be filled to a maximum depth of 4 feet. By measuring the amount of time it took to fill a bucket, the author determined that water leaves his outside faucet at a rate of 9 gallons per minute. How quickly is the depth of the water in the pool changing when the pool is being filled from a hose connected to the faucet? How long will it take to fill the pool?

SOLUTION

We will first convert the dimensions to metric measurements. We know that 1 foot ≈ 30.48 centimeters and 1 gallon ≈ 3785 cubic centimeters.

The radius of the pool is 12(30.48) = 365.76 centimeters. The rate at which the water is leaving the faucet is 9(3785) = 34,065 cubic centimeters per minute.

The pool is cylindrical, so its volume is $V = \pi r^2 h$, where r is the radius and h is the depth of the water in the pool. Since the radius of the pool will not change, we have

$$V = \pi(365.76)^2 h$$

$$= 420{,}283.45h \text{ cubic centimeters}$$

Differentiating with respect to t, we get

$$\frac{dV}{dt} = 420{,}283.45\frac{dh}{dt}\frac{\text{cubic centimeters}}{\text{minute}}$$

But we know that $\frac{dV}{dt} = 34{,}065$ cubic centimeters per minute, so

$$34{,}065 = 420{,}283.45\frac{dh}{dt}$$

$$\frac{dh}{dt} = 0.08105\frac{\text{centimeters}}{\text{minute}}$$

$$= 0.08105\frac{\text{centimeters}}{\text{minute}} \cdot \frac{60 \text{ minutes}}{1 \text{ hour}}$$

$$= 4.863 \text{ centimeters per hour}$$

The pool depth is increasing at a rate of 4.863 centimeters per hour. Since 4 feet = 121.92 centimeters, the amount of time it will take to fill the pool is

$$\frac{121.92 \text{ centimeters}}{4.863 \text{ centimeters per hour}} = 25.07 \text{ hours}$$

It will take about 25 hours to fill the pool.

EXAMPLE 4 Calculating a Change in Wind Chill Temperature

Wind chill temperature is the temperature in calm air that has the same chilling effect on a person as that of a particular combination of temperature and wind. For example, when the temperature is 35° Fahrenheit (°F) and the wind speed is 40 miles per hour, it feels like it is 3°F. The wind chill temperature may be calculated according to the following formula:

$$F = 91.4 - (0.474677 - 0.020425w + 0.303107\sqrt{w})$$

$$(91.4 - T) \text{ degrees Fahrenheit}$$

where w is the wind speed (in mph) and T is the temperature (in °F). (**Source:** National Climatic Data Center)

Shutterstock

If the temperature is currently 20°F and is dropping at a rate of 3°F per hour, and if the wind is currently blowing at 16 mph and is increasing at a rate of 2 mph per hour, how quickly is the wind chill temperature changing?

SOLUTION

We must first differentiate the wind chill equation with respect to time.

$$\frac{d}{dt}(F) = \frac{d}{dt}(91.4 - (0.474677 - 0.020425w$$

$$+ 0.303107\sqrt{w})(91.4 - T))$$

$$\frac{dF}{dt} = 0 - \left(-0.020425\frac{dw}{dt}\right.$$

$$+ 0.5(0.303107w^{-0.5})\frac{dw}{dt}\bigg)(91.4 - T)$$

$$+ \left(-\frac{dT}{dt}\right)(-(0.474677$$

$$-0.020425w + 0.303107\sqrt{w}))$$ Product Rule

$$\frac{dF}{dt} = \left(0.020425\frac{dw}{dt}\right.$$

$$\left. - 0.5(0.303107w^{-0.5})\frac{dw}{dt}\right)(91.4 - T)$$

$$+ \left(\frac{dT}{dt}\right)(0.474677 - 0.020425w$$

$$+ 0.303107\sqrt{w})$$

The current wind speed is 16 mph, so $w = 16$.

$\frac{dw}{dt}$ is the rate of change in the wind and equals 2 mph per hour.

The current temperature is 20°F, so $T = 20$.

$\frac{dT}{dt}$ is the rate of change in the temperature and equals -3°F per hour.

Substituting these values into the derivative equation, we get

$$\frac{dF}{dt} = (0.020425(2)$$

$$- 0.5(0.303107(16)^{-0.5})(2))(91.4 - (20))$$

$$+ (-3)(0.474677 - 0.020425(16)$$

$$+ 0.303107\sqrt{16})$$

$$= (-0.03493)(71.4) - 4.081$$

$$= -2.494 - 4.081$$

$$= -6.575$$

The wind chill index is dropping by about 6.6°F per hour. That is, although the temperature is dropping by only 3°F per hour, it feels like it is dropping by 6.6°F per hour because of the wind.

Because of the number of terms and the complexity of the coefficients, the type of calculation shown in Example 4 is time-consuming and prone to error. For this reason, we recommend using the instructions for using the Equation Solver, which are on your Chapter 5 Tech Card.

······

5.4 Exercises

In Exercises 1–10, differentiate the function with respect to t. Explain the physical meaning of each of the rates. (For each of the problems, A is area, V is volume, r is
radius, h is height, l is length, w is width, b is base, and t is time. Use inches, square inches, cubic inches, and minutes as the units of length, area, volume, and time, respectively.)

1. $A = lw$ (area of a rectangle)

2. $A = 2\pi rh + 2\pi r^2$ (surface area of a cylinder including top and bottom)

3. $V = lwh$ (volume of a box)

4. $V = \frac{4}{3}\pi r^3$ (volume of a sphere)

5. $A = 2lw + 2wh + 2lh$ (surface area of a box)

6. $V = \frac{1}{3}\pi r^2 h$ (volume of a cone)

7. $A = 4\pi r^2$ (surface area of a sphere)

8. $A = \frac{1}{2}bh$ (area of a triangle)

9. $A = \pi r^2$ (area of a circle)

10. $A = 2\pi r^2$ (surface area of a hemisphere)

In Exercises 11–15, solve each related-rate problem using the techniques demonstrated in this section.

11. *Swimming Pool Depth* A circular above ground swimming pool has a 9-foot radius and may be filled to a maximum depth of 4 feet. The pool is being filled with a faucet that releases water at a rate of 10 gallons per minute. How quickly is the depth of the water in the pool changing when the pool is being filled from a hose connected to the faucet? How long will it take to fill the pool?

12. *Water Pressure* A circular above ground swimming pool has a 9-foot radius and may be filled to a maximum depth of 4 feet. If the depth of the water in the pool is rising at a rate of 0.5 inch per minute, at what rate is the water leaving the hose that is filling the pool?

13. *Water Depth* A cylindrical can has a 5-inch radius and an 8-inch height and is to be filled with water. At what constant rate must the depth of the water increase if the can is to be filled in exactly 30 seconds?

14. *Pizza Size* The owner of a pizzeria is contemplating changing the size of his large pizza. He intends to decrease the size of the large pizza from a 16-inch diameter to a 14-inch diameter by gradually

reducing the diameter by $\frac{1}{2}$ inch each month. (He figures that consumers won't notice the change if it is gradual.) At what rate will the surface area of the pizza be changing 2 months into the reduction period?

15. *Oil Spill Spread* The surface area of a circular oil spill is increasing at a rate of 100 square feet per hour. How quickly is the radius increasing when the radius is 157 feet?

In Exercises 16–20, solve each related-rate problem using the Chapter 5 Tech Card. The wind chill temperature may be estimated using the following model equation:

$$F = 91.4 - (0.474677 - 0.020425w$$

$$+ 0.303107\sqrt{w})$$

$$\cdot (91.4 - T) \text{ degrees Fahrenheit}$$

where w is the wind speed (in mph) and T is the temperature (in °F). (Source: National Climatic Data Center)

16. *Wind Chill* If the temperature is currently 0°F and is dropping at a rate of 5°F per hour, and the wind is currently blowing at 18 mph and is increasing at a rate of 2 mph per hour, how quickly is the wind chill temperature changing?

17. *Wind Chill* If the temperature is currently –10°F and is warming at a rate of 8°F per hour, and the wind is currently blowing at 25 mph and is increasing at a rate of 3 mph per hour, how quickly is the wind chill temperature changing?

18. *Wind Chill* If the temperature is currently –2°F and is decreasing at a rate of 4°F per hour, and the wind is currently blowing at 55 mph and is decreasing at a rate of 5 mph per hour, how quickly is the wind chill temperature changing?

19. *Wind Chill* If the temperature is currently 40°F and is increasing at a rate of 4°F per hour, and the wind is currently blowing at 5 mph and is not increasing its speed, how quickly is the wind chill temperature changing?

20. *Wind Chill* If the temperature is currently 32°F and is not changing, and the wind is currently blowing at 25 mph and is increasing its speed by 4 mph per hour, how quickly is the wind chill temperature changing?

ATTENTION
NEED MORE PRACTICE? FIND MORE HERE:
CENGAGEBRAIN.COM

Chapter 6

The Integral

On April 20, 2010, an oil well blowout in the Gulf of Mexico spawned the largest offshore oil spill in American history. The spill continued for weeks as industry experts struggled to cap the oil well 5,000 feet below the ocean surface. Although early estimates were sketchy, the U.S. Geological Survey Director issued an official spill flow rate estimate of 12,000 to 19,000 barrels per day on May 27, 2010. If a function modeling the variable rate at which oil is spilling is known, the Fundamental Theorem of Calculus may be used to calculate the total amount of oil spilled over a specified time period.

6.1 Indefinite Integrals

6.2 Integration by Substitution

6.3 Using Sums to Approximate Area

6.4 The Definite Integral

6.5 The Fundamental Theorem of Calculus

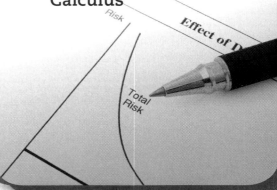

6.1 Indefinite Integrals

A speedometer measures a car's velocity. If we record our car's velocity at 15-second intervals, we will generate a table of data that may be used to find a model for the velocity of the car. Can we determine how far the car traveled from the velocity equation alone? Yes. By using a process called *integration*, we can determine the distance equation if we know the velocity equation.

In this section, we will introduce integration: the process of finding the equation of a function that has a given derivative. We will also define and demonstrate how to use basic rules of integration.

We will begin by looking at three functions: $f(x) = x^3 - x^2 - 8x + 17$, $g(x) = x^3 - x^2 - 8x + 12$, and $h(x) = x^3 - x^2 - 8x + 7$. The equations of these functions look remarkably similar. In fact, their equations differ only by a constant. Similarly, the basic shape of the graphs is the same; they differ only in their vertical placement on the coordinate system (see Figure 6.1).

Let's differentiate each of the functions.

$$f'(x) = 3x^2 - 2x - 8$$

$$g'(x) = 3x^2 - 2x - 8$$

$$h'(x) = 3x^2 - 2x - 8$$

Figure 6.1

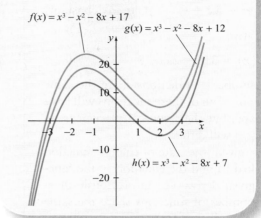

The functions all have the same derivative! In fact, any function of the form

$$f(x) = x^3 - x^2 - 8x + C \qquad (C \text{ is any constant})$$

will have the derivative

$$f'(x) = 3x^2 - 2x - 8$$

Thus, an infinite number of functions share the same derivative. The graphs of all of these functions will have the same basic shape, but their vertical placement on the coordinate system will differ.

Often we are given the rate-of-change function (derivative) and asked to find the equation of the function that has the given derivative. As previously illustrated, an infinite number of functions share the same derivative. To represent all such functions, we will write the letter C in the place of the constant term.

EXAMPLE 1 Finding the Antiderivative of a Function

→ Find the general form of the function that has the derivative $f(x) = 2x$.

SOLUTION

We are looking for a function $F(x)$ such that $\dfrac{d}{dx}(F(x)) = f(x)$. Let's consider the function $F(x) = x^2 + C$.

$$\frac{d}{dx}(F(x)) = \frac{d}{dx}(x^2 + C)$$
$$= 2x + 0$$
$$= 2x$$

Therefore, any function of the form $F(x) = x^2 + C$ has the derivative $f(x) = 2x$. Specific functions include $F(x) = x^2 + 2$, $F(x) = x^2 - 12$, $F(x) = x^2 + 256$, and so on.

The notation $\int f(x)\,dx$ is read "the integral of $f(x)$ with respect to x" or "the antiderivative of $f(x)$ with respect to x." Just as $\dfrac{d}{dx}(f(x))$ means "find the derivative of $f(x)$ with respect to x," $\int f(x)\,dx$ means "find the general form of the function that has derivative $f(x)$ with respect to x." The integral sign $\int$ must always be written together with a *differential* (e.g., dx) to indicate the variable of integration. (We discuss differentials in more detail in the next section.) A function f is said to be **integrable** if there exists a function F such that $F' = f$.

Definition of Terms

Integration: The process of finding the general form of a function that has a given derivative. Also referred to as *antidifferentiation*.

Indefinite integral: The general form of a function that has a given derivative. Also referred to as the *antiderivative*.

Integral sign: The symbol $\int$, which indicates that integration is to be performed.

Integrand: The function written to the right of the integral sign. This is the function that is to be integrated.

EXAMPLE 2 Finding the Antiderivative of a Function

→ Find $\displaystyle\int 3x^2\,dx$.

SOLUTION

We are to find the general equation of the function whose derivative is $f(x) = 3x^2$. Recall that $\dfrac{d}{dx}(x^3) = 3x^2$. Therefore, the indefinite integral of $f(x) = 3x^2$ is $F(x) = x^3 + C$.

We can check our work by differentiating the function F with respect to x.

$$\frac{d}{dx}(F(x)) = \frac{d}{dx}(x^3 + C)$$
$$= 3x^2 + 0$$
$$= 3x^2$$

You may have noticed that we have shifted our notation slightly from our earlier treatment of derivatives. Since we are given the rate-of-change function as the original function, we are using the notation $f(x)$ instead of $f'(x)$ to represent the rate-of-change function. We use the notation $F(x)$ to represent the function whose derivative is $f(x)$. That is, $F'(x) = f(x)$. Admittedly, this shift in notation is a bit confusing. However, since the notation shift is so widely adopted by people who use calculus, we will also use this commonly accepted notation. Making the transition may be difficult at first, but by the end of the chapter, you will be using the new notation with ease.

Basic Integration Rules

As you might expect, the rules for derivatives and the rules for integrals are intimately related. In fact, many of the rules even share the same names.

Power Rule for Integrals

Let $f(x) = x^n$, where x is a variable and n is a constant with $n \neq -1$. Then

$$\int x^n \, dx = \frac{x^{n+1}}{n+1} + C$$

The rule can be easily verified by differentiating the resultant function.

$$\frac{d}{dx}\left(\frac{x^{n+1}}{n+1} + C\right) = \frac{1}{n+1}\frac{d}{dx}(x^{n+1}) + 0$$

Constant Multiple Rule

$$= \frac{n+1}{n+1}x^{n+1-1}$$

Power Rule

$$= x^n$$

Notice that $n \neq -1$. If $n = -1$, the denominator of $\frac{x^{n+1}}{n+1} + C$ is equal to 0, which makes the expression undefined. We will deal with the case of $n = -1$ later.

Constant Multiple Rule for Integrals

Let $f(x) = k \cdot g(x)$, where x is a variable and k is a constant. Then

$$\int (k \cdot g(x)) \, dx = k \cdot \int g(x) \, dx$$
$$= k \cdot G(x) + C$$

where $g(x) = G'(x)$.

EXAMPLE 3 Integrating a Function

Integrate $f(x) = 6x^2$.

SOLUTION

$$\int 6x^2 \, dx = 6\int x^2 \, dx \qquad \text{Constant Multiple Rule}$$

$$= 6\left(\frac{x^3}{3} + C_1\right) \qquad \text{Power Rule}$$

$$= \frac{6x^3}{3} + 6C_1$$

$$= 2x^3 + C$$

Notice that we rewrote $6C_1$ as C. Since C_1 is a constant, multiplying it by 6 will still result in a constant. The same would be true if we were to add two different constants C_1 and C_2. The result would still be a constant, which we could write as C. Therefore, when doing integration, we will write a single constant C to represent a multiple of a constant or a sum of constants.

Remember that the derivative represents the slope of the tangent line. The derivative of a linear function $f(x) = mx + b$ is $f'(x) = m$. That is, for a linear function, the derivative of the function is the slope of the line. Similarly, the antiderivative of a constant function $f(x) = m$ is the linear function $F(x) = mx + b$. This relationship is summarized in the Constant Rule for Integrals.

Constant Rule for Integrals

Let k be a constant. Then

$$\int k \, dx = kx + C$$

The Sum and Difference Rule is another extremely useful rule to use when working with integrals.

Sum and Difference Rule for Integrals

Let $f(x)$ and $g(x)$ be integrable functions of x. Then

$$\int (f(x) \pm g(x))dx = \int f(x)dx \pm \int g(x)dx$$

EXAMPLE 4 Antidifferentiating a Function

Find the antiderivative of $h(x) = 2x + 4$.

SOLUTION

$$\int (2x + 4)\,dx = \int 2x\,dx + \int 4\,dx \quad \text{Sum and Difference Rule}$$

$$= 2\int x\,dx + 4\int 1\,dx \quad \text{Constant Multiple Rule}$$

$$= 2\left(\frac{x^2}{2}\right) + 4(x) + C \quad \text{Power and Constant Rules}$$

$$= x^2 + 4x + C$$

Exponential Rule for Integrals

Let $a > 0$ and $a \neq 1$. Then

$$\int a^x dx = \frac{1}{\ln(a)} a^x + C$$

Special case: $\int e^x dx = e^x + C$

The Exponential Rule may be easily verified by differentiating $y = \frac{1}{\ln(a)}a^x + C$.

$$\frac{d}{dx}\left(\frac{1}{\ln(a)}a^x + C\right) = \frac{d}{dx}\left(\frac{1}{\ln(a)}a^x\right) + \frac{d}{dx}(C) \quad \text{Sum and Difference Rule}$$

$$= \frac{1}{\ln(a)}\left(\frac{d}{dx}(a^x)\right) + 0 \quad \text{Constant and Constant Multiple Rules}$$

$$= \frac{1}{\ln(a)}(\ln(a)a^x) \quad \text{Exponential Rule}$$

$$= \frac{\ln(a)}{\ln(a)}(a^x)$$

$$= a^x$$

The special case occurs because

$$\int e^x dx = \frac{1}{\ln(e)}e^x + C$$

$$= \frac{1}{1}e^x + C \quad \text{Since } \ln(e) = 1$$

$$= e^x + C$$

EXAMPLE 5 Integrating an Exponential Function

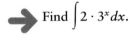 Find $\int 2 \cdot 3^x dx$.

SOLUTION

$$\int 2 \cdot 3^x dx = 2\int 3^x dx \quad \text{Constant Multiple Rule}$$

$$= 2 \cdot \frac{1}{\ln(3)}(3^x) + C \quad \text{Exponential Rule}$$

$$= \frac{2}{\ln(3)}(3^x) + C$$

Recall that $x^{-1} = \frac{1}{x}$. We saw previously that we cannot use the Power Rule for Integrals to integrate $f(x) = x^{-1}$. However, the function is integrable.

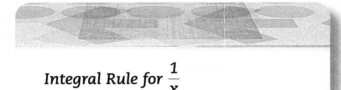

Integral Rule for $\frac{1}{x}$

If $x \neq 0$, then

$$\int \frac{1}{x}\,dx = \ln|x| + C$$

The absolute value of x is necessary to ensure that the two functions $f(x) = \dfrac{1}{x}$ and $F(x) = \ln|x| + C$ have the same domain ($\ln x$ is defined only for $x > 0$, but $\ln|x|$ is defined for all $x \neq 0$). We can convince ourselves that the rule holds true by differentiating $\ln|x| + C$.

$$\frac{d}{dx}(\ln|x| + C) = \frac{d}{dx}(\ln|x|) + \frac{d}{dx}(C) \quad \text{Sum and Difference Rule}$$

$$= \frac{d}{dx}\left(\begin{cases} ln(x) & x > 0 \\ ln(-x) & x < 0 \end{cases}\right) + 0 \quad \begin{array}{l}\text{Definition of absolute value} \\ \text{Constant Rule}\end{array}$$

$$= \begin{cases} \dfrac{1}{x}, x > 0 & \text{Logarithm Rule} \\[2mm] \dfrac{1}{(-x)}(-1), x < 0 & \text{Logarithm and Chain Rules} \end{cases}$$

$$= \begin{cases} \dfrac{1}{x}, x > 0 \\[2mm] \dfrac{1}{x}, x < 0 \end{cases}$$

$$= \frac{1}{x}, x \neq 0$$

We see that for all nonzero values of x, $\dfrac{d}{dx}(\ln|x| + C) = \dfrac{1}{x}$.

EXAMPLE 6 Integrating a Function of the Form $y = \dfrac{c}{x}$

Integrate $\dfrac{3}{x}$ with respect to x.

SOLUTION

$$\int \left(\frac{3}{x}\right) dx = 3\int\left(\frac{1}{x}\right) dx \quad \begin{array}{l}\text{Constant Multiple} \\ \text{Rule for Integrals}\end{array}$$

$$= 3\ln|x| + C \quad \text{Integral Rule for } \frac{1}{x}$$

Indefinite Integral Applications

Recall that the units of $\dfrac{dy}{dx}$ are the units of y over the units of x. Since $\int\left(\dfrac{dy}{dx}\right) dx = y + C$, the units of the antiderivative of $\dfrac{dy}{dx}$ are the units of y. In real-world applications, we are often given enough information to figure out the specific value of C, as demonstrated in Example 7.

EXAMPLE 7 Using Integration to Find a Position Function

Based on 2.5 minutes of data, the velocity of the author's minivan as he drove from a stoplight onto a highway may be modeled by $V(t) = -0.1773t^2 + 0.6798t + 0.5019$ miles per minute, where t is the number of minutes since he left the stoplight. Find the position function $D(t)$ that shows his distance from the stoplight at time t. Find the total distance traveled during the first minute and during the first 2 minutes.

SOLUTION

The position function $D(t)$ is the antiderivative of the velocity function $V(t)$.

$$\int (V(t))\, dt = \int (-0.1773t^2 + 0.6798t + 0.5019)\, dt$$

$$D(t) = -0.1773\frac{t^3}{3} + 0.6798\frac{t^2}{2} + 0.5019t + C$$

$$= -0.0591t^3 + 0.3399t^2 \quad \begin{array}{l}\text{Constant Multiple,} \\ \text{Power, and Sum and}\end{array}$$

$$+ 0.5019t + C \text{ miles} \quad \text{Difference Rules}$$

You may ask, "What is the meaning of C in this context?" Let's see. $D(0)$ represents the distance traveled after 0 minutes. Intuitively, we know that after 0 minutes, the minivan hasn't traveled at all, so $D(0) = 0$.

$$D(0) = -0.0591(0)^3 + 0.3399(0)^2$$
$$+ 0.5019(0) + C$$
$$= C$$

But we know that $D(0) = 0$, so it follows that $C = 0$. $D(t) = -0.0591t^3 + 0.3399t^2 + 0.5019t$ is the distance function for the minivan. According to our distance model, $D(1) = 0.78$ and $D(2) = 1.89$. That is, the total distance traveled during the first minute is 0.78 miles. The total distance traveled during the first two minutes is 1.89 miles.

EXAMPLE 8 Using Antidifferentiation to Find a Cost Function

Have you ever dreamed of being a published author? Lulu (www.lulu.com) publishes books for aspiring authors. In 2010, the marginal cost for printing a 6″ × 9″ soft-cover book with a color cover and black and white interior pages was

$$c(p) = 0.015 \text{ dollars per page}$$

where p is the number of pages in the book. Including the setup costs, a 32-page book has a \$3.71 production cost. Find the book production cost function.

SOLUTION

Recall that marginal cost is the derivative of the cost function. Therefore, the antiderivative of the marginal cost is the cost function.

$$\int c(p) \, dp = \int 0.015 \, dp$$
$$C(p) = 0.015p + k \qquad \text{Constant Rule for Integrals}$$

(*Note:* We used k as a constant, since the function was named C.)

The cost of a 32-page book is \$3.71, so

$$C(32) = 0.015(32) + k$$
$$3.71 = 0.48 + k$$
$$k = 3.23$$

The constant $k = 3.23$ represents the fixed setup cost. The cost to produce a book with p pages is $C(p) = 0.015p + 3.23$ dollars.

....

6.1 Exercises

In Exercises 1–10, find the general antiderivative of the function.

1. $f(x) = 2$

2. $r(t) = t^{-1} - 3t$

3. $v(t) = 0.5t + 20$

4. $s(t) = t^3 - 3t^2 + 3t - 1$

5. $h(x) = \dfrac{2.3}{x} + 1$

6. $p(x) = 3^x - x^3$

7. $f(t) = 4t^{-2} + 2t^{-1} - 1$

8. $h(x) = 2(3^x) - 3(2^x)$

9. $q = \dfrac{2000}{p} - \dfrac{500}{p^2}$

10. $y = 5t^{-2} + 2^t$

In Exercises 11–18, calculate the indefinite integral.

11. $\int (3x - 5) \, dx$

12. $\int \dfrac{t - 2}{t} \, dt$

(*Hint:* Divide each term in the numerator by the denominator before integrating.)

13. $\int (3x^{-2} - 4x^3 + 2) \, dx$

14. $\int (5t^{-2} - 16t^2 - 9) \, dt$

15. $\int \dfrac{5u^2 - 4u + 1}{u^2} \, du$

(*Hint:* Divide each term in the numerator by the denominator before integrating.)

16. $\int \left(\dfrac{400}{x^2} + \dfrac{200}{x} + 50 \right) dx$

17. $\int \left(\dfrac{4}{x} - \dfrac{5}{x^2} \right) dx$

18. $\int (4t + 4^t) \, dt$

In Exercises 19 and 20, apply the concept of integration in solving the real-world applications.

19. *Publishing Costs* Lulu (www.lulu.com) publishes books for aspiring authors. In 2010, the marginal printing cost for a $8\frac{1}{2}$″ × 11″ paperback book was

$$c(p) = 0.018 \text{ dollars per page}$$

where p is the number of pages in the book. Including setup costs, a 300-page book has an \$8.90 production cost. Find the book production cost function.

20. *Minivan Position* Based on 1.5 minutes of data, the velocity of the author's minivan as he drove from a stoplight into a sparsely populated residential area may be modeled by

$$V(t) = -2.133t^3 + 3.999t^2 - 0.1533t + 0.3167$$
$$\text{miles per minute}$$

where t is the number of minutes since he left the stoplight. Find the position function $D(t)$ that shows his distance from the stoplight at time t.

ATTENTION
NEED MORE PRACTICE? FIND MORE HERE:
CENGAGEBRAIN.COM

6.2 Integration by Substitution

The process of *integration by substitution* is a clever way to integrate a function that initially doesn't appear to be integrable. (Recall that a function f is said to be integrable if there exists a function F such that $F' = f$.) Because of the relatively complex nature of the process, we will defer illustrations of how this process may be used in real-life applications to later sections.

In this section, we will introduce the concept of differentials. We will then demonstrate how to do integration by substitution.

Differentials

Recall that $\dfrac{d}{dx}(y) = \dfrac{dy}{dx}$. The derivative $\dfrac{dy}{dx}$ is a single term, not the quotient of dy and dx. However, it would be useful for us to be able to treat $\dfrac{dy}{dx}$ as if it were a fraction. For this purpose, we introduce the concept of *differentials*. We define the **differential** dx to be an arbitrarily small real number and

$$dy = y'dx$$

Observe that dividing both sides of the equation $dy = y'dx$ by dx yields

$$\frac{dy}{dx} = y'$$

which is consistent with our earlier notation for the derivative. Thus, by using differentials, we can treat dy and dx in $\dfrac{dy}{dx}$ as separate entities and use the rules for fractions in manipulating the terms.

EXAMPLE 1 Calculating the Differential dy
Find dy given $y = x^2$.

SOLUTION

$$dy = 2x \ dx$$

EXAMPLE 2 Calculating the Differential dy
Find dy given $y = 2x^3 - 3x$.

SOLUTION

$$dy = (6x^2 - 3)\,dx$$

We may also use differentials with other variables, such as u and t, as demonstrated in the next two examples.

EXAMPLE 3 Calculating the Differential du
Find du given $u = e^{3t} + 2$.

SOLUTION

$$du = 3e^{3t}dt$$

EXAMPLE 4 Calculating the Differential du
Find du given $u = \ln(t)$.

SOLUTION

$$du = \frac{1}{t}dt$$

Integration by Substitution

The technique of **integration by substitution** is used to rewrite a function that doesn't appear to be readily integrable in such a way that it can be integrated using the basic integration rules. Integration by substitution undoes the effects of the chain rule for derivatives. For example, consider the function $F(x) = (x^2 + 4)^{-3}$. Using the Chain Rule, the derivative of the function is $F'(x) = -3(x^2 + 4)^{-4}(2x)$, which may be simplified to $F'(x) = -6x(x^2 + 4)^{-4}$. (The $2x$ term used in the calculation is derivative of the "inside" function $x^2 + 4$.) From the preceding calculations, we know that $\displaystyle\int -6x(x^2 + 4)^{-4}dx = (x^2 + 4)^{-3} + C$; however, the integration process is not obvious. We will demonstrate the process in Example 5 before formalizing the procedure.

EXAMPLE 5 Integrating by Substitution
Find the antiderivative of $f(x) = -6x\,(x^2 + 4)^{-4}$.

SOLUTION

We need to determine $\displaystyle\int -6x(x^2 + 4)^{-4}dx$. The function is a product of two functions that we hope to be able to write in the form $u^n du$. We'll try $u = x^2 + 4$. We have

$$u = x^2 + 4$$

We calculate the differential $du = u' dx$

$$du = 2x\, dx$$

Let's see if we can rewrite the integral in terms of u and du. Recall that $u = x^2 + 4$ and $du = 2x\, dx$.

$$\int -6x(x^2 + 4)^{-4} dx = \int -3(x^2 + 4)^{-4}(2x)dx$$

Since $-6x = -3(2x)$

$$= \int -3u^{-4} du$$

Since $u = x^2 + 4$ and $du = 2x\, dx$

$$= -3\frac{u^{-3}}{-3} + C$$

Power Rule for Integrals

$$= u^{-3} + C$$

Rewriting the result in terms of x, we have

$$\int -6x(x^2 + 4)^{-4} dx = (x^2 + 4)^{-3} + C$$

How To Integrate by Substitution

A function that is the product of two functions can sometimes be integrated by substitution. To integrate by substitution, do the following:

1. Select one function to be u. Often the more complex function or the function in the parentheses is the best choice for u.

2. Calculate the differential du. If $u = f(x)$, then $du = f'(x)dx$.

3. Rewrite the original integrand in terms of u and du. The resultant integrand should not contain any x variables.

4. Integrate the resultant function, if possible. If the resultant function is not integrable, repeat the process but select a different function to be u.

It is not always possible to find a substitution of variables that makes a difficult function integrable. However, if the integral can be rewritten in any of the following forms, it may be integrated using the basic integration rules.

- $\int u^n du$

- $\int e^u du$

- $\int \frac{1}{u} du$

EXAMPLE 6 Integrating by Substitution

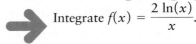 Integrate $f(x) = \dfrac{2\ln(x)}{x}$.

SOLUTION

We must find $\int \dfrac{2\ln(x)}{x} dx$. This function cannot be readily integrated using any of the rules covered in Section 6.1. This is because $f(x) = \dfrac{2\ln(x)}{x}$ is the product of $2\ln(x)$ and $\dfrac{1}{x}$. (Contrary to what you might think, $\int (f(x)g(x))\, dx \neq \int f(x)\, dx \cdot \int g(x)\, dx$.)

We will let $u = \ln(x)$. We use the concept of differentials to find du.

$$u = \ln(x)$$

$$du = \left(\frac{d}{dx}(\ln(x))\right)dx$$

$$du = \frac{1}{x}dx$$

We now rewrite the integral in terms of u.

$$\int 2(\ln(x))\left(\frac{1}{x}dx\right) = \int 2u\, du$$

Since $\ln(x) = u$ and $\frac{1}{x}dx = du$

But we know that

$$\int 2u\, du = u^2 + C$$

(Although the differential du represents an arbitrarily small number and the du of an integral means "with respect to u," the two uses of the notation du may be interchanged. A more in-depth explanation may be obtained from an advanced calculus text.)

Since the variable u was something that we introduced into the problem, we must rewrite the solution in terms of x. Recall that $u = \ln(x)$.

$$u^2 + C = (\ln(x))^2 + C$$

Thus, we have

$$\int \frac{2 \ln(x)}{x} dx = (\ln(x))^2 + C$$

EXAMPLE 7 Integrating by Substitution

Integrate $g(t) = e^{2t-4}$.

SOLUTION

We must find $\int e^{2t-4}\, dt$. We hope to be able to rewrite the integral as $\int e^u du$. We let $u = 2t - 4$. Then

$$du = 2\, dt$$

Looking at the integrand (the function to be integrated), we see that we have dt but not $2\, dt$. Can we solve $du = 2\, dt$ for dt? Yes!

$$2\, dt = du$$
$$dt = \frac{1}{2} du$$

Substituting, we get

$$\int e^{2t-4} dt = \int e^u \left(\frac{1}{2} du\right) \quad \text{Since } u = 2t - 4 \text{ and } dt = \frac{1}{2} du$$

$$= \frac{1}{2} \int e^u\, du \quad \text{Constant Multiple Rule for Integrals}$$

$$= \frac{1}{2} e^u + C \quad \text{Exponential Rule}$$

$$= \frac{1}{2} e^{2t-4} + C \quad \text{Rewrite in terms of } t$$

EXAMPLE 8 Integrating by Substitution

Find the general form of a function that has the derivative $f(x) = \dfrac{2x + 3}{x^2 + 3x + 4}$.

SOLUTION

We must find $\int \dfrac{2x + 3}{x^2 + 3x + 4}\, dx$. We observe that the numerator is the derivative of the denominator.

Consequently, we hope to be able to rewrite the integral in the form $\int \dfrac{1}{u}\, du$. We let $u = x^2 + 3x + 4$. Then

$$du = (2x + 3)dx$$

Rewriting the integral in terms of u and solving, we get

$$\int \frac{2x + 3}{x^2 + 3x + 4}\, dx = \int \frac{(2x + 3)dx}{x^2 + 3x + 4}$$

$$= \int \frac{du}{u} \quad \text{Since } u = x^2 + 3x + 4 \text{ and } du = (2x + 3)\, dx$$

$$= \int \frac{1}{u}\, du$$

$$= \ln|u| + C \quad \text{Integral Rule for } \frac{1}{x}$$

$$= \ln|x^2 + 3x + 4| + C \quad \text{Rewrite in terms of } x$$

All functions of the form $F(x) = \ln|x^2 + 3x + 4| + C$ have the derivative

$$f(x) = \frac{2x + 3}{x^2 + 3x + 4}$$

In each of the previous examples, we selected the "correct" u each time. However, it is very common to attempt integration by substitution with the "wrong" function for u. If you rewrite a function in terms of u and the resulting function isn't integrable, go back and select a different function to be u. This is part of the process of integration by substitution and should not be viewed as an error. As you gain experience with these types of problems, you will more easily recognize what is the best choice for u. In Example 9, we will demonstrate a type of problem that requires an especially clever selection of u.

EXAMPLE 9 Integrating by Substitution a Function Containing a Radical

Find $\int x\sqrt{x - 1}\, dx$.

SOLUTION

We begin by rewriting the integrand with a rational exponent.

$$\int x(x - 1)^{1/2}\, dx$$

Let's pick $u = x - 1$. Then

$$du = 1\, dx$$
$$du = dx$$

We make the substitution.

$$\int x u^{1/2}\, du$$

What do we do with the extra x? The integrand must be written in terms of u. (We can't use the Constant Multiple Rule to move the x to the left-hand side of the integral sign because x is a variable, not a constant.) Recall that $u = x - 1$; therefore, $x = u + 1$. Thus we may rewrite the integral as

$$\int x u^{1/2}\, du = \int (u + 1)u^{1/2}\, du \quad \text{Since } x = u + 1$$

Since we have completely rewritten the integral in terms of u, we may continue. We will multiply out the integrand and then integrate each term.

$$\int (u^{3/2} + u^{1/2})\, du = \int u^{3/2}\, du + \int u^{1/2}\, du$$

$$= \frac{u^{5/2}}{\frac{5}{2}} + \frac{u^{3/2}}{\frac{3}{2}} + C \quad \text{Power Rule for Integrals}$$

$$= \frac{2}{5}u^{5/2} + \frac{2}{3}u^{3/2} + C$$

$$= \frac{2}{5}(x - 1)^{5/2} + \frac{2}{3}(x - 1)^{3/2} + C$$

Thus $\int x\sqrt{x - 1}\, dx = \dfrac{2}{5}(x - 1)^{5/2} + \dfrac{2}{3}(x - 1)^{3/2} + C.$

EXAMPLE 10 Determining When Integration by Substitution Won't Work

➡️ Integrate $f(x) = xe^x$.

SOLUTION

We must find $\int x e^x\, dx$. If we pick $u = x$, then $du = dx$ and the integral becomes

$$\int u e^u\, du$$

Observe that this is identical to the original function written, except that it is written in terms of u. Using the integration techniques we've covered, we don't know how to integrate a function in this form.

Let's try $u = e^x$. Then

$$u = e^x$$
$$\ln(u) = \ln(e^x)$$
$$\ln(u) = x$$
$$x = \ln(u)$$

Since $\dfrac{du}{dx} = e^x$, $du = e^x dx$. The resultant integral is

$$\int x e^x\, dx = \int (\ln(u))\, du \quad \begin{array}{l}\text{Since } x = \ln(u) \\ \text{and } e^x dx = du\end{array}$$

We haven't yet learned how to integrate $\ln(u)$, so we are unable to find the solution using this choice of u.

As a final attempt, we'll try $u = xe^x$. Then $du = (1 \cdot e^x + e^x \cdot x)dx$, by the Product Rule. If we attempt to make a substitution, we'll have a mixture of u and x in the integrand, leaving us with a function that we can't integrate $\left(\int x e^x dx = \int u \dfrac{du}{1 \cdot e^x + e^x \cdot x} \right)$.

The function $f(x) = xe^x$ cannot be integrated using integration by substitution. Nevertheless, there is a method that can be used to integrate $f(x) = xe^x$. In Section 7.1, we will show how to integrate this function using integration by parts.

It is important to note that, despite the many advances in calculus, there are some functions that we still don't know how to integrate. Nevertheless, we must attempt to integrate a function using all known methods before we can conclude that it is not integrable.

···

6.2 Exercises

In Exercises 1–20, integrate the functions. Some integrals will require integration by substitution, while others may be integrated using basic integration rules. All of the functions can be integrated using the techniques we've covered.

1. $f(x) = 2x(x^2 + 3)^5$

2. $h(x) = 2x(3x + 5)$

3. $s(x) = e^x(e^x + 1)$

4. $f(x) = 1.2x^2 - 2.4x + 0.6$

5. $p(x) = (x - 2)e^{x^2 - 4x}$

6. $y = 3xe^{3x}$

7. $g(t) = 4e^{3t}$

8. $s(x) = \dfrac{\ln(x^2)}{x}$

9. $h(t) = \dfrac{2t - 1}{t^2 - t + 2}$

10. $f(x) = \dfrac{6x^2}{x^3 - 9}$

11. $p(x) = \dfrac{1}{x \ln x}$

12. $y = 3x\sqrt{x - 2}$

13. $g(x) = 2x\sqrt{x^2 + 1}$

14. $f(t) = t\sqrt{t^2 - 1}$

15. $f(x) = \dfrac{3x^2 - 2x + 1}{x^3 - x^2 + x - 1}$

16. $f(x) = \dfrac{x}{2x - 1}$

17. $h(x) = \dfrac{2x}{4x^2 - 5}$

18. $a(t) = (4t - 1)(t^2 - 2t)$

19. $q = \dfrac{4p}{2p^2 + 1} - 4p\sqrt{2p^2 + 1}$

20. $s(t) = (4t - 2)(t^2 - t)^{-1}$

6.3 Using Sums to Approximate Area

River rafting or kayaking is an exciting sport that requires both physical skill and an intellectual understanding of the hydraulics of a river. One factor that enthusiasts must consider before navigating a river is the flow rate of the water. The flow rate indicates how much water (in cubic feet) will pass a fixed point in a second. The flow rate fluctuates constantly, ever changing the intensity of the river. For the inex- perienced, flow-rate values mean little. One way to add conceptual meaning to the num- bers is to determine how much water has passed a given point over a period of time. As will be explained in this section, we can do this by estimating this area between a graph of the flow rate and the horizontal axis.

In this section, we will demonstrate how left- and right-hand sums may be used to approximate the area between the graph of a function and the x-axis. This will prepare us to link the notion of area with the con- cept of the definite integral in Section 6.4. The relation- ship between the two concepts is truly remarkable.

Suppose that the flow rate $f(x)$ of a river is a constant 3,000 cubic feet per second. Let's consider the graph of the flow-rate function $f(x) = 3$, shown in Figure 6.2.

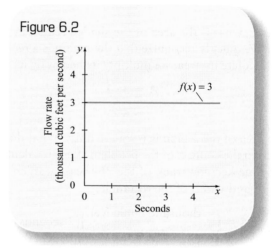

Figure 6.2

The graph is a horizontal line with y-intercept $(0, 3)$. We ask the question, "What is the area of the region between the graph of the function and the x-axis on the interval [1, 3]?"

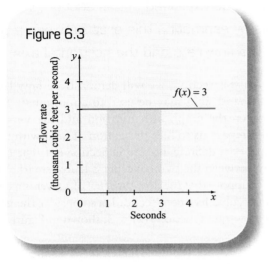

Figure 6.3

That is, what is the area of the shaded region in Figure 6.3? We quickly recognize that the region is a rectangle. To calculate its area, we multiply its height by its width.

$$A = 3 \cdot 2$$
$$= 6$$

The area of the region is 6 square units. What does this mean in the context of the problem? Let's recalculate A, this time keeping track of the physical meaning of the length and width of the rectangle.

$$A = \left(3 \, \frac{\text{thousand cubic feet}}{\text{second}} \right)(2 \text{ seconds})$$
$$= 6000 \text{ cubic feet}$$

Between the first and third second after we started timing, 6,000 cubic feet of water passed the water flow measurement station. In general, if $f(x)$ is a positive rate-of-change function, then the area between the graph of $f(x)$ and the x-axis on the interval $[a, b]$ represents the total change in $F(x)$ (an antiderivative of $f(x)$) from $x = a$ to $x = b$.

When the shaded region is a familiar geometric shape (triangle, rectangle, and so on), it is easy to calculate the area using known formulas. However, what do we do if the shaded region is not a familiar geometric shape? Can we still calculate the area? In Example 1, we will estimate the area of an irregularly shaped region by drawing a series of rectangles. As you will see,

the accuracy of our estimate will increase as we increase the number of rectangles.

EXAMPLE 1 Using Riemann Sums to Estimate an Area

Estimate the area of the region between $f(x) = \sqrt{x} + 2$ and the x-axis on the interval [0, 4].

SOLUTION

We begin by drawing the graph of $f(x) = \sqrt{x} + 2$ (see Figure 6.4).

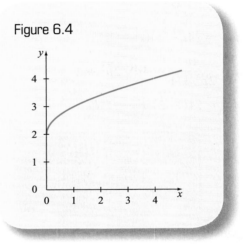

Figure 6.4

We will first estimate the area by drawing a single rectangle whose upper left corner touches the graph of f (Figure 6.5).

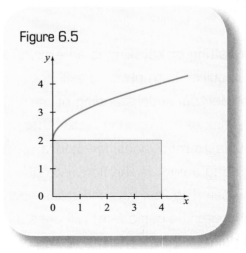

Figure 6.5

The area of the rectangle is $A = 2 \cdot 4 = 8$ square units. Since the entire region below the graph is not completely shaded, 8 square units is an underestimate of the actual area. To improve our accuracy, we will draw two rectangles that are each 2 units wide. The upper left-hand corner of each rectangle touches the graph (see Figure 6.6).

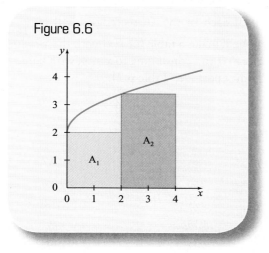

Figure 6.6

The area of the first rectangle is easily calculated: $A_1 = 2 \cdot 2 = 4$ square units. We can easily see that the second rectangle has a width of 2 units, but it is difficult to tell from the graph what the height of the rectangle is. However, since the upper left-hand corner of the rectangle is on the graph of f, we can use the equation of $f(x)$ to calculate the height of the rectangle. Since the rectangle touches the graph when $x = 2$, the height of the rectangle is

$$f(2) = \sqrt{2} + 2$$
$$\approx 1.414 + 2$$
$$\approx 3.414 \text{ units}$$

Consequently, the approximate area of the second rectangle is

$$A_2 = 3.414 \cdot 2$$
$$= 6.828$$

We can estimate the area of the region by adding together the areas of the two rectangles.

$$A = A_1 + A_2$$
$$= 4 + 6.828$$
$$= 10.828 \text{ square units}$$

This estimate is substantially better than our first estimate because it has increased the size of the shaded region below the graph. However, it is still an under-

estimate. How can we increase the accuracy of our estimate? Add more rectangles. This time we use four rectangles of width 1 unit, as shown in Figure 6.7.

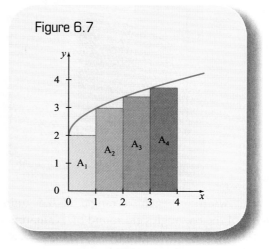

Figure 6.7

The width of each rectangle is 1 unit. Since the upper left-hand corner of each rectangle touches the graph of f, the heights of the rectangles are $f(0)$, $f(1)$, $f(2)$, and $f(3)$, respectively. The areas of the rectangles are given by

$$A_1 = f(0) \cdot 1$$
$$= 2 \cdot 1$$
$$= 2 \text{ square units}$$
$$A_2 = f(1) \cdot 1$$
$$= 3 \cdot 1$$
$$= 3 \text{ square units}$$
$$A_3 = f(2) \cdot 1$$
$$= (\sqrt{2} + 2) \cdot 1$$
$$= 3.414 \text{ square units}$$
$$A_4 = f(3) \cdot 1$$
$$= (\sqrt{3} + 2) \cdot 1$$
$$= 3.732 \text{ square units}$$

The area of the entire shaded region is the sum of the individual areas.

$$A = A_1 + A_2 + A_3 + A_4$$
$$= 2 + 3 + 3.414 + 3.732$$
$$= 12.146 \text{ square units}$$

How can we improve our estimate? Increase the number of rectangles. This time we use 8 rectangles of width 0.5 units (see Figure 6.8).

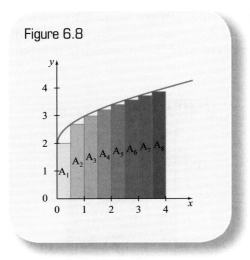

Figure 6.8

The width of each rectangle is 0.5 units. The heights of the rectangles are found by evaluating $f(x)$ at $x = 0.0$, $x = 0.5$, $x = 1.0$, $x = 1.5$, $x = 2.0$, $x = 2.5$, $x = 3.0$, and $x = 3.5$ (see Table 6.1).

Table 6.1

x	$f(x) = \sqrt{x} + 2$
0.0	2.000
0.5	2.707
1.0	3.000
1.5	3.225
2.0	3.414
2.5	3.581
3.0	3.732
3.5	3.871

The area of the shaded region is given by

$$A = A_1 + A_2 + A_3 + A_4 + A_5 + A_6 + A_7 + A_8$$

$$= 2(0.5) + 2.707(0.5) + 3(0.5) + 3.225(0.5)$$
$$+ 3.414(0.5) + 3.581(0.5)$$
$$+ 3.732(0.5) + 3.871(0.5)$$

$$= (2.000 + 2.707 + 3.000 + 3.225 + 3.414$$
$$+ 3.581 + 3.732 + 3.871)(0.5)$$

$$= (25.530)(0.5)$$

$$= 12.765 \text{ square units}$$

We will next estimate the area using 16 rectangles. We know that our area estimate will continue to improve as we increase the number of rectangles; however, as the number of rectangles increases and their width decreases, it becomes increasingly difficult to draw the rectangles. We need a better method.

Since the width of the entire region is 4 units, the width of each of the 16 rectangles will be $\frac{4}{16} = 0.25$ units. We'll make a table for $f(x)$ with the x-values starting at $x = 0$ and continuing to $x = 4$, with values spaced 0.25 unit apart (see Table 6.2).

Table 6.2

x	$f(x) = \sqrt{x} + 2$
0.00	2.000
0.25	2.500
0.50	2.707
0.75	2.866
1.00	3.000
1.25	3.118
1.50	3.225
1.75	3.323
2.00	3.414
2.25	3.500
2.50	3.581
2.75	3.658
3.00	3.732
3.25	3.803
3.50	3.871
3.75	3.936
4.00	4.000

We know that the height of each left-hand rectangle is determined by evaluating $f(x)$ at the x-value on the leftmost edge of the rectangle. The leftmost edge of the first rectangle occurs at $x = 0$. The leftmost edge of the second rectangle occurs at $x = 0.25$. Continuing on down, we see that the leftmost edge of the sixteenth rectangle occurs at $x = 3.75$. The heights of the rectangles are highlighted in the red box in Table 6.3.

Since each rectangle has the same width, we may add up all of the heights of the rectangles and multiply them by the width instead of having to multiply the

Table 6.3

Rectangle Number	x	$f(x) = \sqrt{x} + 2$
1	0.00	2.000
2	0.25	2.500
3	0.50	2.707
4	0.75	2.866
5	1.00	3.000
6	1.25	3.118
7	1.50	3.225
8	1.75	3.323
9	2.00	3.414
10	2.25	3.500
11	2.50	3.581
12	2.75	3.658
13	3.00	3.732
14	3.25	3.803
15	3.50	3.871
16	3.75	3.936
	4.00	4.000

height of each rectangle individually by its width. (Both methods will yield the same result, but the first method requires fewer computations.)

$$A = (2 + 2.5 + 2.707 + 2.866 + 3 + 3.118 + 3.225$$
$$+ 3.323 + 3.414 + 3.5 + 3.581 + 3.658$$
$$+ 3.732 + 3.803 + 3.871 + 3.936)(0.25)$$
$$= (52.234)(0.25)$$
$$\approx 13.06 \text{ square units}$$

This method is easier than drawing the rectangles; however, even it can become cumbersome when there are large numbers of rectangles. As it turns out, the exact area of the region is $13\frac{1}{3}$ square units. Our final estimate was fairly close to the exact value.

Riemann Sums

Each of the estimates in Example 1 is called a **left-hand sum**, since the upper *left* corner of each of the rectangles used in the sum touched the graph of f. The notion of

using sums is credited to the famous mathematician Bernhard Riemann. Consequently, these sums are commonly called **Riemann sums**. The term *Riemann sum* also applies to right-hand sums (sums where the upper right-hand corner of the rectangle touches the graph).

EXAMPLE 2 Using Riemann Sums to Estimate an Area

→ Estimate the area of the region between $f(x) = \dfrac{1}{x}$ and the x-axis on the interval [1, 2] using a left-hand sum with $n = 5$.

SOLUTION

We are asked to estimate the area of the shaded region in Figure 6.9 by using a left-hand sum with five rectangles.

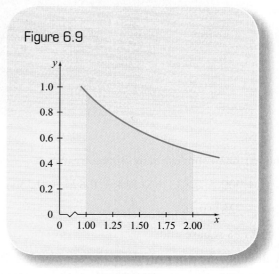

Figure 6.9

The width of each rectangle is $\dfrac{2-1}{5} = 0.2$ units. We will create a table of values for $f(x)$ starting at $x = 1$ and continuing to $x = 2$ with x-values spaced 0.2 units apart (see Table 6.4).

Table 6.4

Rectangle Number	x	$f(x) = \dfrac{1}{x}$
1	1.0	1.000
2	1.2	0.833
3	1.4	0.714
4	1.6	0.625
5	1.8	0.556
	2.0	0.500

Finding the Left-Hand Sum for Small Values of n (n ≤ 10)

Let f be a function defined on the interval $[a, b]$ whose graph lies above the x-axis. The left-hand sum estimate for the area between the graph of f and the x-axis may be found using the following steps.

1. Calculate the width of each rectangle $\Delta x = \dfrac{b - a}{n}$. "$\Delta x$" is read "delta x" and is the distance between consecutive x-values.

2. Create a table of values for $f(x)$ starting at $x = a$ and ending at $x = b$, with intermediate x-values spaced Δx units apart.

3. Add up the first to the penultimate (second to last) values

of $f(x)$ listed in the table. (This is the sum of the heights of the rectangles.)

4. Multiply the result of Step 3 by Δx to get the left-hand sum approximation of the area.

(*Note:* This method may be used for any value of n; however, for values of n larger than 10, the process becomes extremely tedious.)

Shutterstock

The left-hand sum estimate of the area is

$$A = (1.000 + 0.833 + 0.714 + 0.625 + 0.556)(0.2)$$

$$= (3.728)(0.2)$$

$$= 0.746 \text{ square units}$$

By drawing the rectangles as shown in Figure 6.10, we can see that this is an overestimate of the actual area.

EXAMPLE 3 Using a Left-Hand Sum to Estimate an Area

➡ Estimate the area between the graph of $f(x) = -x^2 + 4x$ and the x-axis on the interval $[0, 4]$ using a left-hand sum with 8 rectangles.

SOLUTION

We are asked to estimate the area of the region shown in Figure 6.11.

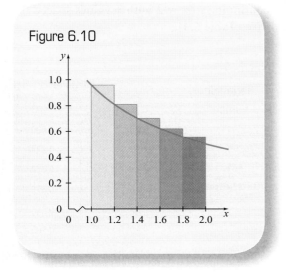

Figure 6.10

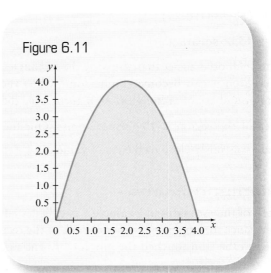

Figure 6.11

The width of each rectangle is

$$\Delta x = \frac{4 - 0}{8}$$

$$= \frac{1}{2}$$

$$= 0.5$$

We will create a table of values starting at $x = 0$ and continuing to $x = 4$ with the x-values spaced 0.5 units apart (see Table 6.5).

Table 6.5

Rectangle Number	x	$f(x) = -x^2 + 4x$
1	0.0	0.00
2	0.5	1.75
3	1.0	3.00
4	1.5	3.75
5	2.0	4.00
6	2.5	3.75
7	3.0	3.00
8	3.5	1.75
	4.0	0.00

The sum of the heights of the rectangles is

$$0 + 1.75 + 3 + 3.75 + 4 + 3.75 + 3 + 1.75 = 21$$

We now multiply the sum of the heights by the width of a rectangle.

$$A = 21(0.5)$$

$$= 10.5 \text{ square units}$$

This left-hand sum estimate is close to the exact area $(10\frac{2}{3}$ square units).

As you may have guessed, we can also calculate a right-hand sum. The right-hand sum is the sum of the areas of rectangles whose upper right-hand corner touches the graph of f. The process for finding the right-hand sum is similar to the process for finding the left-hand sum, as will be shown in Example 4.

EXAMPLE 4 Using a Right-Hand Sum to Estimate an Area

Estimate the area between the graph of $f(x) = x^3 - 3x^2 + 3x$ and the x-axis on the interval $[0, 2]$ using a right-hand sum with 4 rectangles.

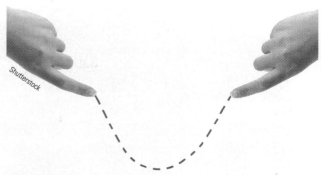

SOLUTION

We are asked to estimate the area of the shaded region in Figure 6.12.

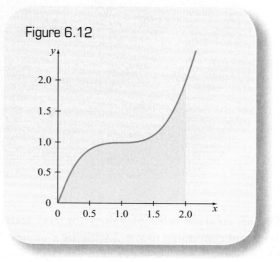

Figure 6.12

For the right-hand sum, the upper right-hand corner of each rectangle will touch the graph of $f(x)$, as shown in Figure 6.13. The width of each rectangle will be

$$\Delta x = \frac{2 - 0}{4}$$

$$= 0.5$$

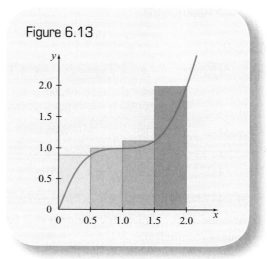

Figure 6.13

Finding the Right-Hand Sum for Small Values of n (n ≤ 10)

Let f be a function defined on the interval [a, b] whose graph lies above the x-axis. The right-hand sum estimate for the area between the graph of f and the x-axis may be found using the following steps.

1. Calculate the width of each rectangle. $\Delta x = \dfrac{b - a}{n}$.

2. Create a table of values for f(x) starting at x = a and ending at x = b, with intermediate x-values spaced Δx units apart.

3. Add up the second to the last values of f(x) listed in the table. (This is the sum of the heights of the rectangles.)

4. Multiply the result of Step 3 by Δx to get the left-hand sum approximation of the area.
(Note: This method may be used for any value of n; however, for values of n larger than 10, the process becomes extremely tedious.)

We will create a table of values for $f(x)$ starting at $x = 0$ and continuing to $x = 2$ with the x-values spaced 0.5 units apart (see Table 6.6). For the right-hand sum, we select the second through the last term in the table. These are the respective heights of the rectangles.

We sum the heights together and multiply them by the width of a rectangle.

$$A = (0.875 + 1.000 + 1.125 + 2.000)(0.5)$$

$$= (5)(0.5)$$

$$= 2.5 \text{ square units}$$

Table 6.6

Rectangle Number	x	$f(x) = x^3 - 3x^2 + 3x$
	0	0.000
1	0.5	0.875
2	1.0	1.000
3	1.5	1.125
4	2.0	2.000

The right-hand sum estimate for the area is 2.5 square units (using 4 rectangles).

The process for finding the right-hand sum is identical to the process for finding the left-hand sum, with the exception of Step 3. For the left-hand sum, we add up the first through the penultimate term in the table. For the right-hand sum, we add up the second through the last term.

The table method for finding the left- and right-hand sums works well for small values of n. However, for large values of n (say $n = 100,000$), the table method is not practical. In Section 6.4, we will demonstrate how summation notation may be used for left- and right-hand sums with large values of n.

You may wonder, "Is the left-hand or the right-hand sum a better estimate of the area?" Actually, it varies from function to function. Often you can get one of the best area estimates by averaging the left- and right-hand sums.

EXAMPLE 5 Using Riemann Sums to Estimate an Area

Estimate the area between the graph of $f(x) = x^3 - 5x^2 + 6x + 1$ and the x-axis on the interval [0, 4] with $n = 4$ using a left-hand sum, a right-hand sum, and the average of the sums.

SOLUTION

We are asked to estimate the area of the shaded region in Figure 6.14.

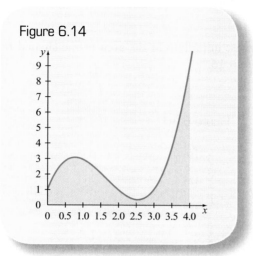

Figure 6.14

We generate the table of values for $f(x)$ given in Table 6.7.

Table 6.7

x	$f(x) = x^3 - 5x^2 + 6x + 1$
0	1
1	3
2	1
3	1
4	9

The left-hand sum is given by

$$LHS = (1 + 3 + 1 + 1)(1)$$

$$= 6 \text{ square units}$$

The right-hand sum is given by

$$RHS = (3 + 1 + 1 + 9)(1)$$

$$= 14 \text{ square units}$$

Averaging the two sums, we get

$$\frac{LHS + RHS}{2} = \frac{6 + 14}{2}$$

$$= 10 \text{ square units}$$

Our estimate is fairly close to the exact area of $9\frac{1}{3}$ square units. (In Section 6.5, we will show you how to calculate the exact area.)

EXAMPLE 6 Using Riemann Sums to Estimate Distance Traveled

The velocity of the author's minivan as he drove from a traffic light onto a highway may be modeled by

$$V(t) = -0.1773t^2 + 0.6798t + 0.5019$$

miles per minute

where t is time in minutes. Estimate the area between the velocity graph and the t-axis on the interval [0.0, 2.5] using 5 rectangles. Then interpret the real-world meaning of the result.

SOLUTION

We are asked to estimate the area of the shaded region in Figure 6.15 using 5 rectangles.

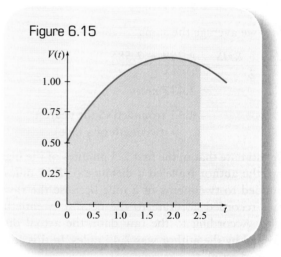

Figure 6.15

We construct a table of values for $V(t)$ starting at $t = 0.0$ and ending at $t = 2.5$ with t-values spaced $\Delta t = \dfrac{2.5 - 0}{5} = 0.5$ minutes apart (see Table 6.8).

Table 6.8

t	$V(t) = -0.1773t^2 + 0.6798t + 0.5019$
0.0	0.502
0.5	0.797
1.0	1.004
1.5	1.123
2.0	1.152
2.5	1.093

Since the height of the rectangles represents velocity in miles per minute and the width of the rectangles represents minutes, the units of the area will be

$$\frac{\text{miles}}{\text{minute}} \cdot \cancel{\text{minute}} = \text{miles}$$

We first find the left-hand sum.

$$LHS = (0.502 + 0.797 + 1.004$$
$$+ 1.123 + 1.152)(0.5)$$

$$= (4.578)(0.5)$$

$$\approx 2.289 \text{ miles}$$

Then we find the right-hand sum.

$$RHS = (0.797 + 1.004 + 1.123$$
$$+ 1.152 + 1.093)(0.5)$$

$$= (5.169)(0.5)$$

$$\approx 2.585 \text{ miles}$$

Now we average the sums.

$$\frac{LHS + RHS}{2} = \frac{2.289 + 2.585}{2}$$

$$= 2.437 \text{ miles}$$

$$\approx 2.45 \text{ (rounded to the nearest}$$
$$\text{twentieth of a mile)}$$

We estimate that in the first 2.5 minutes of the highway trip, the author traveled a distance of 2.45 miles. (We rounded to twentieths of a mile because the raw data were recorded accurate to the nearest twentieth of a mile.) According to the raw data, the actual distance traveled by the author was 2.45 miles. In this case, our rounded estimate was right on!

Although the TI-83 Plus doesn't come preloaded with a Riemann sum program, several excellent programs are available at www.ticalc.org. One outstanding program is *Riemann.8xp,* written by Mike Miller of Corban College. We transferred the downloaded program from our PC to our calculator using the TI-Connect software. In the example on your Chapter 6 Tech Card, we will repeat Example 7 using the Riemann.8xp program. (Users of the TI-83 calculator should download the file Rieman83.83P.)

EXAMPLE 7 Using Riemann Sums to Estimate Water Volume

Based on river flow forecasts for June 8 to June 11, 2010, the flow rate of the Ohio River near Cincinnati, Ohio, may be modeled by

$$f(t) = -6.60t^2 + 13.7t + 118$$
thousand cubic feet per second

where t is the number of 24-hour periods since June 8, 2010. (**Source:** www.erh.noaa.gov)

How much water (in thousands of cubic feet) passed by this point near Cincinnati, Ohio, between June 8 and June 11, 2010?

SOLUTION

We must initially find the area of the shaded region in Figure 6.16.

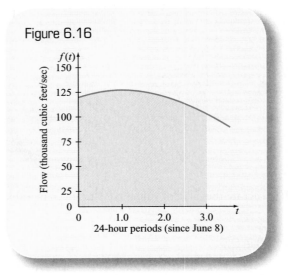

Figure 6.16

We decide to use 6 rectangles and generate the appropriate table of values (Table 6.9). Table values are rounded to the nearest integer.

Table 6.9

t	$f(t) = -6.60t^2 + 13.7t + 118$
0.0	118
0.5	123
1.0	125
1.5	124
2.0	119
2.5	111
3.0	100

This problem is a bit tricky, since the time intervals are 24-hour periods. We'll need to do some unit conversion before reaching our final answer. Since the flow rate is given in thousands of cubic feet per second and t is in terms of 24-hour periods, the units of the area of each rectangle will be

$$\frac{\text{thousand cubic feet}}{\text{second}} \cdot 24\text{-hour periods}$$

We'll convert these to the appropriate units after finding the left- and right-hand sums.

$$LHS = (118 + 123 + 125 + 124 + 119 + 111)(0.5)$$

$$= (720)(0.5)$$

$$= 360$$

$$RHS = (123 + 125 + 124 + 119 + 111 + 100)(0.5)$$

$$= (702)(0.5)$$

$$= 351$$

$$\frac{LHS + RHS}{2} = \frac{360 + 351}{2}$$

$$= 355.5 \, \frac{\text{thousand cubic feet}}{\text{second}} \cdot 24\text{-hour periods}$$

$$= 355.5 \text{ thousand cubic feet} \cdot \frac{24\text{-hour periods}}{\text{second}}$$

How many seconds are there in a 24-hour period?

$$24\text{-hour periods} = 24 \cdot 1 \text{ hour}$$

$$= 24 \cdot 60 \text{ minutes}$$

$$= 1440 \cdot 1 \text{ minute}$$

$$= 1440 \cdot 60 \text{ seconds}$$

$$= 86,400 \text{ seconds}$$

Thus we have

$$= 355.5 \text{ thousand cubic feet} \cdot \frac{86,400 \, \cancel{\text{seconds}}}{\cancel{\text{second}}}$$

$$= 30,715,200 \text{ thousand cubic feet}$$

$$\approx 30.7 \text{ billion cubic feet}$$

Between June 8 and June 11, 2010, we estimate that more than 30.7 billion cubic feet of water passed by the measurement station on the Ohio River near Cincinnati, Ohio.

As noted earlier in the section, if $f(x)$ is a rate-of-change function, then the area between the graph of $f(x)$ and the x-axis on the interval $[a, b]$ represents the total change in $F(x)$ (an antiderivative of $f(x)$) from $x = a$ to $x = b$. We will discuss this in greater detail in Section 6.5.

6.3 Exercises

In Exercises 1–3, draw the rectangles used to calculate the left-hand sum estimate of the area between the graph of the function and the horizontal axis on the specified interval. (In each case, use 4 rectangles.) Then calculate the left-hand sum.

1. $f(x) = 2$; $[1, 5]$

2. $h(x) = 2^x + x$; $[0, 2]$

3. $v(t) = 0.5t + 20$; $[3, 7]$

In Exercises 4–8, use the left-hand sum to estimate the area between the graph of the function and the horizontal axis on the specified interval. For each exercise, calculate the sum with $n = 2$, $n = 4$, and $n = 10$ rectangles.

4. $s(t) = t^3 - 3t^2 + 3t - 1$; $[1, 2]$

5. $h(x) = \frac{2}{x} + 1$; $[1, 5]$

6. $s(x) = \ln(x)$; $[e, e^2]$

7. $y = 4t^2 - 1$; $[2, 4]$

8. $f(t) = \frac{t + 2}{t - 2}$; $[3, 11]$

In Exercises 9–18, use left- and right-hand sums (with $n = 4$) to estimate the area between the graph of the function and the horizontal axis on the specified interval. In each exercise, calculate the left-hand sum, the right-hand sum, and the average of the two sums. The exact area, A, is given so that you can compare your estimates to the actual area.

9. $g(x) = x^2 - 2x + 2$ on $[3, 5]$; $A = 20\frac{2}{3}$

10. $f(x) = -(x - 2)^3 + 2$ on $[1, 2]$; $A = 2.25$

11. $h(x) = x^3 - 3x^2 + 2x + 1$ on $[0, 2]$; $A = 2$

12. $f(x) = x^3 - 9x^2 + 23x$ on $[2, 6]$; $A = 64$

13. $h(x) = x^3 - 9x^2 + 26x$ on $[1, 5]$; $A = 96$

14. $f(x) = 4x - x^4$ on $[0, 1]$; $A = 1.8$

15. $h(x) = e^{2x} - e^x$ on $[0, \ln 2]$; $A = 0.5$

16. $s(t) = \ln(t)$ on $[1, e]$; $A = 1$

17. $y = e^x$ on $[\ln 2, \ln 5]$; $A = 3$

18. $y = 5x^4 - \ln 5(5^x)$ on $[2, 4]$; $A = 392$

In Exercises 19–20, use Riemann sums as a part of the problem-solving process.

19. *River Flow* Between 5:45 A.M. and 7:45 A.M. on November 22, 2002, the stream flow rate of the Columbia River below the Priest Rapids Dam

dropped dramatically. The flow rate during that time period may be modeled by

$$F(t) = 14.50t^4 - 80.15t^3 + 168.3t^2 - 180.8t + 158.9 \text{ cubic feet per second}$$

where t is the number of hours since 5:45 A.M. (**Source:** waterdata.usgs.gov)

(a) Convert the units of the flow-rate function from cubic feet per second to cubic feet per hour.

(b) Use left- and right-hand sums with $n = 8$ to approximate the total amount of water that flowed past the flow-rate measurement station between 5:45 A.M. and 7:45 A.M.

(c) Why do you think the flow rate decreased so substantially?

20. *River Flow* Based on data from November 22 to 24, 2002, the flow rate of the Snake River near Irwin, Idaho, at noon may be modeled by

$$f(t) = -50t^2 + 100t + 1050 \text{ cubic feet per second}$$

where t is the number of 24-hour periods since noon on November 22, 2002. (**Source:** waterdata.usgs.gov)

How much water (in gallons) passed by this point near Irwin, Idaho, between noon on November 22 and noon on November 24, 2002? (Use $n = 8$.)

ATTENTION
NEED MORE PRACTICE? FIND MORE HERE!
CENGAGEBRAIN.COM

6.4 The Definite Integral

Based on data from 2007 to 2009, the marginal revenue of the Coca-Cola Company may be modeled by

$$M = -5.04s + 120 \text{ dollars per unit case}$$

where s is the number of unit cases (in billions). A unit case is equivalent to 24 eight-ounce servings of finished beverage. In 2009, 24.4 billion unit cases were sold. In 2007, 22.7 billion unit cases were sold. (**Source:** Modeled from Coca-Cola Company 2009 Annual Report) By how much did the revenue of the Coca-Cola Company grow as the number of unit cases sold increased from 22.7 billion to 24.4 billion? If you were a financial consultant for Coca-Cola, would you encourage the company to reduce its prices so that it could increase unit case sales? We can address questions such as these using the notion of the definite integral.

In this section, we will continue our discussion of area estimates by introducing *summation notation* and the *definite integral*. We will also show you several helpful definite integral properties.

Summation Notation

We use *summation notation* in order to represent the sum of a large number of terms easily. The notation may feel a bit awkward at first; however, as you become skilled in using it, you will come to appreciate its usefulness. Let's return to the table we introduced in Example 4 of Section 6.3 (Table 6.6), reproduced here as Table 6.10.

Table 6.10

x	$f(x) = x^3 - 3x^2 + 3x$
0	0.000
0.5	0.875
1.0	1.000
1.5	1.125
2.0	2.000

Recall that for the left-hand sum, the sum of the rectangle heights may be written as $f(0) + f(0.5) + f(1.0) + f(1.5)$. Notice that the x-values are $\Delta x = 0.5$ units apart. An alternative way to write this sum is

$$\sum_{i=0}^{3} f(0.5i)$$

The Greek letter *sigma* (Σ) tells us that we are to sum the values of the form $f(0.5i)$. The i is the *index of summation*. It is a variable, and it is always equal to a whole number. The numbers above and below Σ are the *limits of summation*. In this case, the initial value of i is 0 and the ending value is 3.

We will first substitute the initial value of i into $f(0.5i)$.

$$f(0.5(0)) = f(0)$$

We will then increase i from 0 to 1 and substitute $i = 1$ into $f(0.5i)$.

$$f(0.5(1)) = f(0.5)$$

The summation symbol Σ tells us to sum these two values together. So far we have

$$f(0) + f(0.5)$$

However, the summation tells us that we must continue to increase the index until $i = 3$, each time summing the values.

$$f(0.5(2)) = f(1)$$
$$f(0.5(3)) = f(1.5)$$

Therefore, $\sum_{i=0}^{3} f(0.5i) = f(0) + f(0.5) + f(1.0) + f(1.5)$. From Table 6.10, we see that $f(0) = 0.000$, $f(0.5) = 0.875$, $f(1.0) = 1.000$, and $f(1.5) = 1.125$. Therefore,

$$\sum_{i=0}^{3} f(0.5i) = f(0) + f(0.5) + f(1.0) + f(1.5)$$
$$= 0.000 + 0.875 + 1.000 + 1.125$$
$$= 3.000$$

In the next two examples we will temporarily move away from our discussion of heights and areas in order to give you practice using summation notation.

EXAMPLE 1 Using Summation Notation

Calculate $\sum_{i=1}^{10} i^2$.

SOLUTION

We are to sum terms of the form i^2 by substituting in the values 1 through 10 for i.

$$\sum_{i=1}^{10} i^2 = (1)^2 + (2)^2 + (3)^2 + (4)^2 + (5)^2$$
$$+ (6)^2 + (7)^2 + (8)^2 + (9)^2 + (10)^2$$
$$= 1 + 4 + 9 + 16 + 25$$
$$+ 36 + 49 + 64 + 81 + 100$$
$$= 385$$

EXAMPLE 2 Using Summation Notation

Calculate $\sum_{i=0}^{6} (i - 3)$.

SOLUTION

We must sum terms of the form $i - 3$ by substituting in the values 0 through 6 for i.

$$\sum_{i=0}^{6} (i - 3) = (0 - 3) + (1 - 3) + (2 - 3)$$
$$+ (3 - 3) + (4 - 3) + (5 - 3) + (6 - 3)$$
$$= (-3) + (-2) + (-1)$$
$$+ 0 + 1 + 2 + 3$$
$$= 0$$

We can now describe the left- and right-hand sums in terms of summation notation.

Summation Notation for the Left-Hand Sum

Let f be a function defined on the interval $[a, b]$. If the graph of f is above the x-axis, then the area of the region between the graph of f and the x-axis may be approximated by

$$\sum_{i=0}^{n-1} f(x_i)\Delta x = f(x_0)\Delta x + f(x_1)\Delta x + \cdots + f(x_{n-1})\Delta x$$

where $\Delta x = \dfrac{b - a}{n}$ and

$$x_0 = a$$
$$x_1 = a + \Delta x$$
$$\vdots$$
$$x_{n-1} = a + (n - 1)\Delta x$$

$\sum_{i=0}^{n-1} f(x_i)\Delta x$ is the left-hand sum estimate of the area using n rectangles.

Summation Notation for the Right-Hand Sum

Let f be a function defined on the interval $[a, b]$. If the graph of f is above the x-axis, then the area of the region between the graph of f and the x-axis may be approximated by

$$\sum_{i=1}^{n} f(x_i)\Delta x = f(x_1)\Delta x + f(x_2)\Delta x + \cdots + f(x_n)\Delta x$$

where $\Delta x = \dfrac{b-a}{n}$ and

$$x_1 = a + \Delta x$$
$$x_2 = a + 2 \cdot \Delta x$$
$$\vdots$$
$$x_n = b$$

$\sum_{i=1}^{N} f(x_i)\Delta x$ is the right-hand sum estimate of the area using n rectangles.

In general, a Riemann sum does not require that each rectangle have the same width. However, by making the rectangles have the same width, we reduce the number of calculations needed to find the solution.

Let's work an area example using summation notation.

EXAMPLE 3 Using Summation Notation in Calculating a Riemann Sum

Use a left-hand sum with $n = 6$ to estimate the area between the graph of $f(x) = x^2 + 2$ and the x-axis on the interval $[1, 3]$.

SOLUTION

We are asked to estimate the area of the shaded region in Figure 6.17 using a left-hand sum with 6 rectangles.

We first find the width of each rectangle.

$$\Delta x = \frac{3 - 1}{6}$$

$$= \frac{1}{3} \text{ unit}$$

The left-hand sum is given by

$$\sum_{i=0}^{n-1} f(x_i)\Delta x = \sum_{i=0}^{6-1} f(x_i)\left(\frac{1}{3}\right)$$

Figure 6.17

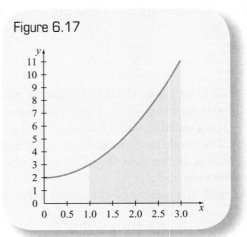

$$= \sum_{i=0}^{5} ((x_i)^2 + 2)\left(\frac{1}{3}\right)$$

$$= \sum_{i=0}^{5} ((1 + i \cdot \Delta x)^2 + 2)\left(\frac{1}{3}\right) \quad \text{Since } x_i = a + i\Delta x \text{ and } a = 1$$

$$= \sum_{i=0}^{5} \left(\left(1 + i\left(\frac{1}{3}\right)\right)^2 + 2\right)\left(\frac{1}{3}\right) \quad \text{Since } \Delta x = \frac{1}{3}$$

$$= \left(\left(1 + 0\left(\frac{1}{3}\right)\right)^2 + 2\right)\left(\frac{1}{3}\right)$$

$$+ \left(\left(1 + 1\left(\frac{1}{3}\right)\right)^2 + 2\right)\left(\frac{1}{3}\right)$$

$$+ \left(\left(1 + 2\left(\frac{1}{3}\right)\right)^2 + 2\right)\left(\frac{1}{3}\right)$$

$$+ \left(\left(1 + 3\left(\frac{1}{3}\right)\right)^2 + 2\right)\left(\frac{1}{3}\right)$$

$$+ \left(\left(1 + 4\left(\frac{1}{3}\right)\right)^2 + 2\right)\left(\frac{1}{3}\right)$$

$$+ \left(\left(1 + 5\left(\frac{1}{3}\right)\right)^2 + 2\right)\left(\frac{1}{3}\right)$$

$$= \left(((1)^2 + 2) + \left(\left(\frac{4}{3}\right)^2 + 2\right)\right.$$

$$+ \left(\left(\frac{5}{3}\right)^2 + 2\right) + \left(\left(\frac{6}{3}\right)^2 + 2\right)$$

$$\left. + \left(\left(\frac{7}{3}\right)^2 + 2\right) + \left(\left(\frac{8}{3}\right)^2 + 2\right)\right)\left(\frac{1}{3}\right)$$

$$= \left(3 + \left(\frac{34}{9}\right) + \left(\frac{43}{9}\right) + (6) + \left(\frac{67}{9}\right)\right.$$

$$\left. + \left(\frac{82}{9}\right)\right)\left(\frac{1}{3}\right)$$

$$\approx 11.37$$

Admittedly, formulating the sum using summation notation was more complicated than the table method. However, as we use an increasingly large number of rectangles, using summation notation will be the easiest way to represent the left- and right-hand sums symbolically.

The Definite Integral

What would happen if we used infinitely many rectangles to estimate the area between the graph of a function and the horizontal axis? Would the estimate be exact? As we have seen, when we increase the number of rectangles, the width of each rectangle is reduced and the area estimate improves. The definite integral captures the idea of infinitely many rectangles.

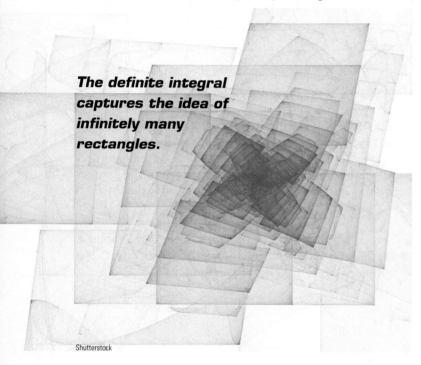

The definite integral captures the idea of infinitely many rectangles.

Shutterstock

The use of infinitely many rectangles may also be represented using summation notation in conjunction with limit notation.

We have $\int_a^b f(x)dx = \lim_{n\to\infty} \sum_{i=1}^n f(x_i)\Delta x$ and $\int_a^b f(x)dx = \lim_{n\to\infty} \sum_{i=0}^{n-1} f(x_i)\Delta x$. That is, when we use infinitely many rectangles, the right- and left-hand sums become equal. When we write the symbol $\lim_{n\to\infty}$, we mean that n is made to be infinitely large.

The Definite Integral

Let f be a continuous function defined on the interval $[a, b]$. The **definite integral of f from a to b** is given by

$$\int_a^b f(x)dx = \lim_{n\to\infty} \left(\left(f(x_1) + f(x_2) + \cdots + f(x_n)\right)\Delta x\right)$$

where $x_1, x_2, \ldots, x_n$ are the right-hand endpoints of subintervals of length

$$\Delta x = \frac{b-a}{n}.$$

For the definite integral $\int_a^b f(x)dx$, a is called the **lower limit of integration** and b is called the **upper limit of integration**. As with indefinite integrals, $f(x)$ is called the **integrand**. If $\int_a^b f(x)dx$ is defined, then f is said to be **integrable** on $[a, b]$.

The connection between the indefinite integral $\int f(x)dx$ and the definite integral $\int_a^b f(x)dx$ is not immediately obvious. For a nonnegative function $f(x)$, $\int f(x)dx$ is a family of functions with the same derivative, while $\int_a^b f(x)dx$ is the sum of the signed areas between the graph of f and the x-axis. However, despite the dramatically different meanings of indefinite and definite integrals, the two are closely related, as will be discussed in Section 6.5.

Consider the graph of $f(x) = x^3 - 4x^2 + 3x$ on the interval $[0, 3]$ (Figure 6.18).

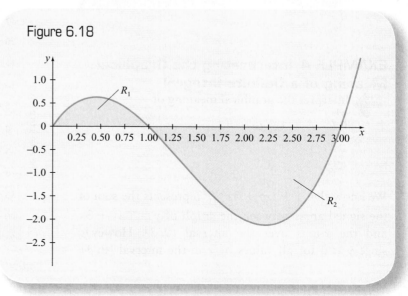

Figure 6.18

The first region, R_1, has area $\dfrac{5}{12}$ square units. The second region, R_2, has area $2\dfrac{2}{3}$ square units. Therefore, $\displaystyle\int_0^3 (x^3 - 4x^2 + 3x)\, dx = \dfrac{5}{12} - 2\dfrac{2}{3} = -2\dfrac{1}{4}.$ The fact that the number is negative indicates that more area lies below the x-axis than above it.

In considering the relationship between $\displaystyle\int_a^b f(x)\, dx$ and the left- and right-hand sums, it is helpful to think of $f(x)$ as the height of an infinitely narrow rectangle and dx as the width of that rectangle. The integral sign $\displaystyle\int$ is a somewhat distorted S, meaning "sum," and tells us to sum the areas of infinitely many adjacent rectangles of width dx between $x = a$ and $x = b$. (From our earlier discussion of differentials, we know that we can choose the value of dx to be as small as we like.) The height of each rectangle varies as x moves from a to b in steps of length dx.

The Meaning of $\displaystyle\int_a^b f(x)dx$ for a Function f

Let f be a function on $[a, b]$. The definite integral

$$\int_a^b f(x)dx$$

is the sum of the areas of the enclosed regions above the x-axis minus the sum of the areas of enclosed regions below the x-axis on the interval $[a, b]$.

In other words, $\displaystyle\int_a^b f(x)dx$ is the **sum of the signed areas.**

EXAMPLE 4 Interpreting the Graphical Meaning of a Definite Integral

Interpret the graphical meaning of

$$\int_0^3 (-x^2 + 3x)\, dx = 4.5.$$

SOLUTION

We know that $\displaystyle\int_0^3 (-x^2 + 3x)\, dx$ represents the sum of the signed areas between the graph of $y = -x^2 + 3x$ and the x-axis over the interval $[0, 3]$. However, since $y \geq 0$ for all values of x in the interval $[0, 3]$,

$\displaystyle\int_0^3 (-x^2 + 3x)\, dx$ gives the area of the shaded region in Figure 6.19.

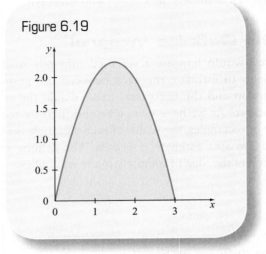

Figure 6.19

The area of the shaded region is 4.5 square units.

EXAMPLE 5 Interpreting the Meaning of a Definite Integral in an Applied Setting

Based on data from 1998 and 2008 and a projection for 2018, the rate of change in the number of jobs in the mining industry may be modeled by

$$r(t) = -2.56t + 28.0 \text{ thousand jobs per year}$$

where t is the number of years since 1998. (**Source:** Modeled from U.S. Bureau of Labor Statistics data) The graph of $r(t)$ is depicted in Figure 6.20.

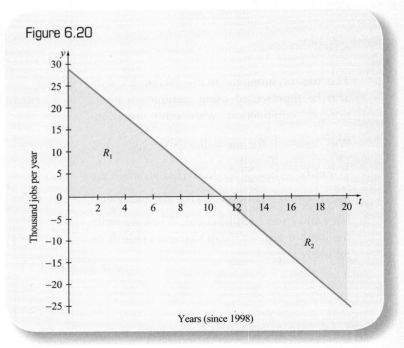

Figure 6.20

Years (since 1998)

The areas of the two shaded regions are 153.125 square units and 105.125 square units, respectively.

Calculate and interpret the meaning of

$$\int_0^{20} (-2.56t + 28.0)\,dt$$

SOLUTION

$\int_0^{20} (-2.56t + 28.0)\,dt$ is the sum of the signed areas. Since region R_2 is below the t-axis, its signed area is negative. Therefore,

$$\int_0^{20} (-2.56t + 28.0)\,dt = 153.125 + (-105.125) = 48$$

Recall that the units of the integral are found by multiplying the units of the independent variable by the units of the dependent variable. That is,

$$\text{years since 1998} \cdot \frac{\text{thousand jobs}}{\text{year since 1998}} = \text{thousand jobs}$$

So from the end of 1998 to the end of 2018, the number of mining jobs increased by a net of 48,000 jobs. That is, taking into account the increases and decreases in the number of mining jobs, there will be 48,000 more mining jobs at the end of 2018 than there were at the end of 1998.

Definite Integral Properties

Some of the definite integral properties are directly related to the indefinite integral rules. You will readily note the similarities.

Definite Integral Properties

For integrable functions f and g, the following properties hold:

- $\int_a^a f(x)\,dx = 0$

- $\int_a^b f(x)\,dx = -\int_b^a f(x)\,dx$

- $\int_a^b k \cdot f(x)\,dx = k \int_a^b f(x)\,dx$ for constant k

- $\int_a^b (f(x) \pm g(x))\,dx = \int_a^b f(x)\,dx \pm \int_a^b g(x)\,dx$

- $\int_a^b f(x)\,dx = \int_a^c f(x)\,dx + \int_c^b f(x)\,dx$ for $a \leq c \leq b$

EXAMPLE 6 Applying Integration Properties

Given $\int_{-2}^0 (2x + 3)\,dx = 2$, determine $\int_0^{-2} (2x + 3)\,dx$.

SOLUTION

Switching the limits of integration changes the sign of the definite integral. Therefore, $\int_0^{-2} (2x + 3)\,dx = -2$.

EXAMPLE 7 Applying Integration Properties

Given $\int_2^4 2x\,dx = 12$ and $\int_2^4 (2x + 5)\,dx = 22$, find $\int_2^4 5\,dx$.

SOLUTION

We know that $\int_2^4 2x\,dx + \int_2^4 5\,dx = \int_2^4 (2x + 5)\,dx$. Therefore,

$$12 + \int_2^4 5\,dx = 22 \qquad \text{Since } \int_2^4 2x\,dx = 12 \text{ and } \int_2^4 (2x + 5)\,dx = 22$$

$$\int_2^4 5\,dx = 10$$

EXAMPLE 8 Applying Integration Properties

Given $\int_{-1}^1 3x^2\,dx = 2$ and $\int_1^3 3x^2\,dx = 26$, find $\int_{-1}^3 3x^2\,dx$.

SOLUTION

The only difference between the three integrals is the limits of integration. Since the upper limit of integration of $\int_{-1}^1 3x^2\,dx$ is the same as the lower limit of integration of $\int_1^3 3x^2\,dx$, the two integrals may be written together as a single integral $\int_{-1}^3 3x^2\,dx$. The value of this integral is found by summing the values of the two integrals from which it was created. That is,

$$\int_{-1}^3 3x^2\,dx = \int_{-1}^1 3x^2\,dx + \int_{-1}^3 3x^2\,dx$$

$$= 2 + 26 \qquad \text{Since } \int_{-1}^1 3x^2\,dx = 2 \text{ and } \int_1^3 3x^2\,dx = 26$$

$$= 28$$

EXAMPLE 9 Interpreting the Meaning of a Definite Integral in an Applied Setting

Based on data from 2000–2004, the rate of change in the percentage of highway accidents resulting in injuries in the United States annually may be modeled by

$$r(t) = -2.502(0.6521)^t \text{ percentage points per year}$$

where t is the number of years since 2000. (**Source:** Modeled from *Statistical Abstract of the United States, 2007*, Table 1047)

The signed area of the shaded region of the graph in Figure 6.21 is $\int_1^4 r(t)dt$ and equals –2.76.

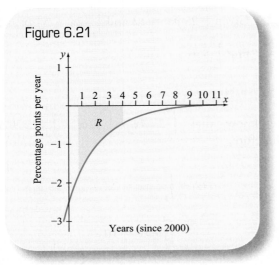

Figure 6.21

Given $\int_1^7 r(t)dt = -3.52$, calculate and interpret the meaning of $\int_4^7 r(t)dt$.

SOLUTION

$\int_4^7 r(t)dt$ is the net change in the percentage of accidents resulting in injuries from the end of 2004 through the end of 2007 in percentage points. We know that $\int_1^4 r(t)dt = -2.76$ and $\int_1^7 r(t)dt = -3.52$, so

$$\int_1^4 r(t)dt + \int_4^7 r(t)dt = \int_1^7 r(t)dt$$

$$-2.76 + \int_4^7 r(t)dt = -3.52$$

$$\int_4^7 r(t)dt = -0.76$$

From the end of 2004 through the end of 2007, the percentage of highway accidents resulting in injuries decreased by 0.76 percentage points. That is, at the end of 2007 the percentage of highway accidents resulting

in injuries was 0.76 percentage points lower than the percentage of accidents resulting in injuries in 2004.

6.4 Exercises

In Exercises 1–5, draw the graph of the integrand function and shade the region corresponding to the definite integral. Then calculate the definite integral using geometric formulas for area.

1. $\int_5^7 1\,dx$

2. $\int_0^4 x\,dx$

3. $\int_1^2 (-2x + 5)\,dx$

4. $\int_{-1}^1 (2x + 2)\,dx$

5. $\int_{-2}^{-1} (-2x + 1)\,dx$

In Exercises 6–10, use the definite integral properties to find the numeric value of each definite integral given that $\int_2^4 f(x)\,dx = 5$, $\int_2^6 f(x)\,dx = 9$, $\int_2^4 g(x)\,dx = 2$, and $\int_2^6 g(x)\,dx = 1$.

6. $\int_2^4 (2f(x))\,dx$

7. $\int_2^6 (f(x) + g(x))\,dx$

8. $\int_4^6 f(x)\,dx$

9. $\int_4^6 (f(x) + g(x))\,dx$

10. $\int_6^2 (3f(x) - 4g(x))\,dx$

In Exercises 11–13, write the definite integral that represents the sum of the signed areas of the shaded regions of the graph.

11. $f(x) = -x^2 + 5x - 6$

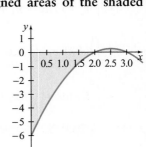

12. $f(x) = -x^3 + 4x + 20$

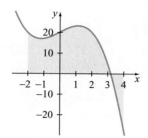

13. $f(x) = -|x - 3| + 3$

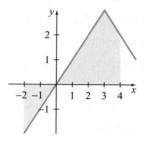

In Exercises 14 and 15, shade the regions of the graph that will be used in calculating the definite integral.

14. $\int_{1}^{3} (-x^2 + 4x - 3)\, dx$

15. $\int_{-1}^{1} e^{-x^2 + 1}\, dx$

In Exercises 16–18, use the properties of definite integrals to find the solution.

16. *Theme Park Ticket Prices* Based on 2007 ticket prices, the rate of change in the cost of an adult Disney Park Hopper Bonus Ticket may be modeled by

$T(d) = -10.71d + 59.04$ dollars per day in park

where d is the number of days that the ticket authorizes entrance into Disneyland and Disney

California Adventure. (**Source:** www.disneyland.com) The graph of $T(d)$ is shown.

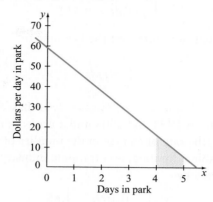

The area of the shaded region of the graph is

$\int_{4}^{5} T(d)\, dd$ and is equal to 10.85. Additionally,

$\int_{2}^{4} T(d)\, dd = 53.82.$

Calculate and interpret the meaning of

$\int_{2}^{5} T(d)\, dd.$

17. *Lean Ground Beef Prices* Based on data from 2002 to 2008, the rate of change in the price of lean ground beef may be modeled by

$$p(t) = 0.060t^2 - 0.340t + 0.436$$
dollars per pound per year

where t is the number of years since 2002. (**Source:** *Statistical Abstract of the United States, 2010*, Table 717)

The graph of $p(t)$ is shown.

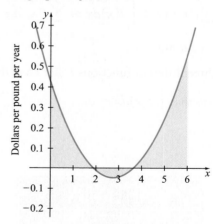

Years (since 2002)

The sum of the signed areas of the shaded regions is

$\int_{0}^{6} p(t)\, dt$

and is equal to 0.816. Additionally,

$$\int_4^6 p(t)\,dt = 0.512$$

Calculate and interpret the meaning of

$$\int_0^4 p(t)\,dt.$$

18. *Fishing, Hunting, and Trapping Jobs* Based on data from 1998 and 2008 and a projection for 2018, the rate of change in the number of jobs in the fishing, hunting, and trapping industry may be modeled by

$$r(t) = 0.109t - 1.63$$
$$\text{thousands of jobs per year}$$

where t is the number of years since 1998. (**Source:** Modeled from U.S. Bureau of Labor Statistics data)

Graph r on the interval $[0, 20]$ and shade the region whose area is given by

$$\int_{10}^{15} (r(t) = 0.109t - 1.63)\,dt.$$ Then calculate $$\int_{10}^{15} (r(t) = 0.109t - 1.63)\,dt$$ and interpret its real-world meaning. (*Hint:* Use the formula for the area of a trapezoid, $A = \left(\dfrac{b_1 + b_2}{2}\right)h$, to calculate the value of the definite integral.)

Exercises 19 and 20 are intended to challenge your understanding of Riemann sums and definite integrals.

19. Suppose that $f(x)$ is *increasing* on the interval $[a, b]$. Is the following statement true for all values of n?

$$\sum_{i=0}^{n-1} f(x_i)\Delta x \le \int_a^b f(x)\,dx \le \sum_{i=1}^{n} f(x_i)\Delta x$$

Justify your answer.

20. Find three different functions f that have the property that $\int_2^4 f(x)\,dx = 0$.

6.5 The Fundamental Theorem of Calculus

Based on online sales projections for 2007 to 2012, the rate of change in the amount of revenue generated by event ticket online sales may be modeled by

$$r(t) = -0.0822t + 1.02 \text{ billion dollars per year}$$

where t is the number of years since 2007. (**Source:** *Statistical Abstract of the United States, 2009*, Table 1015) According to the model, by how much did online event ticket sales revenue increase between the end of 2008 and the end of 2012?

Although we may use the methods previously covered to answer this question, in this section we will demonstrate how the Fundamental Theorem of Calculus can greatly simplify our computations. This theorem is one of the most powerful tools in calculus and gives us a remarkably easy way to calculate the sum of the signed areas between the graph of a function and the x-axis.

In Example 3 of Section 6.4, we used a left-hand sum with 6 rectangles to estimate the area between the graph of $f(x) = x^2 + 2$ and the x-axis on the interval $[1, 3]$ (see Figure 6.22). The left-hand sum approximation of the area was approximately 11.37 square units. This is an underestimate of the actual area.

The right-hand sum approximation of the area between $f(x) = x^2 + 2$ and the x-axis on $[1, 3]$ using 6 rectangles was 14.04 square units. This is an overestimate of the actual area. Averaging the two sums, we estimated that the shaded region had an approximate area of 12.70 square units.

Let's find the antiderivative of the function f.

$$\int f(x)\,dx = \int (x^2 + 2)\,dx$$

$$F(x) = \frac{x^3}{3} + 2x + C$$

Figure 6.22

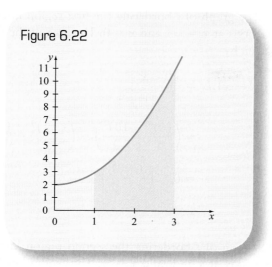

We'll evaluate the function F at the endpoints of the interval.

$$F(3) = \frac{(3)^3}{3} + 2(3) + C$$

$$= 15 + C$$

$$F(1) = \frac{(1)^3}{3} + 2(1) + C$$

$$= 2\frac{1}{3} + C$$

We next find the difference between the two values.

$$F(3) - F(1) = 15 + C - \left(2\frac{1}{3} + C\right)$$

$$= 12\frac{2}{3}$$

$$\approx 12.67$$

Notice that the constant C was eliminated in the computation. This will happen anytime we calculate the difference between two different values of the antiderivative. This value, 12.67, is extremely close to our estimate of 12.70. In fact, the exact area of the region is $12\frac{2}{3}$ square units. Thus, the definite integral $\int_1^3 f(x)\,dx$ not only represents the area of the region between $f(x)$ and the x-axis on the interval [1, 3], it also represents the total change in $F(x)$ from $x = 1$ to $x = 3$, where $F'(x) = f(x)$. This remarkable result is a key component of the Fundamental Theorem of Calculus.

Fundamental Theorem of Calculus

Let f be a continuous function on $[a, b]$. Then

$$\int_a^b f(x)\,dx = F(b) - F(a)$$

where F is any antiderivative of f.

Since F may be *any* antiderivative of f, we will choose F to be the antiderivative of f that has the constant $C = 0$. This choice will make all of our computations easier. We'll work several examples to illustrate the power of the Fundamental Theorem of Calculus.

EXAMPLE 1 Using the Fundamental Theorem of Calculus to Calculate a Definite Integral

Based on online sales projections for 2007 to 2012, the rate of change in the amount of revenue generated by event ticket online sales may be modeled by

$$r(t) = -0.0822t + 1.02 \text{ billion dollars per year}$$

where t is the number of years since 2007. (Source: *Statistical Abstract of the United States, 2009*, Table 1015) Calculate $\int_1^5 (-0.0822t + 1.02)\,dt$. Then interpret the practical meaning of the result.

Shutterstock

SOLUTION

We must first find an antiderivative of $f(t) = -0.0822t + 1.02$ and then evaluate the antiderivative at $x = 5$ and $x = 1$.

$$\int_1^5 (-0.0822t - 1.02)\, dx = (-0.0411t^2 + 1.02t)\Big|_1^5$$

An antiderivative of $f(t) = -0.0822t + 1.02$ is $F(t) = -0.0411t^2 + 1.02t$. The vertical bar tells us that we must evaluate this antiderivative at $t = 5$ and $t = 1$ and then find the difference of the two values.

$$(-0.0411t^2 + 1.02t)\Big|_1^5 = (-0.0411(5)^2 + 1.02(5))$$
$$- (-0.0411(1)^2 + 1.02(1))$$
$$= 4.07 - 0.98$$
$$= 3.09$$

Therefore, $\int_1^5 (-0.0822t + 1.02)\, dt = 3.09$. Since $f(t) = -0.0822t + 1.02$ is positive on the interval $[1, 5]$, the total area between the graph of f and the t-axis on $[1, 5]$ is 3.09. Between the end of 2008 and the end of 2012, event online ticket sales are predicted to increase by \$3.09 billion.

EXAMPLE 2 Using the Fundamental Theorem of Calculus to Find the Area of a Region

Find the area of the region between the graph of the function $f(x) = x^3 - x$ and the x-axis over the interval $[-1, 1]$.

SOLUTION

We are asked to find the combined area of the regions enclosed by the graph of f and the x-axis. We begin by graphing f and shading the enclosed regions (see Figure 6.23).

Figure 6.23

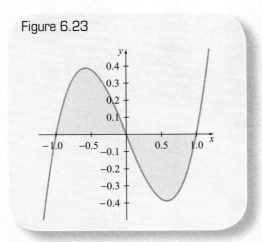

The graph of f bordering the first region intersects the x-axis at $x = -1$ and $x = 0$. The area of the region is given by

$$\int_{-1}^0 (x^3 - x)\, dx = \frac{x^4}{4} - \frac{x^2}{2}\Big|_{-1}^0$$
$$= \left(\frac{(0)^4}{4} - \frac{(0)^2}{2} \right) - \left(\frac{(-1)^4}{4} - \frac{(-1)^2}{2} \right)$$
$$= 0 - \left(\frac{1}{4} - \frac{1}{2} \right)$$
$$= \frac{1}{4}$$

The graph of f bordering the second region intersects the x-axis at $x = 0$ and $x = 1$. The region lies below the x-axis, so $\int_0^1 (x^3 - x)\, dx$ will be a negative number equal to -1 times the area.

$$\int_0^1 (x^3 - x)\, dx = \frac{x^4}{4} - \frac{x^2}{2}\Big|_0^1$$
$$= \left(\frac{(1)^4}{4} - \frac{(1)^2}{2} \right) - \left(\frac{(0)^4}{4} - \frac{(0)^2}{2} \right)$$
$$= \left(\frac{1}{4} - \frac{1}{2} \right) - 0$$
$$= -\frac{1}{4}$$
$$= -1 \cdot \frac{1}{4}$$

The area of the second region is also $\frac{1}{4}$ square unit. Therefore, the area bounded by the graph of f and the x-axis is

$$A = \frac{1}{4} + \frac{1}{4}$$
$$= \frac{1}{2} \text{ square unit}$$

Notice in Example 2 that we used separate definite integrals to calculate the area above the x-axis and the area below the x-axis. Using a single definite integral would have led to an erroneous result. If we calculate $\int_{-1}^1 (x^3 - x)\, dx$, we get the sum of the *signed* areas, not the actual area.

$$\int_{-1}^1 (x^3 - x)\, dx = \frac{1}{4}x^4 - \frac{1}{2}x^2\Big|_{-1}^1$$
$$= \left(\frac{1}{4}(1)^4 - \frac{1}{2}(1)^2 \right) - \left(\frac{1}{4}(-1)^4 - \frac{1}{2}(-1)^2 \right)$$

$$= \left(\frac{1}{4} - \frac{1}{2}\right) - \left(\frac{1}{4} - \frac{1}{2}\right)$$

$$= 0$$

A common error among beginning calculus students is to assume that the definite integral *always* yields the area between the graph of the function and the x-axis. As was just illustrated, this is not the case.

Although we are often given an interval over which to find the area, we are sometimes required to find the interval ourselves. When we see the phrase "Find the area of the region bounded by $f(x)$ and the x-axis," we are being asked to calculate the area of the region(s) enclosed by the graph of f and the x-axis. As long as f is a continuous function, we can determine the area by setting up definite integrals with the limits of integration representing each consecutive pair of x-intercepts, as demonstrated in Example 3.

EXAMPLE 3 Using the Fundamental Theorem of Calculus to Find the Area of a Region

Find the area of the region bounded by $f(x) = 4x^3 - 2x$ and the x-axis.

SOLUTION

We must first determine where f crosses the x-axis. To do this, we set $f(x)$ equal to zero and solve.

$$0 = 4x^3 - 2x$$

$$0 = 4x\left(x^2 - \frac{1}{2}\right)$$

$$0 = 4x\left(x - \sqrt{\frac{1}{2}}\right)\left(x + \sqrt{\frac{1}{2}}\right)$$

The function has x-intercepts at $x = -\sqrt{\frac{1}{2}}$, $x = 0$, and $x = \sqrt{\frac{1}{2}}$. Let's graph the function so that we can see the location of the bounded regions (Figure 6.24).

On the interval $\left(-\infty, -\sqrt{\frac{1}{2}}\right)$, the graph of f and the x-axis do not enclose a region. The first enclosed region occurs on the interval $\left(-\sqrt{\frac{1}{2}}, 0\right)$. The second enclosed region occurs on the interval $\left(0, \sqrt{\frac{1}{2}}\right)$. On the interval $\left(\sqrt{\frac{1}{2}}, \infty\right)$ the graph of f and the x-axis do not enclose a region. The area we're looking for is

$$A = \left|\int_{-\sqrt{1/2}}^{0} (4x^3 - 2x)\, dx\right| + \left|\int_{0}^{\sqrt{1/2}} (4x^3 - 2x)\, dx\right|$$

Notice that we placed absolute values on each of the definite integrals. This will guarantee that the value returned for each region is that region's area instead of its *signed* area.

Figure 6.24

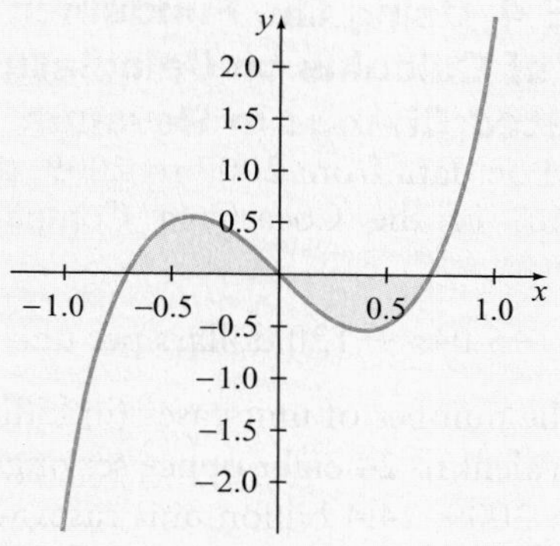

$$A = \left|\int_{-\sqrt{1/2}}^{0} (4x^3 - 2x)\, dx\right| + \left|\int_{0}^{\sqrt{1/2}} (4x^3 - 2x)\, dx\right|$$

$$= \left|(x^4 - x^2)\Big|_{-\sqrt{1/2}}^{0}\right| + \left|(x^4 - x^2)\Big|_{0}^{\sqrt{1/2}}\right|$$

$$= \left|\left((0)^4 - (0)^2\right) - \left(\left(-\sqrt{\frac{1}{2}}\right)^4 - \left(-\sqrt{\frac{1}{2}}\right)^2\right)\right|$$
$$+ \left|\left(\left(\sqrt{\frac{1}{2}}\right)^4 - \left(\sqrt{\frac{1}{2}}\right)^2\right) - \left((0)^4 - (0)^2\right)\right|$$

$$= \left|(0 - 0) - \left(\frac{1}{4} - \frac{1}{2}\right)\right| + \left|\left(\frac{1}{4} - \frac{1}{2}\right) - (0 - 0)\right|$$

$$= \left|0 - \left(-\frac{1}{4}\right)\right| + \left|-\frac{1}{4} - 0\right|$$

$$= \frac{1}{2} \text{ square unit}$$

The area of the region bounded by $f(x) = 4x^3 - 2x$ and the x-axis is $\frac{1}{2}$ square unit.

The relationship between a function and its antiderivative becomes especially meaningful if the function represents a rate of change.

Accumulated Change of a Function

Let f be the rate-of-change function (derivative) of f on $[a, b]$. Then

$$\int_a^b f(x)\, dx = F(b) - F(a)$$

is the accumulated change in F over the interval $[a, b]$.

EXAMPLE 4 Using the Fundamental Theorem of Calculus to Calculate an Accumulated Change in Revenue

Based on data from 2007 to 2009, the marginal revenue of the Coca-Cola Company may be modeled by

$$m = -5.04s + 120 \text{ dollars per unit case}$$

where s is the number of unit cases (in billions). A unit case is equivalent to 24 eight-ounce servings of finished beverage. In 2009, 24.4 billion unit cases were sold. In 2007, 22.7 billion unit cases were sold. (**Source:** Modeled from Coca-Cola Company 2009 Annual Report) By how much did the revenue of the Coca-Cola Company grow as the number of unit cases sold increased from 22.7 billion to 24.4 billion?

SOLUTION

Marginal revenue is the rate of change in the revenue function. The accumulated change in revenue is given by $M = -2.52s^2 + 120s$.

$$\int_{22.7}^{24.4} (-5.04s + 120)\, ds = \left. (-2.52s^2 + 120s) \right|_{22.7}^{24.4}$$

$$= (-2.52(24.4)^2 + 120(24.4))$$

$$- (-2.52(22.7)^2 + 120(22.7))$$

$$= (1427.69) - (1425.47)$$

$$= 2.22$$

Increasing the number of cases sold from 22.7 billion to 24.4 billion increased revenue by $2.22 billion (according to the model).

Shutterstock

Let's look at the functions m and M graphically. The graph of m is a line, as shown in Figure 6.25. On the interval [22.7, 24.4], the shaded region between the graph of m and the horizontal axis is a pair of triangles with one triangle above the horizontal axis and the other triangle below the horizontal axis. The area of the first triangle is 3.10 and the area of the second triangle is 0.88. The sum of the signed areas is $3.10 + (-0.88) = 2.22$.

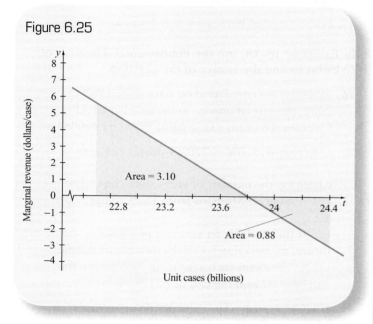

Figure 6.25

The graph of M is a parabola (Figure 6.26).

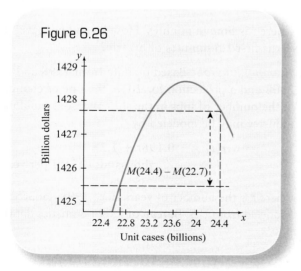

Figure 6.26

The vertical distance between the point (24.4, $M(24.4)$) and the point (22.7, $M(22.7)$) is given by $M(24.4) - M(22.7)$.

$$M(24.4) - M(22.7)$$
$$= (1427.69) - (1425.47)$$
$$= 2.22$$

Amazingly, the sum of the signed areas between the graph of m and the horizontal axis on the interval [22.7, 24.4] is easily calculated by finding the difference $M(24.4) - M(22.7)$, where M is an antiderivative of m.

A common error of beginning calculus students is to assume that the revenue earned by the company was $2.22 billion. The value $2.22 billion is the accumulated *change* in revenue, not the revenue value itself.

Changing Limits of Integration

When the integrand function in a definite integral requires integration by substitution, the limits of integration may be changed along with the variable. This technique will be demonstrated in Example 5.

EXAMPLE 5 Changing the Limits of Integration of a Definite Integral

➡ Calculate $\displaystyle\int_0^1 x^2(x^3 + 1)^3\,dx$.

SOLUTION

Let $u = x^3 + 1$. Then

$$du = 3x^2\,dx$$

$$\frac{du}{3} = x^2\,dx$$

Therefore,

$$\int_0^1 x^2(x^3 + 1)^3\,dx = \int_a^b u^3\left(\frac{du}{3}\right)$$

$$= \frac{1}{3}\int_a^b u^3\,du$$

What are the values of a and b? Recall that the limits of integration of the original function were $x = 0$ and $x = 1$. We will substitute these values of x into the equation relating u and x to find the new limits of integration.

Since $u = x^3 + 1$, the new limits of integration are $a = 0^3 + 1$ and $b = 1^3 + 1$. Therefore,

$$\int_0^1 x^2(x^3 + 1)^3\, dx = \frac{1}{3}\int_1^2 u^3\, du$$

$$= \frac{1}{3}\left(\frac{u^4}{4}\right)\Big|_1^2$$

$$= \frac{2^4}{12} - \frac{1^4}{12}$$

$$= 1.25$$

6.5 Exercises

In Exercises 1–10, use the Fundamental Theorem of Calculus to calculate the definite integral.

1. $\displaystyle\int_5^7 1\, dx$

2. $\displaystyle\int_0^4 x\, dx$

3. $\displaystyle\int_1^2 (-3x^2 + 5)\, dx$

4. $\displaystyle\int_{-1}^1 (8x^3 + 2x)\, dx$

5. $\displaystyle\int_1^3 \left(\frac{1}{x}\right) dx$

6. $\displaystyle\int_2^5 (2^t)\, dt$

7. $\displaystyle\int_{-2}^4 (5 \cdot 2^x)\, dx$

8. $\displaystyle\int_0^1 \left(\frac{2x}{x^2 + 1}\right) dx$

9. $\displaystyle\int_0^2 (x^2 - 1)(x^3 - 3x)^3\, dx$

10. $\displaystyle\int_2^1 (-2x + 1)\, dx$

In Exercises 11–15, calculate the area of the region bounded by the graph of the function and the x-axis. You may find it helpful to graph the function before attempting each exercise.

11. $h(x) = x^3 + x^2 - 12x$

12. $f(x) = x^3 - x^2$

13. $f(x) = x^3 - 1$ between $x = -1$ and $x = 1$

14. $g(x) = 1 - 3^x$ between $x = 0$ and $x = 2$

15. $h(x) = x - x^2$ between $x = -1$ and $x = 2$

In Exercises 16–18, use the Fundamental Theorem of Calculus to find the answer to the question.

16. *AP Calculus Exam* Based on data from 1969 to 2002, the rate of change in the number of AP Calculus AB exam participants may be modeled by

$$R(t) = 315.70t - 770.64 \text{ people per year}$$

where t is the number of years since the end of 1969. (**Source:** Modeled from The College Board data)

If the model is an accurate predictor of the future, by how much will the number of people taking the exam each year increase between the end of 2002 and the end of 2012?

17. *Minivan Position* The velocity of the author's minivan as he drove from a traffic light onto a highway may be modeled by

$$V(t) = -0.1773t^2 + 0.6798t + 0.5019$$
$$\text{miles per minute}$$

where t is time in minutes. How far did he travel in the first 1.5 minutes of his trip?

18. *Ship Building Jobs* Based on data from 1998 and 2008 and a projection for 2018, the rate of change in the number of jobs in the ship and boat building industry may be modeled by

$$r(t) = -0.198t + 1.28$$
$$\text{thousands of jobs per year}$$

where t is the number of years since 1998. (**Source:** Modeled from U.S. Bureau of Labor Statistics data)

Between the end of 1998 and the end of 2018, by how much is the number of jobs in the ship and boat building industry expected to change?

Exercises 19 and 20 are intended to challenge your understanding of the Fundamental Theorem of Calculus.

19. A function $F(x)$ has the properties that $F'(x) = \dfrac{1}{x}$ and $F(1) = 1 + \ln(4)$. Find $F(4)$.

20. Calculate the area of the region bounded by the graph of $g(x) = 2^x$ and the graph of $h(x) = x^2$. (*Hint:* The graphs of g and h intersect in exactly three places.)

Chapter 7

Advanced Integration
Techniques and Applications

Fashion trends come and go. When a new product is first introduced to the market, sales are often slow. However, as the popularity of the product increases, sales increase rapidly. After a period of time, the market becomes saturated with the product, and sales taper off. The cumulative sales of a product that demonstrate the type of sales growth just described may often be modeled by a logistic function.

7.1 Integration by Parts

7.2 Area Between Two Curves

7.3 Differential Equations and Applications

7.4 Differential Equations: Limited Growth and Logistic Growth Models

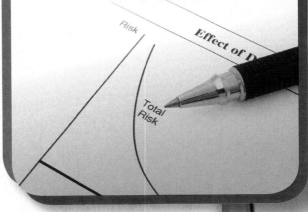

7.1 Integration by Parts

As we saw in Chapter 6, many derivative rules have a corresponding integral rule. Although there is no product rule for integrals, the *integration by parts method* is directly related to the product rule for derivatives.

In this section, we will demonstrate how to use the method of integration by parts to integrate a function. This technique is among the more challenging methods to use, so we will use it only after basic integration methods and the substitution method have failed.

Recall that the Product Rule for derivatives states that

$$\frac{d}{dx}(u(x) \cdot v(x)) = u'(x) \cdot v(x) + v'(x) \cdot u(x)$$

where u and v are differentiable functions. If we integrate both sides of the equation with respect to x, we get

$$\int\left(\frac{d}{dx}(u(x)\cdot v(x))\right)dx = \int(u'(x)\cdot v(x) + v'(x)\cdot u(x))\,dx$$

$$u(x)\cdot v(x) = \int(u'(x)\cdot v(x))\,dx + \int(v'(x)\cdot u(x))\,dx$$

$$= \int\left(\frac{du}{dx}\cdot v(x)\right)dx + \int\left(\frac{dv}{dx}\cdot u(x)\right)dx$$

$$u(x)\cdot v(x) = \int v(x)\,du + \int u(x)\,dv$$

For notational ease, we will write $u(x)$ as u and $v(x)$ as v, keeping in mind that both u and v are functions of x.

$$uv = \int v\,du + \int u\,dv$$

We'll solve this equation for $\int u\,dv$.

$$\int u\,dv = uv - \int v\,du$$

This equation is the basis of the integration by parts method.

> **Integration by Parts**
> Let u and v be differentiable functions of x. Then
> $$\int u\,dv = uv - \int v\,du$$

Consider the integral $\int (xe^x)\,dx$. We attempted to integrate this function by substitution in Section 6.2, but we were unable to do so. We will now attempt to integrate the function using integration by parts. In order to use this method, we must first identify a function u and a function v' such that $uv' = xe^x$. We typically will pick the function with the simplest derivative to be u or the most complex function that may be easily integrated to be v'. In this case, we pick

$$u(x) = x \text{ and } v'(x) = e^x$$

Next we'll differentiate u with respect to x and, using the concept of differentials, solve for du.

$$\frac{d}{dx}(u) = \frac{d}{dx}(x)$$
$$\frac{du}{dx} = 1$$
$$du = 1\,dx$$

Next we'll integrate both sides of the equation $v'(x) = e^x$ with respect to x.

$$v'(x) = e^x$$
$$\int v'(x)\,dx = \int e^x dx$$
$$v(x) = e^x$$

We will add in the constant C at the end of the problem.

We'll now return to the formula $\int u\,dv = uv - \int v\,du$ and plug in the corresponding pieces. (Recall that, by definition, $dv = v'(x)\,dx$. In this case, $dv = e^x dx$.)

$$\int u\,dv = uv - \int v\,du$$
$$\int (x)(e^x dx) = (x)\cdot(e^x) - \int (e^x)(dx)$$

Since $u = x$, $v = e^x$, and $du = dx$.

$$= xe^x - \int e^x dx$$
$$= xe^x - e^x + C$$

(In this example, as in previous examples, the constant C represents the sum of all constants generated throughout the integration process.)

Let's check our work by differentiating $F(x) = xe^x - e^x + C$.

$$\frac{d}{dx}(F(x)) = \frac{d}{dx}(xe^x - e^x + C)$$
$$f(x) = (1\cdot e^x + (e^x)\cdot(x)) - e^x + 0$$
$$= e^x + xe^x - e^x$$
$$= xe^x$$

The result checks out, so we did the problem correctly.

Now that we've demonstrated the basic theory that surrounds the integration by parts method, we will refine the process to increase our efficiency. We'll demonstrate the streamlined process in Example 1 and then detail the steps of the process.

EXAMPLE 1 Integrating by Parts

Integrate $f(x) = 3x\,(2^x)$ with respect to x.

SOLUTION

We are asked to find $\int 3x(2^x)\,dx$. Since $\frac{d}{dx}(3x)$ is a constant, we will select $u = 3x$. Consequently, $dv = 2^x dx$. Observe that $\int 3x(2^x)\,dx$ may be rewritten as $\int u\,dv$. We are now ready to do integration by parts.

		Choose u and dv
$u = 3x$	$dv = 2^x\,dx$	
$du = 3\,dx$	$v = \dfrac{2^x}{\ln(2)}$	Differentiate u and integrate dv with respect to x

$$\int 3x(2^x)\,dx = \int u\,dv$$
$$= uv - \int v\,du$$
$$= (3x)\left(\frac{2^x}{\ln(2)}\right) - \int \frac{2^x}{\ln(2)}(3\,dx)$$
$$= \frac{3x(2^x)}{\ln(2)} - \frac{3}{\ln(2)}\int 2^x dx \qquad \text{Constant Multiple Rule}$$
$$= \frac{3x(2^x)}{\ln(2)} - \frac{3}{\ln(2)}\left(\frac{2^x}{\ln(2)}\right) + C \qquad \text{Exponential Rule}$$
$$= \left(\frac{2^x}{\ln(2)}\right)\left(3x - \frac{3}{\ln(2)}\right) + C \qquad \text{Factor out } \frac{2^x}{\ln(2)}$$

Therefore, $\int 3x\,(2^x)\,dx = \left(\dfrac{2^x}{\ln(2)}\right)\left(3x - \dfrac{3}{\ln(2)}\right) + C$.

$$= \frac{1}{2}x^2 \ln(x) - \frac{1}{2}\left(\frac{1}{2}x^2\right) + C \quad \text{Power Rule}$$

$$= \frac{1}{2}x^2\left(\ln(x) - \frac{1}{2}\right) + C \quad \text{Factor out } \frac{1}{2}x^2$$

Therefore, $\displaystyle\int x \ln(x)\,dx = \frac{1}{2}x^2\left(\ln(x) - \frac{1}{2}\right) + C.$

EXAMPLE 3 Integrating by Parts

➡ Simplify $\displaystyle\int (2x + 1)(x - 3)^3\,dx.$

SOLUTION

We'll pick $u = 2x + 1$ and $dv = (x - 3)^3\,dx.$

$u = 2x + 1$	$dv = (x - 3)^3 dx$	
$du = 2\,dx$	$v = \dfrac{(x - 3)^4}{4}$	By the substitution method

$$uv - \int v\,du = (2x + 1)\left(\frac{(x-3)^4}{4}\right) - \int\left(\frac{(x-3)^4}{4}\right)(2\,dx)$$

$$= \frac{(2x+1)(x-3)^4}{4} - \frac{2}{4}\int (x-3)^4\,dx \quad \text{Constant Multiple Rule}$$

$$= \frac{1}{4}(2x+1)(x-3)^4 - \frac{1}{2}\int (x-3)^4\,dx \quad \text{Simplify fractions}$$

$$= \frac{1}{4}(2x+1)(x-3)^4 - \frac{1}{2}\left(\frac{(x-3)^5}{5}\right) + C \quad \text{Power Rule}$$

$$= \frac{1}{4}(2x+1)(x-3)^4 - \frac{1}{10}(x-3)^5 + C \quad \text{Simplify fractions}$$

$$= \frac{5}{20}(2x+1)(x-3)^4 - \frac{2}{20}(x-3)^5 + C \quad \text{Rewrite fractions}$$

$$= \frac{1}{20}(x-3)^4(5(2x+1) - 2(x-3)) + C \quad \text{Factor out } \frac{1}{20}(x-3)^4$$

$$= \frac{1}{20}(x-3)^4(10x + 5 - 2x + 6) + C \quad \text{Simplify}$$

$$= \frac{1}{20}(x-3)^4(8x + 11) + C \quad \text{Group like terms}$$

Therefore,

$$\int (2x + 1)(x - 3)^3\,dx = \frac{1}{20}(x-3)^4(8x + 11) + C.$$

Shutterstock

We'll demonstrate these steps in the next several examples.

EXAMPLE 2 Integrating by Parts

➡ Integrate $f(x) = x \ln(x).$

SOLUTION

We must find $\displaystyle\int x \ln(x)\,dx.$ We'll pick $u = \ln(x)$ and $dv = x\,dx.$

$u = \ln(x)$	$dv = x\,dx$	Choose u and dv
$du = \dfrac{1}{x}\,dx$	$v = \dfrac{x^2}{2}$	Differentiate u and integrate dv with respect to x

$$uv - \int v\,du = \ln(x) \cdot \frac{x^2}{2} - \int\left(\frac{x^2}{2} \cdot \frac{1}{x}\right)dx$$

$$= \frac{1}{2}x^2 \ln(x) - \frac{1}{2}\int x\,dx \quad \text{Constant Multiple Rule}$$

EXAMPLE 4 Integrating by Parts

Simplify $\int (x^2)(x+1)^3 dx$.

SOLUTION

Let $u = x^2$ and $dv = (x+1)^3\,dx$.

$u = x^2$	$dv = (x+1)^3 dx$
$du = 2x\,dx$	$v = \dfrac{(x+1)^4}{4}$

By the substitution method

$$uv - \int v\,du = (x^2)\left(\frac{(x+1)^4}{4}\right) - \int\left(\frac{(x+1)^4}{4}\right)(2x\,dx)$$

$$= \frac{1}{4}x^2(x+1)^4 - \frac{1}{2}\int x(x+1)^4\,dx$$

Constant Multiple Rule

Observe that $\int x(x+1)^4 dx$ is not readily integrable. We'll integrate this piece separately using integration by substitution and then substitute the result back into the equation.

Let $w = x+1$. Then $dw = dx$ and $x = w - 1$. Thus

$$\int x(x+1)^4 dx = \int (w-1)(w^4)\,dw$$

$$= \int (w^5 - w^4)\,dw$$

$$= \frac{w^6}{6} - \frac{w^5}{5} + C$$

Power Rule

$$= \frac{5w^6}{30} - \frac{6w^5}{30} + C$$

Get a common denominator

$$= \frac{1}{30}w^5(5w - 6) + C$$

Factor out $\frac{1}{30}w^5$

$$= \frac{1}{30}(x+1)^5(5(x+1) - 6) + C$$

Replace w with $x+1$

$$= \frac{1}{30}(x+1)^5(5x + 5 - 6) + C$$

Simplify

$$= \frac{1}{30}(x+1)^5(5x - 1) + C$$

Group like terms

We'll substitute this result back into the integration by parts equation.

$$\frac{1}{4}x^2(x+1)^4 - \frac{1}{2}\int x(x+1)^4\,dx$$

$$= \frac{1}{4}x^2(x+1)^4 - \frac{1}{2}\left(\frac{1}{30}(x+1)^5(5x - 1)\right) + C$$

$$= \frac{15}{60}x^2(x+1)^4 - \frac{1}{60}(x+1)^5(5x - 1) + C$$

$$= \frac{1}{60}(x+1)^4(15x^2 - (x+1)(5x - 1)) + C$$

$$= \frac{1}{60}(x+1)^4(15x^2 - (5x^2 + 4x - 1)) + C$$

$$= \frac{1}{60}(x+1)^4(10x^2 - 4x + 1) + C$$

Therefore,

$$\int (x^2)(x+1)^3 dx = \frac{1}{60}(x+1)^4(10x^2 - 4x + 1) + C.$$

Phew!

Although we used the substitution method to integrate the second integral in Example 4, we could have used integration by parts to integrate the second integral. In Example 5, we will apply the integration by parts method multiple times.

EXAMPLE 5 Integrating by Parts

Integrate $f(x) = x^2 e^x$.

SOLUTION

We're asked to find $\int x^2 e^x dx$. We'll pick $u = x^2$ and $dv = e^x dx$.

$u = x^2$	$dv = e^x dx$
$du = 2x\,dx$	$v = e^x$

$$uv - \int v\,du = (x^2)(e^x) - \int 2xe^x dx$$

$$= x^2 e^x - 2\int xe^x dx$$

The remaining integral, $\int xe^x dx$, is not readily integrable. We will integrate this function using integration by parts. We will select new values for u and dv; however, it is important to note that although the variables are the same, the functions are not equal to the u and v identified previously.

$u = x$	$dv = e^x dx$
$du = dx$	$v = e^x$

$$x^2 e^x - 2\int xe^x \, dx = x^2 e^x - 2\left(xe^x - \int e^x \, dx\right)$$

$$= x^2 e^x - 2(xe^x - e^x) + C$$

$$= x^2 e^x - 2xe^x + 2e^x + C$$

$$= e^x(x^2 - 2x + 2) + C$$

Integration by parts can also be used to integrate functions that formerly did not appear integrable. In Example 6, we will integrate the natural log function.

EXAMPLE 6 Integrating the Natural Logarithm Function

 Simplify $\int \ln(x) \, dx$.

SOLUTION

Let $u = \ln(x)$ and $dv = dx$.

$u = \ln(x)$	$dv = dx$
$du = \dfrac{1}{x} dx$	$v = x$

$$uv - \int v \, du = (\ln(x))(x) - \int x\left(\frac{1}{x} dx\right)$$

$$= x \ln(x) - \int 1 \, dx$$

$$= x \ln(x) - x + C$$

$$= x(\ln(x) - 1) + C$$

Therefore, $\int \ln(x) \, dx = x(\ln(x) - 1) + C$.

EXAMPLE 7 Using Integration to Predict Cumulative Revenue

Based on online sales projections for 2007 to 2012, the annual amount of revenue generated by online event ticket sales may be modeled by

Shutterstock

$$s(t) = -9.05 + 7.56 \ln t \text{ billion dollars}$$

where t is the number of years since 2000. (**Source:** *Statistical Abstract of the United States, 2009,* Table 1015) According to the model, what will be the cumulative amount of revenue generated from the end of 2007 to the end of 2012?

SOLUTION

Since $s(t)$ is the annual sales revenue, we must calculate $\int_{7}^{12} (-9.05 + 7.56 \ln t) \, dt$ to determine the cumulative amount of revenue generated between 2007 and 2012.

$$\text{Cumulative Revenue} = \int_{7}^{12} (-9.05 + 7.56 \ln t) \, dt$$

$$= -9.05t \Big|_{7}^{12} + 7.56 \int_{7}^{12} \ln(t) \, dt \quad \text{Power, Constant Multiple Rules}$$

$$= -9.05t \Big|_{7}^{12} + 7.56t(\ln t - 1) \Big|_{7}^{12} \quad \text{Integral of natural log function}$$

$$= -9.05(12) + 7.56(12)(\ln(12) - 1) - (-9.05(7)$$

$$+ 7.56(7)(\ln(7) - 1)) \quad \text{Fundamental Theorem of Calculus}$$

$$= 26.1 - (-13.3)$$

$$= 39.4$$

According to the model, the cumulative revenue from online ticket sales between 2007 and 2012 will be $39.4 billion.

7.1 Exercises

In Exercises 1–10, integrate the function using integration by parts.

1. $f(x) = xe^{-x}$

2. $g(x) = 5x \ln(x)$

3. $h(t) = (3t - 5)\ln(t)$

4. $f(t) = \dfrac{6}{t} + \ln(t)$

5. $h(t) = t^3 \ln(t)$

6. $h(t) = \dfrac{\ln(t)}{t} - 3$

7. $f(x) = 4xe^{2x+1}$

8. $f(x) = \ln(x^2)$

9. $h(x) = e^{2x}(2x - 1)$

10. $f(t) = (3t + 4)e^{2t}$

In Exercises 11–18, integrate the function using the simplest possible method.

11. $g(t) = \dfrac{3^t}{3^t - 1}$

12. $f(x) = 3x(e^x - x)$

13. $h(x) = 2x(x^2 - 9)$

14. $g(t) = (3t^2)(t^3 - 1)$

15. $g(x) = 3^x(x^2 - 3)$

16. $h(x) = 3xe^{3x}$

17. $f(t) = e^2 t^2$

18. $f(x) = \dfrac{2^{\ln x}}{x}$

In Exercises 19 and 20, use the integration by parts method to determine the answer to the question.

19. **Real Networks Licensing Revenue** Based on data from 1999 to 2003, the annual net revenue from software licensing fees for Real Networks, Inc., may be modeled by

$$R(t) = 1208(t - 0.88)e^{-1.34t} + 56.7 \text{ million dollars}$$

where t is the number of years since the end of 1998. (**Source:** Modeled from Real Networks, Inc., 2003 Report, p. 14)

According to the model, what was the accumulated net revenue from software licensing fees between the end of 2000 and the end of 2003?

20. **Electronic Arts Profit** Based on data from 2000 to 2004, the annual gross profit for Electronic Arts, Inc., may be modeled by

$$P(t) = 274(t - 0.6)e^{-0.232t} + 596 \text{ million dollars}$$

where t is the number of years since the end of 1998. (**Source:** Modeled from Electronic Arts 2004 Annual Report, p. 19)

According to the model, what was the accumulated profit between the end of 2000 and the end of 2004?

ATTENTION

NEED MORE PRACTICE? FIND MORE HERE:
CENGAGEBRAIN.COM

7.2 Area Between Two Curves

Electronic Arts, Inc., is arguably the most dominant force in the electronic gaming software market. In fiscal year 2010, the company had 27 titles that sold over 1 million copies, and five titles sold in excess of 4 million copies each, including *The Sims 3, Madden NFL 10, Need for Speed: Shift, FIFA 10*, and *Battlefield: Bad Company 2*.

Based on data from fiscal years 2006 to 2010, the *rate of change* in the annual net revenue of Electronic Arts, Inc., may be modeled by

$$r(t) = -384t^2 - 1336t - 444$$
$$\text{million dollars per year}$$

and the *rate of change* in the annual cost of goods sold may be modeled by

$$c(t) = -286.2t^2 - 1022t - 348$$

million dollars per year

where t is the number of years since 2006. (**Source:** Electronic Arts Annual Report, 2010) By how much did the company's gross profits increase between the end of fiscal year 2006 and the end of fiscal year 2010? This problem may be interpreted graphically by calculating the area of the region between the graph of r and the graph of c.

In this section, we will discuss how to calculate the area of the bounded region between two graphs. We will also explain the meaning of the definite integral of the difference between two rate-of-change functions.

> **Area Between Two Curves**
>
> If the graph of f lies above the graph of g on an interval $[a, b]$, then the area of the region between the two graphs from $x = a$ to $x = b$ is given by
>
> $$\int_a^b (f(x) - g(x))\, dx$$

This result is relatively easy to verify graphically. Consider the graphs of the functions f and g shown in Figure 7.1. We see that on the interval $[0, 4]$, $f(x) > g(x)$. That is, the graph of f lies above the graph of g.

Shutterstock

Figure 7.1

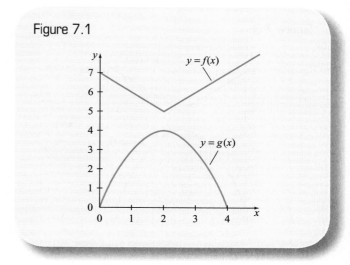

Consider $A = \int_1^3 (f(x) - g(x))\, dx$. By the Sum and Difference Rule for integrals, we know that this equation is equivalent to

$$A = \int_1^3 f(x)\, dx - \int_1^3 g(x)\, dx$$
$$= (\text{area between } f \text{ and } x\text{-axis on } [1, 3])$$
$$- (\text{area between } g \text{ and } x\text{-axis on } [1, 3])$$

The shaded region in Figure 7.2 is the area between the graph of f and the x-axis on the interval $[1, 3]$.

Figure 7.2

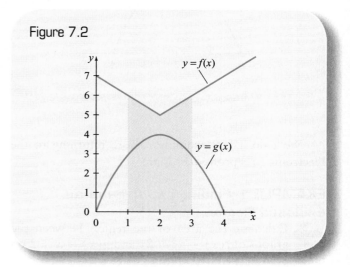

The shaded region in Figure 7.3 is the area between the graph of g and the x-axis on the interval $[1, 3]$.

Figure 7.3

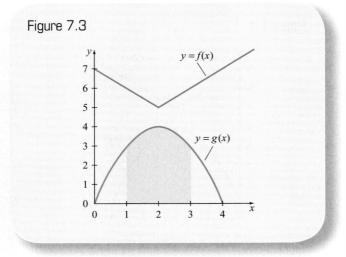

Figure 7.5

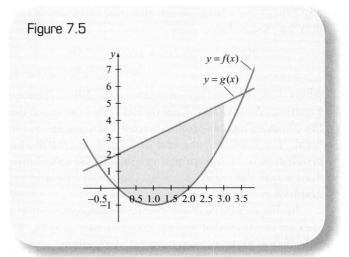

Subtracting the second area from the first area results in the area between the graphs of *f* and *g* on the interval $[1, 3]$ (Figure 7.4).

We see that the graph of $g(x) = x + 2$ lies above the graph of $f(x) = x^2 - 2x$ on the interval $[0, 3]$. We will calculate the area of the region bounded by the two graphs.

$$\int_0^3 (g(x) - f(x))\, dx = \int_0^3 ((x + 2) - (x^2 - 2x))\, dx$$

$$= \int_0^3 (-x^2 + 3x + 2)\, dx$$

$$= \left(-\frac{1}{3}x^3 + \frac{3}{2}x^2 + 2x\right)\Big|_0^3$$

$$= \left(-\frac{1}{3}(3)^3 + \frac{3}{2}(3)^2 + 2(3)\right) - (0)$$

$$= -9 + \frac{27}{2} + 6$$

$$= \frac{21}{2}$$

$$= 10.5$$

The area of the region between the two graphs is 10.5 square units.

Figure 7.4

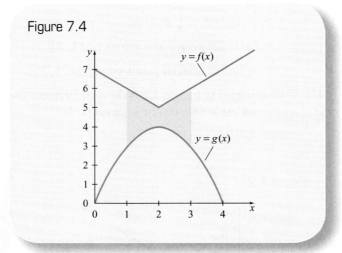

We'll do a few examples before returning to the Electronic Arts profit question.

EXAMPLE 1 Finding the Area of an Enclosed Region

Calculate the area of the region between the graphs of $f(x) = x^2 - 2x$ and $g(x) = x + 2$ on the interval $[0, 3]$.

SOLUTION

We begin by graphing the functions together and shading the appropriate region (see Figure 7.5).

EXAMPLE 2 Finding the Area of an Enclosed Region

Calculate the area of the region bounded by the graphs of $f(x) = -x^2 + 2x + 1$ and $g(x) = x^2 - 2x + 1$.

SOLUTION

We begin by graphing the functions together and shading the bounded region (Figure 7.6).

It appears that the graphs intersect at $x = 0$ and $x = 2$; however, we must confirm this algebraically.

Figure 7.6

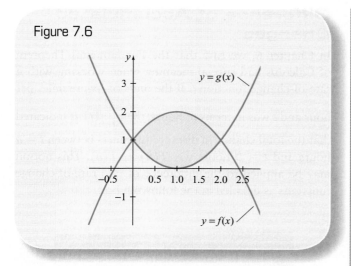

$$f(x) = g(x)$$

$$-x^2 + 2x + 1 = x^2 - 2x + 1$$

$$0 = 2x^2 - 4x$$

$$2x(x - 2) = 0$$

$$x = 0, \; x = 2$$

Our algebraic solution confirms our graphical observation.

We see that the graph of f lies above the graph of g on the interval $[0, 2]$. We will calculate the area of the "eye-shaped" region between the two graphs.

$$A = \int_0^2 (f(x) - g(x)) \, dx$$

$$= \int_0^2 ((-x^2 + 2x + 1) - (x^2 - 2x + 1)) \, dx$$

$$= \int_0^2 (-x^2 + 2x + 1 - x^2 + 2x - 1) \, dx$$

$$= \int_0^2 (-2x^2 + 4x) \, dx$$

$$= \left(-\frac{2}{3}x^3 + 2x^2 \right)\Big|_0^2$$

$$= \left(-\frac{2}{3}(2)^3 + 2(2)^2 \right) - 0$$

$$= \frac{8}{3}$$

The area of the region between the two graphs is $2\frac{2}{3}$ square units.

EXAMPLE 3 Finding the Area of an Enclosed Region

Find the area of the bounded region between $f(x) = x^3 - 3x^2 + 3x$ and $g(x) = -x^2 + 4x - 2$.

SOLUTION

We begin by graphing the functions together and shading the bounded regions (see Figure 7.7).

Figure 7.7

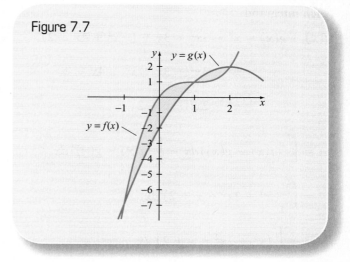

The graphs intersect in three different places: $x = -1$, $x = 1$, and $x = 2$. We can confirm our graphical observation by evaluating both f and g at these points and ensuring that the function values are equal at these x-values.

Table 7.1

x	$f(x) = x^3 - 3x^2 + 3x$	$g(x) = -x^2 + 4x - 2$
-1	-7	-7
1	1	1
2	2	2

Our computational results, shown in Table 7.1, confirm our graphical observation.

There are two bounded regions. The first bounded region occurs on the interval $[-1, 1]$. On this interval, the graph of f lies above the graph of g. The second bounded region occurs on the interval $[1, 2]$. On this interval, the graph of g lies above the graph of f. The combined area of the regions is given by

$$A = \int_{-1}^{1} (f(x) - g(x))\,dx + \int_{1}^{2} (g(x) - f(x))\,dx$$

$$= \int_{-1}^{1} (f(x) - g(x))\,dx - \int_{1}^{2} (f(x) - g(x))\,dx$$

Factor out a −1 to make the integrands equal

Since both integrands are equal, we will simplify the expression $f(x) - g(x)$ and substitute the result into each integrand.

$$f(x) - g(x) = x^3 - 3x^2 + 3x - (-x^2 + 4x - 2)$$

$$= x^3 - 3x^2 + 3x + x^2 - 4x + 2$$

$$= x^3 - 2x^2 - x + 2$$

$$A = \int_{-1}^{1} (f(x) - g(x))\,dx - \int_{1}^{2} (f(x) - g(x))\,dx$$

$$= \int_{-1}^{1} (x^3 - 2x^2 - x + 2)\,dx$$

$$- \int_{1}^{2} (x^3 - 2x^2 - x + 2)\,dx$$

$$A = \left(\frac{1}{4}x^4 - \frac{2}{3}x^3 - \frac{1}{2}x^2 + 2x \right)\Big|_{-1}^{1}$$

$$- \left(\frac{1}{4}x^4 - \frac{2}{3}x^3 - \frac{1}{2}x^2 + 2x \right)\Big|_{1}^{2}$$

$$= \left(\left(\frac{1}{4}(1)^4 - \frac{2}{3}(1)^3 - \frac{1}{2}(1)^2 + 2(1) \right) \right.$$

$$\left. - \left(\frac{1}{4}(-1)^4 - \frac{2}{3}(-1)^3 - \frac{1}{2}(-1)^2 + 2(-1) \right) \right)$$

$$- \left(\left(\frac{1}{4}(2)^4 - \frac{2}{3}(2)^3 - \frac{1}{2}(2)^2 + 2(2) \right) \right.$$

$$\left. - \left(\frac{1}{4}(1)^4 - \frac{2}{3}(1)^3 - \frac{1}{2}(1)^2 + 2(1) \right) \right)$$

$$= \left(\left(\frac{13}{12} \right) - \left(-\frac{19}{12} \right) \right) - \left(\left(\frac{8}{12} \right) - \left(\frac{13}{12} \right) \right)$$

$$= \frac{37}{12} = 3\frac{1}{12}$$

The combined area of the bounded regions between the two graphs is $3\frac{1}{12}$ square units.

Difference of Accumulated Changes

In Chapter 6, we saw that the Fundamental Theorem of Calculus had special meaning when working with a rate-of-change function f. If the units of f were miles per hour and t was in terms of hours, then $\int_{a}^{b} f(t)\,dt$ indicated that the total change in distance (in miles) between $t = a$ hours and $t = b$ hours was $F(b) - F(a)$. This notion may be applied to the difference of two rate-of-change functions as detailed in the following box.

> **Difference of Two Accumulated Changes**
>
> If f and g are continuous rate-of-change functions defined on the interval $[a, b]$, then the difference of the accumulated change of f from $x = a$ to $x = b$ and the accumulated change of g from $x = a$ to $x = b$ is given by
>
> $$\int_{a}^{b} (f(x) - g(x))\,dx$$

EXAMPLE 4 Calculating an Accumulated Change

Based on data from fiscal years 2006 to 2010, the *rate of change* in the annual net revenue of Electronic Arts, Inc., may be modeled by

$$r(t) = -384t^2 + 1336t - 444$$
million dollars per year

and the *rate of change* in the annual cost of goods sold may be modeled by

$$c(t) = -286t^2 + 1022t - 348$$
million dollars per year

where t is the number of years since 2006. (**Source:** Electronic Arts Annual Report, 2010)

By how much did the company's gross profits increase between the end of fiscal year 2006 and the end of fiscal year 2010?

SOLUTION

The total change in the annual net revenues of Electronic Arts, Inc., between the end of 2006 and the end of 2010 may be determined by calculating the sum of the signed areas of the shaded regions between the graph of r and the horizontal axis on the interval $[0, 4]$ (see Figure 7.8).

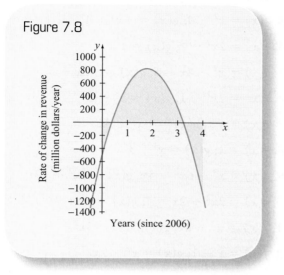

Figure 7.8

In other words, the change in the annual net revenues is given by

$$\int_0^4 r(t)\,dt = \int_0^4 (-384t^2 + 1336t - 444)\,dt$$

$$= (-128t^3 + 668t^2 - 444t)\Big|_0^4$$

$$= (-128(4)^3 + 668(4)^2 - 444(4)) - (0)$$

$$\approx 720 \text{ million dollars}$$

Between the end of fiscal year 2006 and the end of fiscal year 2010 annual net revenues increased by about $720 million.

The total change in the annual cost of goods sold between the end of fiscal year 2006 and the end of fiscal year 2010 may be determined by calculating the sum of the signed areas of the shaded regions between the graph of c and the horizontal axis on the interval $[0, 4]$ (see Figure 7.9).

In other words, the total change in the annual cost of goods sold is given by

$$\int_0^4 c(t)\,dt = \int_0^4 (-286t^2 + 1022t - 348)\,dt$$

$$= (-95.33t^3 + 511t^2 - 348t)\Big|_0^4$$

$$= (-95.33(4)^3 + 511(4)^2 - 348(4)) - (0)$$

$$\approx 683 \text{ million dollars}$$

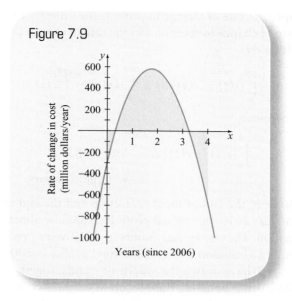

Figure 7.9

Between the end of fiscal year 2006 and the end of fiscal year 2010, the annual cost of goods sold increased by about $683 million.

The change in annual gross profit between the end of fiscal year 2006 and the end of fiscal year 2010 is given by

$$\int_0^4 (r(t) - c(t))\,dt$$

since the rate of change in profit is the difference of the rate of change in revenue and the rate of change in cost. However,

$$\int_0^4 (r(t) - c(t))\, dt = \int_0^4 r(t)\, dt - \int_0^4 c(t)\, dt$$

Therefore,

$$\int_0^4 (r(t) - c(t))\, dt = 720 - 683$$
$$= 37 \text{ million dollars}$$

Between the end of fiscal year 2006 and the end of fiscal year 2010, the annual profit increased by about \$37 million. That is, annual profits in 2010 were predicted to be \$37 million more than annual profits in 2006.

We have drawn the graphs of r and c together and shaded the bounded regions between the two graphs, as shown in Figure 7.10. These regions represent the accumulated change in annual profit.

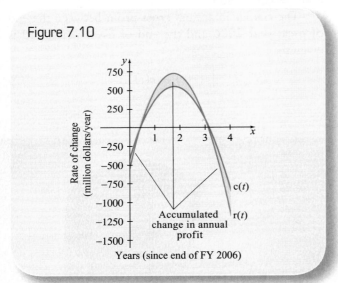

Figure 7.10

Shaded regions above the horizontal axis represent an increase in annual profits. Shaded regions below the horizontal axis represent a decrease in annual profits.

7.2 Exercises

In Exercises 1–20, calculate the combined area of the region(s) bounded by the graphs of the two functions. (On some of the exercises, you may find it helpful to use your graphing calculator to locate the intersection points of the graphs.)

1. $f(x) = x^2$; $g(x) = x$

2. $f(x) = x^2 - 3x + 2$; $g(x) = 2$

3. $f(x) = x^2 - 4$; $g(x) = -x^2 + 4$

4. $f(x) = 2x^2 - 1$; $g(x) = 1$

5. $f(x) = x^2 - 4x - 5$; $g(x) = -3x - 5$

6. $f(x) = 2x^2 - 1$; $g(x) = x^3 - 1$

7. $f(x) = -x$; $g(x) = -x^2 + 3x$

8. $f(x) = x$; $g(x) = x^3 - x$

9. $f(x) = x^3 - 6x^2 + 5x$; $g(x) = 6x^2 + 50x$

10. $f(x) = 2x^2 + 2x + 4$; $g(x) = x^3 + 2x + 4$

11. $f(x) = x^2 - 2x$; $g(x) = x^3 - 3x^2$

12. $f(x) = x - 1$; $g(x) = x^2 - 2x$

13. $f(x) = x^3 - x$; $g(x) = -x^3 + x$

14. $f(x) = 2^x + x$; $g(x) = 3^x - x$

15. $f(x) = 3^x$; $g(x) = x^3$

16. $f(x) = 2^x$; $g(x) = x^3 + 1$

17. $f(x) = 2^x$; $g(x) = x^2$

18. $f(x) = 3\ln(x)$; $g(x) = 2x - 4$

19. $f(x) = 2 + \ln(x)$; $g(x) = 5x^2$

20. $f(x) = x^2 - 4$; $g(x) = \dfrac{\ln(x)}{x}$

In Exercises 21–23, graph the functions together on the same axes. Then determine the solution by using the integral techniques demonstrated in this section. (You may integrate each function separately and then calculate the difference, or, if you prefer, you may calculate the difference between the functions and then integrate the result.)

21. ***Company Cost and Profit*** Based on data from fiscal years 2005 to 2009, the *rate of change* in the annual revenue of Microsoft Corporation may be modeled by

$$S(t) = -3408t^2 + 12{,}130t - 1424$$
$$\text{million dollars per year}$$

and the *rate of change* in the annual operating expenses (cost) may be modeled by

$$C(t) = -1908t^2 + 6940t - 487.6$$
million dollars per year

where *t* is the number of years since 2005. (**Source:** Microsoft Annual Report, 2009)

According to the models, what was the total accumulated change in annual profit between 2005 and 2009?

22. *Mining and Ship Building Jobs* Based on data from 1998 and 2008 and a projection for 2018, the rate of change in the number of jobs in the mining industry may be modeled by

$$m(t) = -2.56t + 28.0$$
thousand jobs per year

and the rate of change in the number of jobs in the ship and boat building industry may be modeled by

$$b(t) = -0.198t + 1.28$$
thousand jobs per year

where *t* is the number of years since 1998. (**Source:** Modeled from U.S. Bureau of Labor Statistics data)

Calculate the accumulated change in the number of jobs in each of the employment categories between 1998 and 2018. Then calculate the difference in the accumulated change between the two groups and interpret the practical meaning of the result.

23. *Laundry and Fishing Jobs* Based on data from 1998 and 2008 and a projection for 2018, the rate of change in the number of jobs in the laundry and dry cleaning services industry may be modeled by

$$d(t) = 0.614t - 7.90$$
thousand jobs per year

and the rate of change in the number of jobs in the fishing, hunting, and trapping industry may be modeled by

$$f(t) = 0.109t - 1.63$$
thousand jobs per year

where *t* is the number of years since 1998. (**Source:** Modeled from U.S. Bureau of Labor Statistics data)

Calculate the accumulated change in the number of jobs in each of the employment categories between 1998 and 2018. Then calculate the difference in the accumulated change between the two groups and interpret the practical meaning of the result.

Exercises 24 and 25 are intended to challenge your understanding of calculating the area between curves. In each exercise, we are interested in the region that is bordered by all of the given functions.

24. What is the combined area of the regions bordered by the three functions $f(x) = 2^x$, $g(x) = 2^{-x}$, and $h(x) = -x + 4$?

25. What is the area of the region bordered by the three functions $f(x) = x^2$, $g(x) = 5x - 6$, and $h(x) = -x^2$?

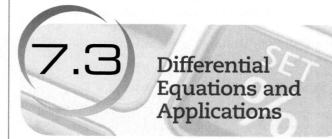

7.3 Differential Equations and Applications

In February 1992, a 79-year-old woman sustained third-degree burns on more than 6 percent of her body when she spilled a cup of McDonald's coffee on herself. She requested that McDonald's pay the associated medical expenses, and when the company denied her request, she filed suit. During the trial, a McDonald's quality assurance manager testified that the company required that coffee in the pot be held at between 180 and 190 degrees Fahrenheit (°F). Liquids at 180 degrees will cause third-degree burns in two to seven seconds. If the temperature of the coffee had been 25 degrees cooler, the woman might have avoided serious injury. The jury initially awarded the woman $2.7 million in damages; however, a judge later reduced this amount to $480,000. Ultimately, the woman and McDonald's settled the case for an undisclosed amount. (**Sources:** www.lect law.com; www.consumerrights.net)

How long does it take a 186-degree cup of coffee to cool to 155 degrees? Questions such as these may be answered using differential equations.

In this section, we will introduce differential equations and show how integral calculus may be used to solve them. Additionally, we'll discuss Newton's Law of Heating and Cooling and other applications of differential equations.

A **first-order differential equation** is an equation that relates an unknown function and its first derivative. In the first few examples, we will look at some simple first-order differential equations and their solutions.

EXAMPLE 1 Solving a Separable Differential Equation

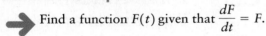 Find a function $F(t)$ given that $\dfrac{dF}{dt} = F$.

SOLUTION

We are looking for a function that is its own derivative.

$$\frac{dF}{dt} = F$$

$$\frac{dF}{F} = dt \qquad \text{Multiply by } dt \text{ and divide by } F$$

$$\frac{1}{F}\,dF = dt \qquad \text{Rewrite } \frac{dF}{F} \text{ as } \frac{1}{F}dF$$

$$\int \frac{1}{F}\,dF = \int dt \qquad \text{Take the integral of both sides}$$

$$\ln|F| = t + C \qquad \text{Integrate}$$

$$e^{t+C} = |F| \qquad \text{Recall } \ln|y| = x \text{ is equivalent to } e^x = |y|$$

$$F = \pm\, e^t \cdot e^C \qquad \text{Rules of exponents and definition of absolute value}$$

$$= \pm\, e^C e^t$$

Letting $k = e^C$, we get the *general solution* of the differential equation:

$$F(t) = \pm\, ke^t$$

Any function of this form is a solution to the differential equation $\dfrac{dF}{dt} = F$. For example, $F(t) = 2e^t$, $F(t) = -3e^t$, and $F(t) = e^t$ are all *particular solutions* to the differential equation.

EXAMPLE 2 Finding a Particular Solution for a Separable Differential Equation

Find the particular solution of the differential equation $\dfrac{dF}{dt} = F^2$ given $F(0) = 4$.

SOLUTION

$$\frac{dF}{dt} = F^2$$

$$\frac{dF}{F^2} = dt \qquad \text{Divide by } F^2 \text{ and multiply by } dt$$

$$\frac{1}{F^2}\,dF = dt \qquad \text{Rewrite } \frac{dF}{F^2} \text{ as } \frac{1}{F^2}dF$$

$$F^{-2}\,dF = dt \qquad \text{Rewrite } \frac{1}{F}\,dF \text{ as } F^{-2}\,dF$$

$$\int F^{-2}\,dF = \int dt \qquad \text{Integrate both sides}$$

$$\frac{F^{-1}}{-1} = t + C \qquad \text{Integrate with Power Rule}$$

$$F^{-1} = -t - C \qquad \text{Multiply both sides by } -1$$

$$\frac{1}{F} = -t - C \qquad \text{Rewrite } F^{-1} \text{ as } \frac{1}{F}$$

$$F = \frac{1}{-t - C} \qquad \text{Divide by } -t - C \text{ and multiply by } F$$

This is the *general solution* to the differential equation. However, since we know that $F(0) = 4$, we will be able to find the *particular solution*.

$$F = \frac{1}{-t - C}$$

$$4 = \frac{1}{-0 - C} \qquad \text{Substitute } F = 4 \text{ and } t = 0$$

$$4 = -\frac{1}{C}$$

$$C = -\frac{1}{4}$$

Therefore, the particular solution is $F(t) = \dfrac{1}{-t + \dfrac{1}{4}}$. We can verify the accuracy of our work by differentiating this function.

$$\frac{d}{dt}(F(t)) = \frac{d}{dt}\left(\frac{1}{-t + \dfrac{1}{4}}\right)$$

$$\frac{dF}{dt} = \frac{d}{dt}\left(-t + \frac{1}{4}\right)^{-1}$$

$$= -1\left(-t + \frac{1}{4}\right)^{-2}(-1)$$

$$= \left(-t + \frac{1}{4}\right)^{-2}$$

$$= \frac{1}{\left(-t + \dfrac{1}{4}\right)^{2}}$$

$$= \left(\frac{1}{-t + \dfrac{1}{4}}\right)^{2}$$

$$= F^{2}$$

Since the result is equivalent to the differential equation we were given at the start of the problem, we believe that our calculated particular solution, $F(t) = \dfrac{1}{-t + \dfrac{1}{4}}$, is correct. To verify, we check to see that $F(0) = 4$.

$$F(0) = \frac{1}{-(0) + \dfrac{1}{4}}$$

$$= 4$$

We used the **separation of variables** process to solve the differential equations in Examples 1 and 2. We will now formally describe the process.

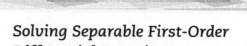

Solving Separable First-Order Differential Equations

A first-order differential equation is said to be **separable** if it may be written as

$$\frac{dy}{dx} = \frac{f(x)}{g(y)}$$

for some functions $f(x)$ and $g(y)$. The solution to the differential equation is obtained by moving the x and y variables to opposite sides of the equal sign and integrating. That is,

$$\int g(y)\,dy = \int f(x)\,dx$$

It is essential that all x-terms be grouped with the dx and all y-terms be grouped with the dy.

EXAMPLE 3 Finding a General Solution to a Separable Differential Equation

The rate of change of an account earning compound interest may be given by the differential equation $\dfrac{dA}{dt} = kA$, where k is a positive constant. Find the general solution to the differential equation.

SOLUTION

We solve using the separation of variables process.

$$\frac{dA}{dt} = kA$$

$$\frac{dA}{A} = k\,dt \qquad \text{Separate the variables}$$

$$\int \frac{1}{A}\,dA = \int k\,dt \qquad \text{Take the integral of both sides}$$

$$\ln A = kt + C \qquad \begin{array}{l}\text{A, the account value, is}\\ \text{assumed to be positive}\end{array}$$

$$A = e^{kt + C} \qquad \begin{array}{l}\text{Since, in general, }\ln A = b\\ \text{is equivalent to }A = e^{b}\end{array}$$

$$= e^{kt} \cdot e^{C} \qquad \text{Rules of exponents}$$

$$= Pe^{kt} \text{ where } P = e^{C}$$

The general solution is $A = Pe^{kt}$.

We immediately recognize the result from Example 3 as the continuous compound interest formula. The standard compound interest formula $A = P\left(1 + \dfrac{r}{n}\right)^{nt}$ may be converted to the form $A = Pe^{kt}$ by letting

$$\left(1 + \frac{r}{n}\right)^{n} = e^{k}$$

$$k = \ln\!\left(\left(1 + \frac{r}{n}\right)^{n}\right)$$

A Differential Equation for Continuous Compound Interest

The value of an investment account earning continuous compound interest is

$$A = Pe^{kt} \text{ dollars}$$

where A is the value of the investment after t years, P is the initial value of the investment (in dollars), and k is the continuous interest rate.

The rate of change in the value of the investment is given by

$$\frac{dA}{dt} = kA\frac{\text{dollars}}{\text{year}}$$

EXAMPLE 4 Finding a Particular Solution to a Differential Equation

The rate of change of an investment account earning compound interest is given by $\dfrac{dA}{dt} = kA$, where k is a positive constant. The initial account value was $1,000. At the end of the third year, the account value is $1,120. Find the particular solution to the differential equation.

SOLUTION

The general solution to the differential equation $\dfrac{dA}{dt} = kA$ is $A = Pe^{kt}$. We also know that $A(0) = 1000$ and $A(3) = 1120$.

$$A = Pe^{kt}$$
$$1000 = Pe^{k(0)}$$
$$1000 = P \cdot 1$$
$$P = 1000$$

Thus $A = 1000e^{kt}$. We will find the value of k by substituting in the second point, $(3, 1120)$.

$$A = 1000e^{kt}$$
$$1120 = 1000e^{k(3)}$$
$$1.12 = e^{3k}$$
$$\ln(1.12) = \ln(e^{3k})$$
$$0.1133 = 3k$$
$$k = 0.03778$$

Therefore, $A = 1000e^{0.03778t}$ is the particular solution to the differential equation. This equation may also be written as $A = 1000(1.0385)^{t}$, since $e^{0.03778} = 1.0385$. The account is earning interest at a rate of 3.778 percent compounded continuously or a rate of 3.85 percent compounded annually.

Newton's Law of Heating and Cooling

The world-renowned scientist and scholar Sir Isaac Newton determined that the rate of change in an object's temperature is proportional to the difference between the constant temperature of the environment surrounding the object and the object's temperature. This observation is summarized in **Newton's Law of Heating and Cooling**.

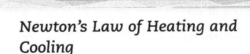

Newton's Law of Heating and Cooling

Let T be the temperature of an object at time t and A be the temperature of the environment surrounding the object (ambient temperature). Then

$$\frac{dT}{dt} = k(T - A)$$

where k is a constant that varies depending on the physical properties of the object.

According to Newton's Law of Heating and Cooling,

$$\frac{dT}{dt} = k(T - A)$$

We will use substitution to rewrite the equation.

We define the function $y = T - A$ and differentiate this function with respect to t.

$$\frac{d}{dt}(y) = \frac{d}{dt}(T - A)$$

$$\frac{dy}{dt} = \frac{d}{dt}(T) - \frac{d}{dt}(A)$$

$$= \frac{dT}{dt} - \frac{dA}{dt}$$

$$= \frac{dT}{dt} - 0 \quad \frac{dA}{dt} = 0 \text{ since } A \text{ is a constant}$$

$$= \frac{dT}{dt}$$

So $\frac{dy}{dt} = \frac{dT}{dt}$. Rewriting the equation, $\frac{dT}{dt} = k(T - A)$, in terms of y, we get

$$\frac{dy}{dt} = ky$$

since $\frac{dT}{dt} = \frac{dy}{dt}$ and $T - A = y$. Separating the variables, we rewrite the equation as

$$\frac{1}{y}dy = k\,dt$$

Integrating both sides of the equation, we get

$$\int \frac{1}{y}dy = \int (k\,dt)$$

$$\ln|y| = kt + C$$

Recall that if $x = \ln|y|$, then $e^x = |y|$. Consequently,

$$\ln|y| = kt + C$$

$$|y| = e^{kt + C}$$

$$y = \pm (e^{kt} \cdot e^C)$$

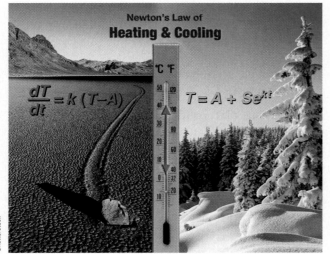

Newton's Law of
Heating & Cooling

$\frac{dT}{dt} = k\,(T-A)$ $T = A + Se^{kt}$

Shutterstock

Now we'll back substitute $y = T - A$.

$$T - A = \pm (e^{kt} \cdot e^C)$$

$$T = A \pm (e^{kt} \cdot e^C)$$

$$T = A + Se^{kt} \text{ where } S = \pm e^C$$

Thus the general solution to the differential equation $\frac{dT}{dt} = k(T - A)$ is $T = A + Se^{kt}$.

EXAMPLE 5 Using a Differential Equation to Model Water Temperature

To test Newton's Law of Heating and Cooling, the author heated a cup of water to 186°F and placed it in a 76°F room. The temperature of the water was recorded at 2-minute time intervals, as shown in Table 7.2.

Table 7.2

Hot Water Cooling in a 76°F Room	
Minutes (t)	Temperature (degrees Fahrenheit) (T)
0	186
2	175
4	167
6	159
8	153
10	147
12	142
14	138

Use Newton's Law of Heating and Cooling to find a model for the water temperature.

SOLUTION

We know that the general solution to the differential equation given in Newton's Law of Heating and Cooling is $T = A + Se^{kt}$. We'll substitute in values from the table to determine the numeric values of k and C. We know that $A = 76$, since the ambient temperature is 76°F. Additionally, we know that at time $t = 0$, $T = 186$.

$$T = 76 + Se^{kt}$$

$$186 = 76 + Se^{k(0)}$$

$$110 = Se^0$$

$$110 = S$$

Our equation $T = A + Se^{kt}$ may now be rewritten as

$$T(t) = 76 + 110e^{kt}$$

We'll use the table values $t = 14$ and $T = 138$ to find the value of k. (We can select any pair of table values. Since we're developing a model, selecting a different pair of table values would slightly alter the model equation.)

$$138 = 76 + 110e^{k(14)}$$

$$62 = 110e^{14k}$$

$$0.5636 = e^{14k}$$

$$\ln(0.5636) = 14k \qquad \text{Rewrite as logarithm}$$

$$k = -0.04095$$

Since k is negative, the temperature of the water is dropping at a continuous rate of 4.095 percent per minute. The temperature of the water may be modeled by

$$T(t) = 76 + 110e^{-0.04095t} \text{ degrees Fahrenheit}$$

Let's see how well this function fits the table of data (Table 7.3).

Table 7.3

	Hot Water Cooling in a 76°F Room	
Minutes (t)	Actual Temperature (degrees Fahrenheit) (T)	Model Temperature (degrees Fahrenheit) (M)
0	186	186
2	175	177
4	167	169
6	159	162
8	153	155
10	147	149
12	142	143
14	138	138

Newton's Law of Heating and Cooling gave a remarkably good estimate for the measured temperature. The discrepancies between the two may be due to measurement device error or other physical phenomena. As with any mathematical model, some error is to be expected.

EXAMPLE 6 Using a Differential Equation to Calculate Cooling Rates

According to the model developed in Example 5, at what rate was the water cooling 4, 8, and 12 minutes into the cooling period?

SOLUTION

The model for the water temperature was given by

$$T(t) = 76 + 110e^{-0.04095t} \text{ degrees}$$

t seconds after the temperature of the water was measured to be 186 degrees. The rate of change in the temperature is given by the differential equation

$$\frac{dT}{dt} = k(T - A)$$

$$= -0.04095(T - 76)$$

We're asked to evaluate this function at $t = 4$, $t = 8$, and $t = 12$.

$$\left.\frac{dT}{dt}\right|_{t=4} = -0.04095(T(4) - 76)$$

$$= -0.04095(169.4 - 76)$$

$$= -3.82$$

Four minutes into the cooling period, the water was cooling at a rate of 3.82 degrees per minute.

$$\left.\frac{dT}{dt}\right|_{t=8} = -0.04095(T(8) - 76)$$

$$= -0.04095(155.3 - 76)$$

$$= -3.25$$

Eight minutes into the cooling period, the water was cooling at a rate of 3.25 degrees per minute.

$$\left.\frac{dT}{dt}\right|_{t=12} = -0.04095(T(12) - 76)$$

$$= -0.04095(143.3 - 76)$$

$$= -2.76$$

Twelve minutes into the cooling period, the water was cooling at a rate of 2.76 degrees per minute.

Assuming that coffee cools at the same rate as water, we estimate that it would take a cup of McDonald's coffee 8 minutes to cool from 186 degrees to 155 degrees Fahrenheit. However, the size and shape of the coffee cup, the addition of cream and sugar, and other environmental variables could dramatically affect the accuracy of our estimate.

7.3 Exercises

In Exercises 1–10, find the general solution of the separable differential equation.

1. $\dfrac{dy}{dx} = \dfrac{x}{y}$

2. $\dfrac{dA}{dt} = 3A^2$

3. $\dfrac{dy}{dx} = \sqrt{y}$

4. $\dfrac{dA}{dt} = \dfrac{3A}{t}$

5. $\dfrac{dy}{dx} = \dfrac{y^2}{x^2}$

6. $\dfrac{dA}{dt} = 4(20 - A)$

7. $\dfrac{dy}{dx} = 0.25(y - 4)$

8. $\dfrac{dy}{dx} = 4y$

9. $\dfrac{dy}{dx} = 4y - y^2$

$\left(\text{Hint: } \displaystyle\int \dfrac{dy}{y(M - y)} = \dfrac{1}{M} \ln \left| \dfrac{y}{M - y} \right|.\right)$

10. $\dfrac{dA}{dt} = 6 - 3A$

In Exercises 11–15, find the particular solution for the differential equations.

$\left(\text{Hint: } \displaystyle\int \dfrac{dy}{y(M - y)} = \dfrac{1}{M} \ln \left| \dfrac{y}{M - y} \right|.\right)$

11. $\dfrac{dy}{dx} = \dfrac{2x}{y}; \, y(1) = 4$

12. $\dfrac{dy}{dx} = 3y; \, y(0) = -1$

13. $\dfrac{dy}{dx} = 25(y - 5); \, y(0) = 0$

14. $\dfrac{dy}{dx} = y(4 - y); \, y(0) = \dfrac{1}{2}$

15. $\dfrac{dy}{dx} = 2y - y^2; \, y(0) = 100$

In Exercises 16–19, use Newton's Law of Heating and Cooling to find the equation of the model that best fits the data. Then answer the given questions.

16. *Water Temperature* The author placed a cup of ice water in a 76°F room and recorded the temperature of the water at 10-minute intervals as shown in the table.

Cold Water Warming in a 76°F Room	
Minutes (*t*)	Temperature (degrees Fahrenheit) (*T*)
0	34.3
10	40.3
20	43.8
30	48.2
40	51.8
50	54.7

(a) Find a model for the temperature of the water at time *t* minutes.

(b) Determine the rate of change in the water's temperature at 10 minutes and 50 minutes into the warming period.

17. *Cake Temperature* Cakes are typically baked at 350°F. If a pan filled with cake batter is moved from a 75°F kitchen into a 350°F oven, how long will it take for the cake batter to heat up to 200°F? (Since the result will vary depending on the quantity of the batter and the shape of the pan, assume that for the container in question, the batter temperature will be 100°F at the end of the third minute.)

18. *Time of Death* When the forensic investigator arrives upon the scene of a homicide, one of the most important things to determine is the time of death of the victim. Suppose that the victim's internal body temperature was 84.5°F when the body was discovered in a 70°F room. Thirty minutes later, the person's body temperature had dropped to 83.6°F. If the person's internal body temperature was 98°F at the time of death, how long had the person been dead when the forensic investigator arrived on the scene?

19. *Time of Death* Suppose that a homicide victim's internal body temperature was 59.1°F when the body was discovered in a 30°F room. Thirty minutes later, the person's body temperature had dropped to 56.9°F. If the person's internal body temperature was 98°F at the time of death, how long had the person been dead when the forensic investigator arrived on the scene?

In Exercise 20, model the value of the given investment.

20. *Investment Value* A $1,000 investment has a continuous interest rate of 10 percent. Write an equation for the value of the investment after t years.

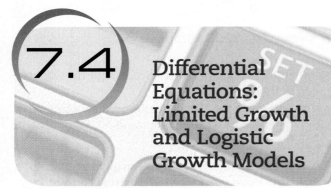

7.4 Differential Equations: Limited Growth and Logistic Growth Models

On September 11, 2001, many of us woke up to the horrific news of the terrorist attacks on the World Trade Center and Pentagon. News bulletins announcing the attack flooded radio and television channels. Initially, the news of the attack spread very rapidly; however, as the day progressed, the news spread more slowly, since most people had already received word of the attack. The spread of information, such as the news of the September 11 attack, may often be modeled mathematically. In this section, we will continue our discussion of differential equa-

tions and show how they can be used to find limited growth and logistic models.

Limited Growth Model

Sociologists often assume that the rate at which news spreads via mass media is proportional to the number of people who have not yet heard the news. This type of growth is called **limited growth**. For a limited growth function y with a maximum value of M, the rate of change, $\frac{dy}{dt}$, is proportional to the difference between the present value of y and M. That is, $\frac{dy}{dt} = k(M - y)$.

EXAMPLE 1 Using Differential Equations to Forecast the Spread of Information

→ At 8:46 A.M. on September 11, 2001, the hijacked American Airlines Flight 11 slammed into the North Tower of the World Trade Center in New York City. Eighteen minutes later, United Airlines Flight 175 crashed into the South Tower of the World Trade Center. By the end of the day, thousands of people were dead as a result of the deadliest terrorist attack in American history.

Suppose that in a community of 10,000 people, 3,000 people were watching television or listening to the radio when the news broke on September 11, 2001. For the purpose of the model, we'll assume that all 3,000 people heard the news simultaneously 5 minutes after the first attack. (It takes time for news crews to reach the site of any newsworthy event.) Find the equation for the

Shutterstock

limited growth model that models the spread of the news of the attack. Then determine how long it took for 90 percent of the community to hear the news and at what rate people were hearing the news at that time.

SOLUTION

The maximum number of people that can be informed is 10,000, so $M = 10{,}000$. Additionally, we know that when $t = 5$, $y = 3000$. We have

$$\frac{dy}{dt} = k(M - y)$$

$$\frac{dy}{dt} = k(10{,}000 - y)$$

We can use the separation of variables process to find the equation of the general solution and then use the point $(5, 3000)$ to find the particular solution.

$$\frac{dy}{dt} = k(10{,}000 - y)$$

$$\frac{dy}{(10{,}000 - y)} = k\,dt$$

Let $u = 10{,}000 - y$; then $du = -dy$. Rewriting the left-hand side of the equation in terms of u, we get

$$\frac{-du}{u} = k\,dt$$

$$\int -\frac{1}{u}\,du = \int k\,dt$$

$$-\ln |u| = kt + C$$

$$-\ln u = kt + C \qquad \text{Since } u > 0$$

$$\ln u = -kt - C$$

$$u = e^{-kt - C}$$

$$10{,}000 - y = e^{-kt}e^{-C} \qquad \text{Since } u = 10{,}000 - y$$

$$y = 10{,}000 - Se^{-kt} \qquad \text{Where } S = e^{-C}$$

At $t = 0$, nobody had heard the news of the attack because the information hadn't yet been broadcasted to the community. (It wasn't until 5 minutes after the crash that the first 3,000 people in the community heard the news.) We can use this information to find the value of S.

$$y(0) = 10{,}000 - Se^{-k(0)}$$

$$0 = 10{,}000 - Se^{0}$$

$$S = 10{,}000$$

A general solution to the differential equation is

$$y = 10{,}000 - 10{,}000e^{-kt}$$

$$= 10{,}000(1 - e^{-kt})$$

We'll now use the point $(5, 3000)$ to find the particular solution.

$$3000 = 10{,}000(1 - e^{-k(5)})$$

$$0.3 = 1 - e^{-5k}$$

$$-0.7 = -e^{-5k}$$

$$0.7 = e^{-5k}$$

$$\ln(0.7) = -5k$$

$$k = \frac{\ln(0.7)}{-5}$$

$$= 0.07133$$

The limited growth model that fits the data is given by

$$y = 10{,}000(1 - e^{-0.07133t})$$

Since 90 percent of 10,000 is 9,000, we need to determine when 9,000 people had heard the news.

$$9000 = 10{,}000(1 - e^{-0.07133t})$$

$$0.9 = 1 - e^{-0.07133t}$$

$$-0.1 = -e^{-0.07133t}$$

$$0.1 = e^{-0.07133t}$$

$$\ln(0.1) = -0.07133t$$

$$t = \frac{\ln(0.1)}{-0.07133}$$

$$= 32.28$$

We estimate that 32 minutes after the attack, 90 percent of the people in the community had heard the news. We'll now determine at what rate the word was spreading.

$$\frac{dy}{dt} = k(10{,}000 - y)$$

$$= 0.07133(10{,}000 - y)$$

$$\left.\frac{dy}{dt}\right|_{t=32.28} = 0.07133(10{,}000 - 9000) \qquad y(32.28) = 9000$$

$$= 71.33$$

Thirty-two minutes after the attack, the news was spreading at a rate of roughly 71 people per minute.

In Example 1, we solved the given differential equation by using the separation of variables method. Using this same approach, we can derive the general solution to the differential equation $\dfrac{dy}{dt} = k(M - y)$.

Limited Growth

Assume that the rate of growth of a function y with maximum value M is proportional to the difference between the present value of y and M. That is,

$$\frac{dy}{dt} = k(M - y)$$

The solution to the differential equation with initial condition

$$y(0) = 0$$

is given by

$$y = M(1 - e^{-kt})$$

and has the graph shown in Figure 7.11.

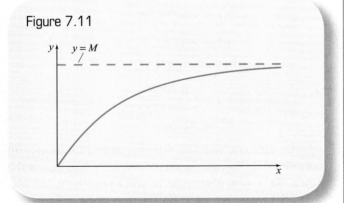

Figure 7.11

As you can deduce, limited growth models can be extremely useful. For example, a limited growth model may be used to model the cumulative number of people infected by a disease during a bioterrorist attack once quarantine efforts are underway. These types of models also have commercial applications.

Commercial fish growers seek to create a fish habitat that will maximize the size of each fish in the least amount of time. The Von Bertalanffy Limited Growth model is widely used in the fish industry. The model assumes that there is a maximal length, L_∞, that a fish will attain under optimal conditions and that the rate of change in the length of the fish, $\frac{dL}{dt}$, is proportional

to the difference between the maximal length and the current length. That is, $\frac{dL}{dt} = k(L_\infty - L)$. Solving this differential equation, we can find the general form of the Von Bertalanffy Limited Growth model.

$$\frac{dL}{dt} = k(L_\infty - L)$$

$$\frac{dL}{L_\infty - L} = k\,dt \qquad \text{Separation of variables}$$

$$\int \frac{1}{L_\infty - L}dL = \int k\,dt \qquad \text{Take the integral of both sides}$$

$$-\ln|L_\infty - L| = kt + C \qquad \text{Since } \frac{d}{dL}(-\ln(L_\infty - L)) = \frac{1}{L_\infty - L}$$

$$\ln(L_\infty - L) = -kt - C \qquad \text{Since } L_\infty - L > 0$$

$$L_\infty - L = e^{-k(t + C)}$$

$$L = L_\infty - e^{-k(t + C)}$$

The solution to the differential equation is commonly written as $L = L_\infty(1 - e^{-k(t - t_0)})$, where t_0 is a negative constant that varies from species to species. The constant t_0 has the property that $L(t_0) = 0$. Our solution may be converted from $L = L_\infty - e^{-k(t + C)}$ to $L = L_\infty(1 - e^{-k(t - t_0)})$ by letting $C = \frac{-\ln(L_\infty)}{k} - t_0$. (No matter what the value of C is, we can always find a t_0 that makes this equality true.)

EXAMPLE 2 Forecasting Fish Size with a Von Bertalanffy Limited Growth Model

Turkish scientists conducted a study on the Eastern Black Sea between 1991 and 1996 and determined the values of k, L_∞, and t_0 for economically important fish species in the region. (**Source:** www.tagem.gov.tr) The whiting fish species has the Von Bertalanffy growing constants given in Table 7.4.

Table 7.4

Whiting	k	L_∞	t_0
Female	0.11	43.3 cm	−1.91 years
Male	0.136	34.2 cm	−2.02 years

Source: www.fishbase.org

(a) According to the model, how long is a 2-year-old female whiting?

(b) According to the model, how long is a 2-year-old male whiting?

(c) Does the 2-year-old female whiting grow faster than the 2-year-old male?

SOLUTION

(a) We know that $L(t) = L_\infty(1 - e^{-k(t - t_0)})$. We have $L_\infty = 43.3$, $k = 0.11$, and $t_0 = -1.91$. Substituting these values into the Von Bertalanffy Limited Growth model yields $L(t) = 43.3(1 - e^{-0.11(t - (-1.91))})$. We determine the length of a 2-year-old female whiting by evaluating this function at $t = 2$.

$$L(2) = 43.3(1 - e^{-0.11(2 - (-1.91))})$$

$$= 43.3(1 - e^{-0.11(3.91)})$$

$$= 43.3(0.3496)$$

$$\approx 15.1$$

We estimate the length of a 2-year-old female whiting to be about 15.1 centimeters.

(b) We know that $L(t) = L_\infty(1 - e^{-k(t - t_0)})$. We have $L_\infty = 34.2$, $k = 0.136$, and $t_0 = -2.02$. Substituting these values into the Von Bertalanffy Limited Growth model yields $L(t) = 34.2(1 - e^{-0.136(t - (-2.02))})$. We determine the length of a 2-year-old male whiting by evaluating this function at $t = 2$.

$$L(2) = 34.2(1 - e^{-0.136(2 - (-2.02))})$$

$$= 34.2(1 - e^{-0.136(4.02)})$$

$$= 34.2(0.4212)$$

$$\approx 14.4$$

We estimate the length of a 2-year-old male whiting to be about 14.4 centimeters.

(c) Finally, we need to evaluate each of the differential equations at $t = 2$. Recall that the general form of the differential equation is

$$\frac{dL}{dt} = k(L_\infty - L).$$

Female:

$$\left.\frac{dL}{dt}\right|_{t=2} = 0.11(43.3 - L(2))$$

$$= 0.11(43.3 - 15.1)$$

$$= 3.1$$

A 2-year-old female whiting grows at a rate of 3.1 centimeters per year.

Male:

$$\left.\frac{dL}{dt}\right|_{t=2} = 0.136(34.2 - L(2))$$

$$= 0.136(34.2 - 14.4)$$

$$= 2.7$$

A 2-year-old male whiting grows at a rate of 2.7 centimeters per year.

From our calculations we conclude that the 2-year-old female whiting grows faster than the 2-year-old male whiting.

Logistic Growth Model

When a successful product is introduced to the market, cumulative product sales often increase very slowly initially; however, as the popularity of the product increases, sales increase rapidly. Then, as the market becomes saturated with the product, cumulative sales growth again slows. This type of growth is called **logistic growth**.

Logistic Growth

Assume that the rate of growth of a function y with maximum value M is proportional to the product of the present value of y and the difference between the present value of y and M. That is,

$$\frac{dy}{dt} = ky(M - y)$$

The solution to this differential equation with initial condition

$$y(0) = \frac{M}{1 + S}$$

is given by

$$y = \frac{M}{1 + Se^{-kMt}}$$

The graph of the solution is an S-shaped (sigmoidal) graph (see Figure 7.12).

Shutterstock

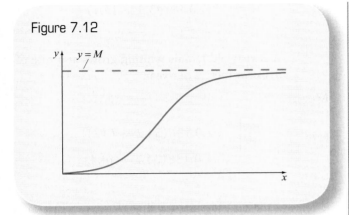

Figure 7.12

For example, consider Blu-ray Disc shipments in the United States and Canada. The Blu-ray Disc was introduced into the market in the fourth quarter of 2006. By the end of 2009, more than 177 million Blu-ray Discs had been shipped. (**Source:** DVD Entertainment Group) Looking at a scatter plot of the Blu-ray Disc shipments (Figure 7.13), it initially looks as if the annual number of discs shipped is increasing exponentially.

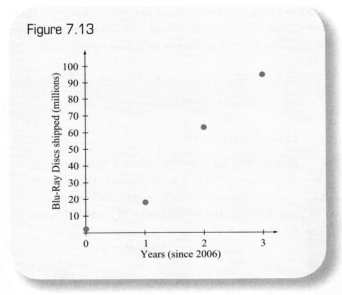

Figure 7.13

However, when we attempt to model the data with an exponential function (Figure 7.14), we see that the exponential function ends up growing much more rapidly than the data.

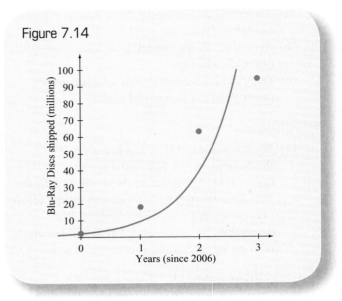

Figure 7.14

However, if we model the data with a logistic function (Figure 7.15), we see that Blu-ray Disc software shipments appear to exhibit logistic growth behavior.

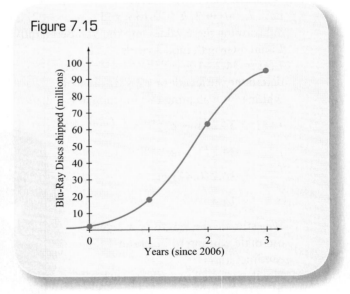

Figure 7.15

Using logistic regression on the TI-84 Plus, we determine that the logistic model for the given Blu-ray Disc data is given by

$$D(t) = \frac{102.0}{1 + 38.24e^{-2.068t}} \text{ million Blu-ray Discs shipped}$$

where t is the number of years since the end of 2006.

Plotting the graph of D together with the scatter plot of the data (Figure 7.16), we see that the solution fits the data remarkably well.

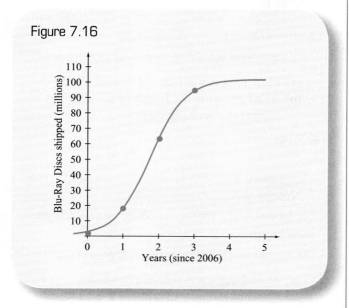

Figure 7.16

How rapidly were Blu-ray Disc shipments increasing at the end of 2010? We could differentiate $D(t) = \dfrac{102.0}{1 + 38.24e^{-2.068t}}$ and evaluate the derivative at $t = 4$. However, it may be easier to calculate $D'(4)$ using the differential equation.

$$\frac{dD}{dt} = kD(M - D)$$

Since $-2.068t = -Mkt$, $k = \dfrac{2.068}{M}$. Since $M = 102.0$, $k = \dfrac{2.068}{102.0}$. The differential equation may now be written as follows:

$$\frac{dD}{dt} = \left(\frac{2.068}{102.0}\right) D(102.0 - D)$$

$$\left.\frac{dD}{dt}\right|_{t=4} = \left(\frac{2.068}{102.0}\right) D(4)(102.0 - D(4))$$

$$D'(4) = \left(\frac{2.068}{102.0}\right) 101.0(102.0 - 101.0)$$

$$\approx 2.0$$

At the end of 2010, the number of Blu-ray Discs shipped annually was increasing at a rate of 2.0 million discs per year. That is, we predict that 2.0 million more discs were shipped in 2011 than in 2010.

How rapidly will Blu-Ray Disc shipments be increasing at the end of 2011? That is, what is $D'(5)$?

$$\left.\frac{dD}{dt}\right|_{t=5} = \left(\frac{2.068}{102.0}\right) D(5)(102.0 - D(5))$$

$$D'(5) = \left(\frac{2.068}{102.0}\right) 101.9(102.0 - 101.9)$$

$$\approx 0.2$$

At the end of 2011, we forecast that the number of Blu-ray Discs shipped annually will be increasing at a rate of 0.2 million discs per year. That is, we predict that 0.2 million more discs will be shipped in 2012 than in 2011.

In the past several examples, we have demonstrated the use of both limited and logistic growth models. Both types of models approach a constant value as the domain values grow large; however, near the origin the models behave quite differently. A limited growth model grows very rapidly at first, whereas a logistic model grows very slowly initially. Consequently, when choosing a model to use, it is important to consider the expected growth behavior of the function at the start of the time period in addition to looking at the shape of the scatter plot.

....

7.4 Exercises

In Exercises 1–6, determine the equation of the limited growth model for each species of fish and then answer the associated questions. You may find it helpful to refer to Example 2 in this section.

1. *Red Mullet*

Red Mullet	k	L_∞	t_0
Female	0.23	25.6 cm	−1.38 years
Male	0.21	22.2 cm	−2.08 years

Source: www.fishbase.org

(a) How long is a 3-year-old male red mullet?
(b) How long is a 3-year-old female red mullet?
(c) Is the 3-year-old male or the 3-year-old female red mullet growing faster?

2. *Turbot*

Turbot	k	L_∞	t_0
Female	0.115	103.4 cm	−0.93 year
Male	0.143	77.3 cm	−1.22 years

Source: www.fishbase.org

(a) At what rate does a 50.0-centimeter-long male turbot grow?
(b) At what rate does a 50.0-centimeter-long female turbot grow?
(c) Which is older, a 50.0-centimeter-long male turbot or a 50.0-centimeter-long female turbot? Explain.

3. *Flounder*

Flounder	k	L_∞	t_0
Female	0.197	44.98 cm	−1.43 years
Male	0.22	33.0 cm	−2.24 years

Source: www.fishbase.org

(a) At what rate does a 20.0-centimeter-long male flounder grow?
(b) At what rate does a 20.0-centimeter-long female flounder grow?
(c) Which is older, a 20-centimeter-long male flounder or a 20-centimeter-long female flounder? Explain.

4. *Pickerel*

Pickerel	k	L_∞	t_0
Female	0.232	17.6 cm	−1.15 years
Male	0.793	20.6 cm	0.08 years

Source: www.fishbase.org

(a) How long is a 4-year-old male pickerel?
(b) How long is a 4-year-old female pickerel?
(c) Is the 4-year-old male or the 4-year-old female pickerel growing faster?

5. *Copper Shark*

Copper Shark	k	L_∞	t_0
	0.038	385.0 cm	−3.477 years

Source: www.fishbase.org

(a) How long is a 3-year-old copper shark?
(b) At what rate is the length of a 3-year-old copper shark changing?

6. *Cowcod*

Cowcod	k	L_∞	t_0
Female	0.052	86.9 cm	−1.94 year

Source: www.fishbase.org

(a) How long is a 2-year-old cowcod?
(b) At what rate is the length of a 2-year-old cowcod changing?

In Exercises 7–12, use the logistic growth model and associated differential equation to answer the questions.

7. *Pediatric AIDS* Based on data from 1992 to 2001, the estimated number of new pediatric AIDS cases in the United States in year t may be modeled by

$$P(t) = \frac{919.9}{1 + 0.0533e^{0.07758t}} + 90 \text{ cases}$$

where t is the number of years since the end of 1992. (**Source:** Modeled from Centers for Disease Control and Prevention data)

The rate of change of the pediatric AIDS function P is the same as the rate of change of the function

$$N(t) = \frac{919.9}{1 + 0.0533e^{0.07758t}}$$

(a) How many new pediatric AIDS cases are estimated to have occurred in 2000?
(b) What is the differential equation whose solution is $N(t)$?
(c) At what rate was the estimated number of new pediatric AIDS cases changing at the end of 2000?
(d) Based on the results of part (c) and the graph of $P(t)$, does it look like efforts to reduce pediatric AIDS in the United States are generating positive results? Explain.

8. ***Adult and Adolescent AIDS*** Based on data from 1981 to 1995, the number of adult and adolescent deaths due to AIDS in a given year in the United States may be modeled by

$$P = \frac{53,955}{1 + 38.834e^{-0.45127t}} \text{ deaths}$$

where t is the number of years since the end of 1981. (**Source:** Modeled from Centers for Disease Control and Prevention data)

(a) According to the model, how many adult and adolescent AIDS deaths occurred in the United States in 1995?

(b) According to the model, at what rate were adult and adolescent AIDS deaths increasing at the end of 1995?

(c) The number of adult and adolescent AIDS deaths in the United States has decreased every year since 1995, in part because of the increased availability of drug treatments and AIDS prevention efforts. In 2001, there were 8,063 deaths. If the number of AIDS deaths had followed the logistic model, how many deaths would have occurred in 2002?

9. ***School Internet Access*** Based on data from 1994 to 2000, the rate of change in the percentage of public-school classrooms with Internet access may be modeled by

$$\frac{dy}{dt} = 0.01077y(85.88 - y) \text{ percentage points per year}$$

where t is the number of years since the end of 1994. In 2000, 77 percent of public-school classrooms had Internet access. (**Source:** Modeled from *Statistical Abstract of the United States, 2001*, Table 243)

(a) Find the particular solution to the differential equation and interpret its real-world meaning.

(b) At what rate was the percentage of public schools with Internet access changing at the end of 2000?

10. ***National Internet Usage*** Based on data from 1995 to 1999 and U.S. Census Bureau projections for 2000 to 2004, the rate of change in annual per capita Internet usage may be modeled by

$$\frac{dy}{dt} = 0.002882y(232.2 - y) \text{ hours per year}$$

where t is the number of years since the end of 1995. (**Source:** *Statistical Abstract of the United*

States, 2001, Table 1125) In 1999, the annual per capita Internet usage was 99 hours.

(a) Find the particular solution to the differential equation and interpret its real-world meaning.

(b) At what rate was the annual per capita Internet usage changing at the end of 2000?

11. ***Stay-at-Home Mothers*** In recent years, the number of married mothers who choose to stay at home to care for their families and bypass an outside career has increased. The number of married couple families with stay-at-home mothers (in thousands) may be modeled by

$$M(t) = \frac{948.2}{1 + 27.23e^{-1.110t}} + 4700$$

where t is the number of years since the end of 1999. (**Source:** Modeled from *Statistical Abstract of the United States, 2006*, Table 59)

The rate of change of the stay-at-home mother function M is the same as the rate of change of the function

$$N(t) = \frac{948.2}{1 + 27.23e^{-1.110t}}$$

(a) Find the differential equation whose solution is N.

(b) Determine at what rate the number of stay-at-home mothers was changing at the end of 2005 and at the end of 2008.

12. ***Game Console Sales*** Based on data from fiscal years 2003 to 2008, the cumulative sales of the Nintendo Game Boy Advance game console may be modeled by

$$C(t) = \frac{82.61}{1 + 1.452e^{-0.8937t}} \text{ million units}$$

where t is the number of years since fiscal year 2003. (**Source:** Modeled from Nintendo Annual Reports, 2003–2009)

(a) Find the differential equation whose solution is C.

(b) Determine at what rate the cumulative Game Boy Advance Sales were changing at the end of fiscal year 2008 and fiscal year 2010.

In Exercises 13 and 14, use a limited growth function to model the data. Then answer the given questions.

13. ***Spread of Information*** Suppose that in a city of 350,000 people, 40,000 people were watching television or listening to the radio when the news

of a bioterrorist attack was first broadcast. For the purpose of the model, assume that all 40,000 people heard the news simultaneously 5 minutes after the attack. Model the spread of the news of the attack. How long did it take for 75 percent of the city to hear the news, and at what rate was the news spreading at that time?

14. *Video Game Sales* Nintendo released the engaging Wii game *Wii Sports* on November 19, 2006. By December 31, 2009, 60.69 million copies of the game had been sold worldwide, making the game the best-selling Wii game in history. (**Source:** www .nintendo.co.jp)

Model the *Wii Sports* game sales, assuming that at the end of the 1,139th day after the game was released (December 31, 2009), 60.69 million copies of the game had been sold. Furthermore, assume that a total of 80 million copies of the game will be sold over the life of the game. According to the model, how long did it take for 50 million copies to be sold, and at what rate were sales increasing at that time?

In Exercises 15–20, use logistic regression to find the logistic model for the data. Then answer the given questions.

15. *Deadly Fights over Money Due to Arguments over Money or Property*

Years (since 1990) (*t*)	Homicides (*H*)
1	520
2	483
3	445
4	387
5	338
6	328
7	287
8	241
9	213
10	206

Source: *Crime in the United States 1995,* 2000 Uniform Crime Report, FBI

According to the model, how many money-related homicides were there in 2008 and at what rate was the number of homicides changing? (According to the 2008 FBI Uniform Crime Report, there were 192 money-related homicides in 2008.)

16. *Game Console Unit Sales*

Fiscal Years (since Fiscal Year 2003) (*t*)	Nintendo Game Cube Cumulative Sales (million units) (*C*)
0	9.6
1	14.6
2	18.5
3	20.9
4	21.6
5	21.7

According to the model, in what year will Nintendo Game Cube Sales reach 22.0 million units? At what rate will Game Cube sales be changing at that time?

17. *Movie Box Office Sales* The movie *Avatar* debuted on December 18, 2009, and rose to be the best-selling movie of all time, having generated $749,620,438 in box office sales in the United States as of July 8, 2010. The following table shows the cumulative box office sales in millions of dollars at various weeks after the movie's release. (**Source:** www.boxofficemojo.com)

	Avatar
Week #	Cumulative Gross Box Office Sales (millions of dollars)
1	137
2	284
3	381
5	517
8	638
12	724
16	743
20	748
22	749
24	749
25	749

Source: www.boxofficemojo.com

(a) According to the model, were cumulative box office sales for *Avatar* increasing at a higher rate in Week 10 or Week 15?

(b) 20th Century Fox was the distributor for *Avatar*. (**Source:** www.boxofficemojo.com) If you were a marketing consultant to 20th Century Fox, what would you tell the company about forecasted box office sales beyond Week 25?

18. *Resource Value*

Value of Fabricated Metals Shipments	
Years (since 1992) (*t*)	Shipment Value (millions of dollars) (*V*)
0	170,403
1	177,967
2	194,113
3	212,444
4	222,995
5	242,812
6	253,720
7	256,900
8	258,960

Source: *Statistical Abstract of the United States, 2001,* Table 982

(*Hint:* Before creating the model, align the data by subtracting 170,000 from each value of *V*. After doing logistic regression, add back the 170,000 to the resultant model equation.)

According to the model, was the value of fabricated metals shipments increasing more rapidly in 1995 or in 1997?

19. *TV Homes with Cable*

Years (since 1970) (*t*)	TV Homes with Cable (percent) (*C*)
0	6.7
5	12.6
10	19.9
15	42.8
16	45.6
17	47.7
18	49.4
19	52.8
20	56.4
21	58.9

continued

continued

Years (since 1970) (*t*)	TV Homes with Cable (percent) (*C*)
22	60.2
23	61.4
24	62.4
25	63.4
26	65.3
27	66.5
28	67.2
29	67.5

Source: *Statistical Abstract of the United States, 2001,* Table 1126

According to the model, what percentage of TV homes will have cable in 2006, and at what rate will that percentage be increasing?

20. *U.S. Army Personnel*

Years (since 1980) (*t*)	Personnel (thousands) (*P*)
0	777
2	780
4	780
6	781
8	772
10	732
12	610
14	541
16	491
18	484
20	482

Source: *Statistical Abstract of the United States, 2001,* Table 500

(*Hint:* Before creating the model, align the data by subtracting 480 from each value of *P*. After doing logistic regression, add back the 480 to the resultant model equation.)

According to the model, how many people were in the U.S. Army in 2005, and at what rate was that number changing?

Chapter 8

Multivariable
Functions and Partial Derivatives

The amount of moisture in the air affects how hot it feels to a person. If the air temperature is 96°F and the relative humidity is 75%, it feels like it is 132°F! The apparent temperature (what it feels like) is a multivariable function of the actual temperature and the relative humidity. Partial derivatives may be used to determine what effect a 1-degree increase in the actual temperature will have on the apparent temperature if the relative humidity remains constant.

8.1 Multivariable Functions

8.2 Partial Derivatives

8.3 Multivariable Maxima and Minima

8.4 Constrained Maxima and Minima and Applications

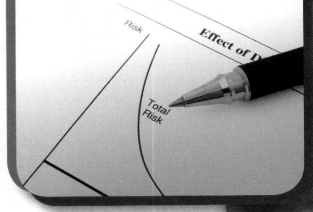

8.1 Multivariable Functions

Businesses seek to manage their labor and capital so as to maximize their production efficiency. Since financial resources are limited, executives have to determine how much money to invest in their employees and how much money to invest in equipment and buildings. Economists Charles Cobb and Paul Douglas formulated the multivariable Cobb-Douglas production function to model the output of small- and large-scale economies.

In this section, we will discuss how to evaluate and graph *multivariable functions*. Additionally, we'll look at several applications of multivariable functions, including the Cobb-Douglas production function.

Functions requiring two or more inputs to generate a single output are called **multivariable functions**. For example, the heat index is a function of the temperature and the humidity. The wind chill factor is a function of temperature and wind speed. The volume of a cylindrical can is a function of its radius and its height. The market share of a business is a function of the market shares of each of its individual competitors. The value of a savings account is a function of the amount invested, the interest rate, the compounding frequency, and the amount of time the money is invested. Functions with two input variables may be represented using a three-dimensional graph.

Function of Two Variables
A **function f of two variables** is a rule that associates each ordered pair of inputs (x, y) with a single output $z = f(x, y)$. The graph of a function of two variables is a three-dimensional **surface**.

calc

In this context, both x and y are **independent variables** and z is the **dependent variable**. The **domain** of the function is the set of all ordered pairs (x, y). The **range** of the function is the set of corresponding values of $f(x, y)$.

Consider the volume of a cylindrical can. The equation for the volume is given by $V = \pi r^2 h$, where r is the radius and h is the height of the can. We may alternatively write $f(h, r) = \pi r^2 h$, where $V = f(h, r)$. What is the volume of a can with a 5-inch height and a 2-inch radius?

$$f(h, r) = \pi r^2 h$$

$$f(5, 2) = \pi(2)^2(5)$$

$$= 20\pi$$

$$\approx 62.83 \text{ cubic inches}$$

The volume of a cylindrical can with a 5-inch height and a 2-inch radius is approximately 62.83 cubic inches.

We may generate a table of values for V by constructing a table with a column for each of the variables (see Table 8.1).

Table 8.1

h	r	$V = \pi r^2 h$
1	1	π
1	2	4π
1	3	9π
2	1	2π
2	2	8π
2	3	18π

Notice that each value of h may be paired with multiple values of r. A more concise way to record the information is to draw a table with the rows containing the values of h and the columns containing the values of r (see Table 8.2). The value of V is contained in the body of the table.

Table 8.2

h ↓	$r \rightarrow$		
	1	2	3
1	π	4π	9π
2	2π	8π	18π

To graph V, we must use a three-dimensional grid with an axis for each of the variables. One of the easiest ways to visualize this is to look at an interior corner of a classroom. The edge between one wall and the floor represents the h-axis. The edge between the other wall and the floor represents the r-axis. The edge joining the two walls represents the V-axis. The origin, $(0, 0, 0)$, is the point of intersection of the three axes (see Figure 8.1).

Figure 8.1

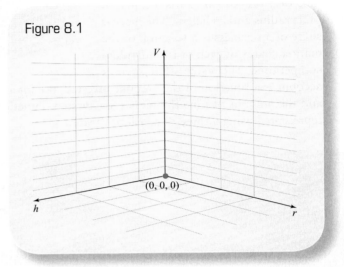

The hV plane is the "wall" formed by the h-axis and the V-axis. The rV plane is the "wall" formed by the r-axis and the V-axis. The hr plane is the "floor" formed by the h-axis and the r-axis. (Using the standard variables x, y, and z, the xz plane and the yz plane form the "walls" and the xy plane is the "floor.")

A point in this coordinate system is an ordered triple of the form (h, r, V). To plot the point $(2, 1, 2\pi)$, we move 2 units from the origin along the h-axis, move 1 unit from the origin in the direction of the r-axis, and finally move 2π units up in the direction of the V-axis, as shown in Figure 8.2.

Figure 8.2

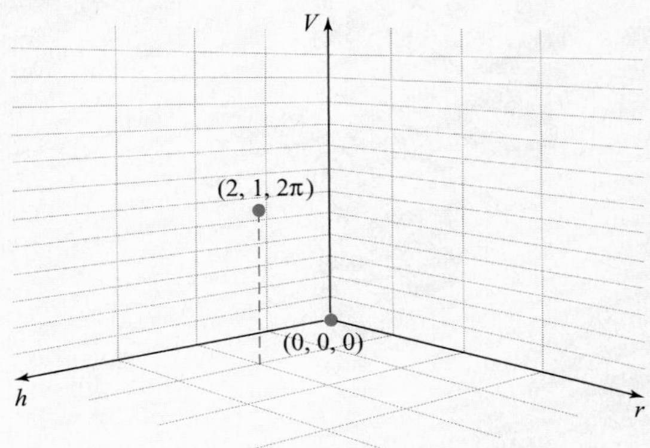

To plot the point $(1, 2, 4\pi)$, we move 1 unit from the origin along the h-axis, move 2 units from the origin in the direction of the r-axis, and finally move 4π units up in the direction of the V-axis, as shown in Figure 8.3.

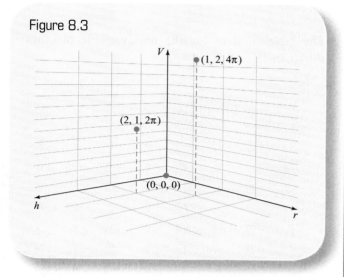

Figure 8.3

(If h, r, or V were negative, we would move in the opposite direction. However, since the equation represents volume as a function of lengths, all variables will be nonnegative.)

Visualizing what the entire surface will look like is often difficult, since we're trying to represent a three-dimensional object on a two-dimensional face of a sheet of paper. However, many technologies and techniques have been developed to make it easier to do so. An Internet search on the keywords "3D Graphing Freeware" will reveal a variety of free graphing utilities that may be used to draw the surface. Additionally, purchased graphing programs such as Maple, Mathematica, and Wolfram Alpha can create beautiful 3D surfaces.

To help you better visualize the surface, the graph of the cylinder volume function is shown from two different angles in Figure 8.4.

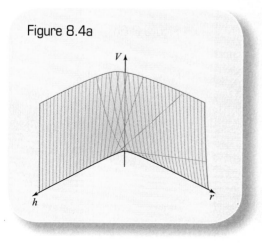

Figure 8.4a

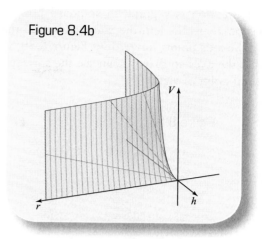

Figure 8.4b

EXAMPLE 1 Graphing a Multivariable Function

Generate a table of data for the multivariable function $f(x, y) = x^2 + y^2$ and use a graphing utility to draw the surface.

SOLUTION

To calculate the values in Table 8.3, we substitute each ordered pair (x, y) into the function f. For example,

$$f(x, y) = x^2 + y^2$$
$$f(-3, -2) = (-3)^2 + (-2)^2$$
$$= 9 + 4$$
$$= 13$$

Table 8.3

x ↓ y →	−3	−2	−1	0	1	2	3
−3	18	13	10	9	10	13	18
−2	13	8	5	4	5	8	13
−1	10	5	2	1	2	5	10
0	9	4	1	0	1	4	9
1	10	5	2	1	2	5	10
2	13	8	5	4	5	8	13
3	18	13	10	9	10	13	18

Notice that the low point of this surface occurs at the origin, $(0, 0, 0)$. As we move away from the origin in any direction, the value of z increases. Two different views of the graph of the surface are shown in Figure 8.5.

Figure 8.5a is a graph in standard position. The x-axis points to the left, the y-axis points to the right, and the z-axis points upward. In Figure 8.5b, we have rotated the axes so that we can see the surface from a different point of view.

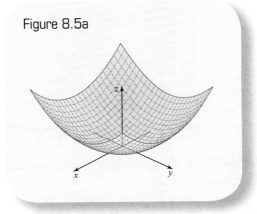

Figure 8.5a

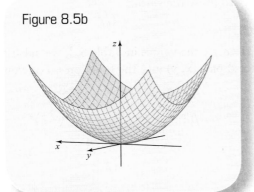

Figure 8.5b

Shutterstock

EXAMPLE 2 Graphing a Multivariable Function

Generate a table of data for the multivariable function $f(x, y) = y^3 - yx^2$ and use a graphing utility to draw the surface.

SOLUTION

The table of data for the multivariable function is shown in Table 8.4.

Table 8.4

x $\downarrow$	$y \rightarrow$						
	−3	−2	−1	0	1	2	3
−3	0	10	8	0	−8	−10	0
−2	−15	0	3	0	−3	0	15
−1	−24	−6	0	0	0	6	24
0	−27	−8	−1	0	1	8	27
1	−24	−6	0	0	0	6	24
2	−15	0	3	0	−3	0	15
3	0	10	8	0	−8	−10	0

The graph of the function on the specified domain is shown in Figure 8.6 in standard position (a) and with the axes rotated (b).

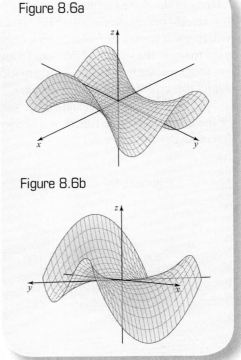

Figure 8.6a

Figure 8.6b

EXAMPLE 3 Using a Multivariable Function to Forecast the Value of an Investment

A $1,000 investment is to be made into an account paying 6% compound interest. The value of the investment compounded n times per year after t years is given by

$$A(n, t) = 1000\left(1 + \frac{0.06}{n}\right)^{nt}$$

Generate a table of data for the multivariable function using $n = 1, 2, 4, 12, 365$ and $t = 0, 1, 2, 3$. Graph the function and interpret the meaning of $(12, 2, 1127.16)$.

SOLUTION

The table of data for the multivariable function is shown in Table 8.5.

Table 8.5

n $\downarrow$	$t \rightarrow$			
	0	1	2	3
1	1000.00	1060.00	1123.60	1191.02
2	1000.00	1060.90	1125.51	1194.05
4	1000.00	1061.36	1126.49	1195.62
12	1000.00	1061.68	1127.16	1196.68
365	1000.00	1061.83	1127.49	1197.20

When we graph the function using a window with $0 \le A \le 1200$, the graph looks like a plane parallel to the nt plane (Figure 8.7).

Figure 8.7

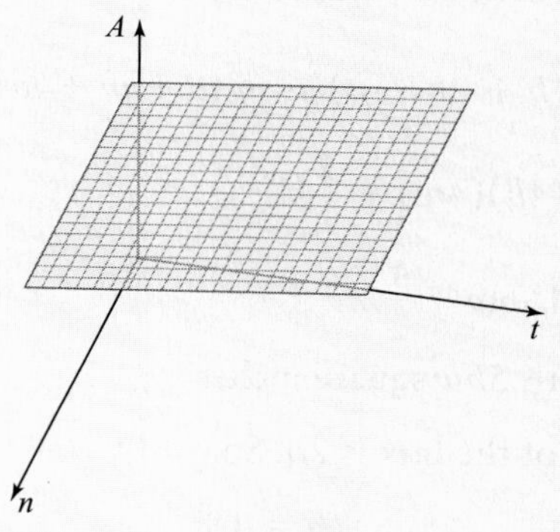

However, when we use the window $1000 \le A \le 1200$, we get a better look at what is happening (Figure 8.8).

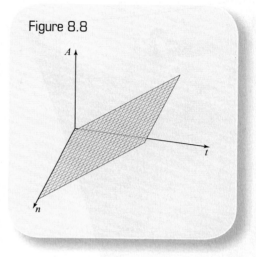

Figure 8.8

The compounding frequency, n, has a greater effect on the value of the investment as t increases. The meaning of $(12, 2, 1127.16)$ is that a $1,000 investment earning 6% interest compounded monthly ($t = 12$) at the end of two years ($n = 2$) will have a value of $1,127.16.

EXAMPLE 4 Using a Cobb-Douglas Production Model

The **Cobb-Douglas production function**,

$$f(L, C) = kL^m C^n$$

is widely used in economics. The model assumes that the output of a company, industry, or country is a function of its labor L and its capital C. (Capital may be interpreted as the dollar value of the money invested in the company, including equipment, material, and building costs.) The model requires that m, n, and k are positive constants and $m + n = 1$.

Based on data from 2008 and 2009, a Cobb-Douglas production model for the Coca-Cola Company is given by

$$P(L, C) = 0.01488L^{0.1135}C^{0.8865}$$

billion unit cases of liquid beverage

where L is the labor cost (including sales, payroll, and other taxes) in millions of dollars and C is the cost of capital expenditures (property, plant, and equipment) in millions of dollars. One unit case of beverage equals 24 eight-ounce servings. (**Source:** Modeled from Coca-Cola Company 2009 Annual Report) According to the model, would it be better to spend $1,900 million on

capital expenditures and $500 million on labor costs or $2,000 million on capital expenditures and $400 million on labor costs? Explain.

SOLUTION

We are asked to calculate $P(500, 1900)$ and $P(400, 2000)$.

$$P(500, 1900) = 0.01488(500)^{0.1135}(1900)^{0.8865}$$

$$= 24.3$$

$$P(400, 2000) = 0.01488(400)^{0.1135}(2000)^{0.8865}$$

$$= 24.8$$

Spending $500 million on labor and $1,900 million on capital expenditures is expected to produce 24.3 billion unit cases of beverage, while spending $400 million on labor and $2,000 million on capital expenditures is expected to produce 24.8 billion unit cases of beverage. The model suggests that it may be better to spend $2,000 million on capital expenditures and $400 million on labor; however, there may be additional constraints to take into consideration. For example, if the Coca-Cola Company were to purchase additional equipment with its capital, it might need to hire additional people to operate the new equipment. Hiring additional people would result in a corresponding increase in the labor cost.

EXAMPLE 5 Using a Multivariable Function to Find the Surface Area of a Box

A box manufacturer designed a closeable box with a square base, as shown in Figure 8.9. Find the surface area equation for the unassembled box. How many square inches of material are required to construct a box 6 inches wide, 6 inches long, and 18 inches high?

Figure 8.9

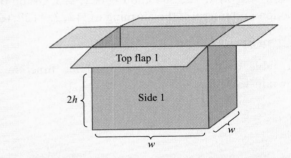

SOLUTION

The surface area of the unassembled box is given by

$$A(h, w) = (h + 2h + h)(w + w + w + w) + \frac{1}{4}w(2h)$$

$$= (4h)(4w) + \frac{1}{2}hw$$

$$= 16hw + \frac{1}{2}hw$$

$$= 16.5hw \text{ square inches}$$

The height of the box is $2h$. So

$$2h = 18$$

$$h = 9$$

The width and length of the box are both 6 inches, so $w = 6$.

$$A(h, w) = 16.5(9)(6)$$
$$= 891$$

The surface area of the unassembled box is 891 square inches.

....

8.1 Exercises

In Exercises 1–10, generate a table of data for the multivariable function using $x = -2, 0, 2$ and $y = -1, 0, 1$. Then use a graphing utility to graph the function. Compare the coordinates of the points in the table to the coordinates of the points on the graph of the function.

1. $z = 3x - 2y$

2. $z = xy - y^2$

3. $z = 4xy + 4$

4. $(x, y) = x^2 + 2xy + y^2$

5. $f(x, y) = e^{-xy}$

6. $f(x, y) = x^2 - 2xy + y^2$

7. $f(x, y) = \dfrac{x^2 + 1}{y^2 + 1}$

8. $f(x, y) = \dfrac{xy}{y^2 + 1}$

9. $z = (x + y)^3$

10. $z = \ln(x^2 y^2 + 1)$

In Exercises 11–20, use the multivariable function to answer the given question.

11. *Cobb-Douglas Function* Based on data from 1999 and 2000, a Cobb-Douglas model for Ford Motor Company is given by

    $$P(L, C) = 0.5319L^{0.4558}C^{0.5442} \text{ thousand vehicles}$$

    where L is the total labor cost and C is the capital expenditure cost, both in millions of dollars. In 2000, Ford Motor Company spent $25,783 million on labor and produced 7,424 thousand vehicles. (**Source:** Modeled from Ford Motor Company 2001 Annual Report)
    According to the model, if Ford spends $20,000 million on labor and $9,000 million on capital expenditures, how many vehicles will it produce?

12. *Housing Costs* The U.S. Department of Housing and Urban Development (HUD) published a research study in January 2002 comparing the cost of building a steel-frame home to the cost of building a wood-frame home. Two homes with identical floor plans were built side by side in Beaufort, South Carolina.
    The material and labor costs of constructing each home are given in the table.

	Material Cost	Labor Cost
Steel Frame	$97,447.52	$32,228.94
Wood Frame	$91,589.38	$25,236.83

Source: HUD

Find an equation for the total cost of constructing w identical wood-frame homes and s identical steel-frame homes in Beaufort, South Carolina. How much will it cost to construct 10 wood-frame and 5 steel-frame homes?

13. *iPad Apps* Chillingo Ltd. is a developer of games for the Apple iPad. As of July 2010, the company offered five paid products including *Angry Birds* HD ($4.99), *Parking Mania* HD ($2.99), *Minigore* HD ($4.99), *Cogs* HD ($4.99), and *Robert Rodriguez presents PREDATORS* ($2.99). (**Source:** Apple App Store)
    Find the formula for the Chillingo game revenue function. How much revenue will Chillingo generate if it sells 20 copies of *Angry Birds*, 50 copies of *Parking Mania*, 30 copies of *Minigore*, 10 copies of *Cogs*, and 15 copies of *PREDATORS*?

14. *Future Value of Investment* Find the equation for the future value of an investment of P dollars into an account paying 6% interest compounded monthly t years after the investment is made. What is the value of a $2,500 investment after 10 years?

15. *Surface Area* Referring to Example 5 in the section, find a formula for the exterior surface area of the assembled box with a square base. (Assume that the top and bottom of the box are closed.) What is the exterior surface area of the box with a 6-inch width, 6-inch length, and 18-inch height?

16. *Area Analysis* In Example 5 we calculated the surface area of an unassembled box, and in Exercise 15 you calculated the exterior surface

area of the assembled box. Of what practical value are each of these results?

17. *Body Mass Index* Body mass index (BMI) is a helpful indicator of obesity or underweight in adults. A high BMI is predictive of death from cardiovascular disease. The Centers for Disease Control and Prevention indicate that a healthy BMI for adults is between 18.5 and 24.9. They use the following guidelines.

Underweight	BMI less than 18.5
Overweight	BMI of 25.0 to 29.9
Obese	BMI of 30.0 or more

The following formula is used to calculate body mass index.

$$BMI = \frac{703 \cdot W}{H^2}$$

where W is your weight in pounds and H is your height in inches. What is your BMI?

18. *Relative Humidity* The relative humidity of the air alters our perception of the temperature. As the humidity increases, the air feels hotter to us. The heat index (apparent temperature) is a function of air temperature and relative humidity, as shown in the table.

When exposed to apparent temperatures exceeding 105°F for prolonged periods of time, people are at high risk of sunstroke, a potentially fatal condition. The heat index values were devised for shady, light wind conditions. Exposure to full sunshine can increase apparent temperature by up to 15°F. What is the apparent temperature when the air temperature is 90°F and the relative humidity is 80%?

19. *Wind Chill* Wind alters our perception of the temperature. As the wind strength increases, the air feels colder to us. This phenomenon is referred to as *wind chill*. Wind chill temperature is a function of wind speed and air temperature and is modeled by

$$W(T, V) = 35.74 + 0.6215T$$
$$- 35.75V^{0.16} + 0.4275TV^{0.16}$$
$$\text{degrees Fahrenheit}$$

where T is the actual air temperature (in degrees Fahrenheit) and V is the velocity of the wind (in miles per hour). (**Source:** National Climatic Data Center) When the air is 10°F and the wind is blowing at 40 miles per hour, what is the wind chill temperature?

20. *Ice Cream Cones* The GE-500 Semi-Automatic is a machine that is designed to create a variety of

	Relative Humidity (%)												
	40	45	50	55	60	65	70	75	80	85	90	95	100
110	136												
108	130	137											
106	124	130	137										
104	119	124	131	137									
102	114	119	124	130	137								
100	109	114	118	124	129	136							
98	105	109	113	117	123	128	134						
96	101	104	108	112	116	121	126	132					
94	97	100	103	106	110	114	119	124	129	135			
92	94	96	99	101	105	108	112	116	121	126	131		
90	91	93	95	97	100	103	106	109	113	117	122	127	132
88	88	89	91	93	95	98	100	103	106	110	113	117	121
86	85	87	88	89	91	93	95	97	100	102	105	108	112
84	83	84	85	86	88	89	90	92	94	96	98	100	103
82	81	82	83	84	84	85	86	88	89	90	91	93	95
80	80	80	81	81	82	82	83	84	84	85	86	86	87

Air Temperature (°F)

Heat Index (Apparent Temperature)

Source: National Weather Service

sizes of ice cream cones, ranging from a diameter of 20 millimeters (mm) to a diameter of 56 mm. The number of cones that can be made in an hour depends upon the diameter of the cone. The production rate of 20-mm cones is 4,400 per hour, while the production rate of 42-mm cones is 2,000 per hour. (**Source:** www.maneklalexports.com)

Let s be the number of hours spent producing 20-mm cones and l be the number of hours spent producing 42-mm cones. Find a formula for the cumulative number of cones produced as a function of s and l. How many cones are produced if 1.5 hours are spent making 20-mm cones and 3 hours are spent making 42-mm cones?

8.2   Partial Derivatives

The wind chill temperature is a function of the air temperature and the wind velocity. If the wind speed is increasing at a rate of 2 mph per hour, how quickly is the wind chill temperature changing? Questions such as these may be answered by using *partial derivatives*.

In this section, we will demonstrate how to calculate partial derivatives. These partial rates of change allow us to determine the effect of a change in one of the input variables on the output.

Recall that the operator $\dfrac{d}{dx}$ means "take the derivative with respect to x." The operator $\dfrac{\partial}{\partial x}$ means "take the partial derivative with respect to x." When calculating a partial derivative, all variables except the variable of differentiation are treated as constants. Consider the function $z = x^2 - xy$.

$$\frac{\partial}{\partial x}(z) = \frac{\partial}{\partial x}(x^2 - xy)$$

$$\frac{\partial z}{\partial x} = 2x - y$$

Observe that we treated the y in xy as if it were a constant. That is, when differentiating $x^2 - xy$, we treated the expression as if it were $x^2 - cx$ (where c is a constant). Since $\dfrac{d}{dx}(x^2 - cx) = 2x - c$, $\dfrac{\partial z}{\partial x} = 2x - y$. Let's now take the partial derivative of the function with respect to y.

$$\frac{\partial}{\partial y}(z) = \frac{\partial}{\partial y}(x^2 - xy)$$

$$\frac{\partial z}{\partial y} = -x$$

Since the variable of differentiation is y, we treated x as a constant. That is, when differentiating $x^2 - xy$, we treated the expression as if it were $c^2 - cy$ (where c is a constant). Since $\dfrac{d}{dy}(c^2 - cy) = -c$, $\dfrac{\partial}{\partial y}(x^2 - xy) = -x$. Observe that $\dfrac{\partial z}{\partial x} \neq \dfrac{\partial z}{\partial y}$. The partial derivative of a function varies based on the variable of differentiation.

Partial Derivatives

The **partial derivatives** of a function $f(x, y)$ are given by

$$\frac{\partial f}{\partial x}$$

(read "the partial of f with respect to x")

$$\frac{\partial f}{\partial y}$$

(read "the partial of f with respect to y")

To calculate $\dfrac{\partial f}{\partial x}$, differentiate the function, treating the y variable as a constant.

To calculate $\dfrac{\partial f}{\partial y}$, differentiate the function, treating the x variable as a constant.

EXAMPLE 1 Finding Partial Derivatives

Find the partial derivatives of the function $f(x, y) = 3xy + y^2 + 5$.

SOLUTION

$$\frac{\partial f}{\partial x} = \frac{\partial}{\partial x}(3xy + y^2 + 5)$$

$$= \frac{d}{dx}(3cx + c^2 + 5) \quad \text{c is a constant representing } y$$

$$= 3c$$

$$= 3y \quad\quad\quad \text{Replace c with } y$$

Replacing the variable y with the constant c is an optional step used to help you apply the partial differentiation rules correctly. As you become skilled at applying the rules, the step will become unnecessary.

$$\frac{\partial f}{\partial y} = \frac{\partial}{\partial y}(3xy + y^2 + 5)$$

$$= \frac{d}{dy}(3cy + y^2 + 5) \quad \text{c is a constant representing } x$$

$$= 3c + 2y$$

$$= 3x + 2y \quad\quad \text{Replace c with } x$$

**Partial Derivatives
(Alternative Notation)**
The **partial derivatives** of a function $f(x, y)$ may alternatively be written as

$$f_x$$

(read "the partial of f with respect to x")

$$f_y$$

(read "the partial of f with respect to y")

EXAMPLE 2 Finding Partial Derivatives

Find the partial derivatives of $f(x, y) = x^2y^2$.

SOLUTION

$$f_x = \frac{\partial}{\partial x}(x^2y^2)$$

$$= \frac{d}{dx}(x^2c^2) \quad \text{c is a constant representing } y$$

$$= 2xc^2$$

$$= 2xy^2 \quad\quad \text{Replace c with } y$$

$$f_y = \frac{\partial}{\partial y}(x^2y^2)$$

$$= \frac{d}{dy}(c^2y^2) \quad \text{c is a constant representing } x$$

$$= 2c^2y$$

$$= 2x^2y \quad\quad \text{Replace c with } x$$

The units of $\frac{\partial f}{\partial x}$ are the units of f divided by the units of x. Similarly, the units of $\frac{\partial f}{\partial y}$ are the units of f divided by the units of y.

EXAMPLE 3 Interpreting the Meaning of the Partial Derivatives of the Compound Interest Function

Find the partial derivatives of the compound interest function $A(r, t) = 1000\left(1 + \frac{r}{4}\right)^{4t}$. Then evaluate the partial derivatives at the point $(0.12, 3, 1425.76)$ and interpret the meaning of the result.

SOLUTION

We will first find the partial derivatives.

$$\frac{\partial}{\partial t}(A) = \frac{\partial}{\partial t}\left(1000\left(1 + \frac{r}{4}\right)^{4t}\right)$$

$$\frac{\partial A}{\partial t} = 1000 \cdot \ln\left(1 + \frac{r}{4}\right) \cdot \left(1 + \frac{r}{4}\right)^{4t} \cdot 4$$

Exponential, Constant Multiple, and Chain Rules

$$A_t = 4000 \ln\left(1 + \frac{r}{4}\right) \cdot \left(1 + \frac{r}{4}\right)^{4t}$$

$$\frac{\partial A}{\partial r} = 1000 \cdot 4t\left(1 + \frac{r}{4}\right)^{4t-1} \cdot \left(\frac{1}{4}\right)$$

Power, Constant Multiple, and Chain Rules

$$A_r = 1000t\left(1 + \frac{r}{4}\right)^{4t-1}$$

Next, we will evaluate each of the partial derivatives at the point $(0.12, 3, 1425.76)$.

$$A_t(0.12, 3) = 4000 \ln\left(1 + \frac{0.12}{4}\right) \cdot \left(1 + \frac{0.12}{4}\right)^{4(3)}$$

$$= 4000 \cdot \ln(1.03)(1.03)^{12}$$

$$= 168.58$$

The units of the partial derivative are the units of the output (dollars) divided by the units of t (years). An investment account earning 12% interest (compounded quarterly) with an initial investment of $1,000 is increasing in value at a rate of $168.58 per year at the end of the third year. That is, from the end of the third year to the end of the fourth year, the account value will increase by about $168.58.

$$A_r = 1000t\left(1 + \frac{r}{4}\right)^{4t-1}$$

$$A_r(0.12, 3) = 1000(3)\left(1 + \frac{0.12}{4}\right)^{4(3)-1}$$

$$= 3000(1.03)^{11}$$

$$= 4152.70$$

The units of the partial derivative are the units of the output (dollars) divided by the units of r (100 percentage points). (Observe that $0.12 \cdot 100$ percentage points $= 12\%$.)

$$4152.70\,\frac{\text{dollars}}{100 \text{ percentage points}}$$

$$= 41.527 \text{ dollars per percentage point}$$

An investment account earning 12% interest per year (compounded quarterly) with an initial investment of $1,000 is increasing in value at a rate of $41.53 per percentage point increase in the interest rate at the end of the third year. That is, increasing the interest rate from 12% per year to 13% per year at the end of the third year will increase the account value at the end of the fourth year by approximately $41.53 over what it would have been without the interest rate increase.

EXAMPLE 4 Interpreting the Meaning of Partial Derivatives

The distance of a cue ball from the eight ball in the game of billiards may be expressed in terms of two perpendicular components parallel to the side or end of the pool table, respectively, as shown in Figure 8.10.

By the Pythagorean Theorem, $d = \sqrt{x^2 + y^2}$. Find the partial derivative of the distance function. Then evaluate each partial derivative at the point $(3, 4, 5)$ and interpret the meaning of the result.

SOLUTION

We will first determine the equations of the partial derivatives.

$$d = \sqrt{x^2 + y^2}$$

$$= (x^2 + y^2)^{1/2}$$

Figure 8.10

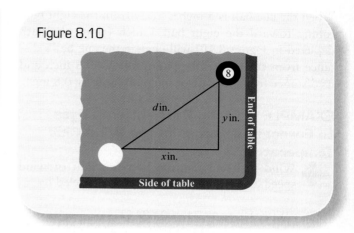

$$d_x = \frac{1}{2}(x^2 + y^2)^{-1/2}(2x)$$

Chain Rule treating y as a constant

$$= x(x^2 + y^2)^{-1/2}$$

$$d_y = \frac{1}{2}(x^2 + y^2)^{-1/2}(2y)$$

Chain Rule treating x as a constant

$$= y(x^2 + y^2)^{-1/2}$$

Evaluating these functions at $(3, 4, 5)$, we determine

$$d_x(3, 4) = (3)(3^2 + 4^2)^{-1/2}$$

$$= \frac{3}{5}$$

$$= 0.6$$

$$d_y(3, 4) = (4)(3^2 + 4^2)^{-1/2}$$

$$= \frac{4}{5}$$

$$= 0.8$$

When the cue ball is 5 inches away from the eight ball, rolling toward the eight ball 1 inch (along the path depicted in Figure 8.10) will change the cue ball's distance from the end of the pool table by 0.6 inch and its distance from the side of the pool table by 0.8 inch.

EXAMPLE 5 Using Partial Derivatives to Forecast Changes in the Wind Chill Temperature

Wind chill temperature is a function of wind speed and air temperature and is modeled by

$$W(T, V) = 35.74 + 0.6215T$$
$$- 35.75V^{0.16} + 0.4275TV^{0.16}$$

degrees Fahrenheit

where T is the actual air temperature (in degrees Fahrenheit) and V is the velocity of the wind (in miles per hour). (**Source:** National Climatic Data Center) If the wind speed is 20 miles per hour, how much will a 1-mile-per-hour change in wind speed alter the wind chill temperature if the current temperature remains at 30 degrees? If the current temperature is 30 degrees, how much will a 1-degree increase in temperature alter the wind chill temperature if the current wind speed remains at 20 miles per hour?

SOLUTION

We begin by calculating the partial derivatives of W.

$$W(T, V) = 35.74 + 0.6215T$$
$$- 35.75V^{0.16} + 0.4275TV^{0.16}$$

$$\frac{\partial W}{\partial T} = 0.6215 + 0.4275V^{0.16}$$

$$\frac{\text{degree of wind chill temperature}}{\text{degree of actual temperature}}$$

$$\frac{\partial W}{\partial V} = 0.16(-35.75)V^{-0.84} + 0.16(0.4275TV^{-0.84})$$

$$= -5.72V^{-0.84} + 0.0684TV^{-0.84}$$

$$= 0.0684V^{-0.84}(-83.63 + T)$$

$$\frac{\text{degree of wind chill temperature}}{\text{miles per hour}}$$

We will now evaluate each of the partial derivatives at (30, 20).

$$\frac{\partial W}{\partial T}(30, 20) = 0.6215 + 0.4275(20)^{0.16}$$

$$= 1.312 \frac{\text{degree of wind chill temperature}}{\text{degree of actual temperature}}$$

Increasing the temperature from 30 to 31 degrees will increase the wind chill temperature by about 1.3 degrees if the wind remains at 20 miles per hour. (Since the partial derivative does not contain the variable T, the wind chill temperature will increase by about 1.3 degrees for every 1 degree increase in temperature regardless of the initial temperature.)

$$\frac{\partial W}{\partial V}(30, 20) = 0.0684V^{-0.84}(-83.63 + T)$$

$$= 0.0684(20)^{-0.84}(-83.63 + 30)$$

$$= -0.2962 \frac{\text{degree of wind chill temperature}}{\text{miles per hour}}$$

Increasing the wind speed from 20 to 21 miles per hour will decrease the wind chill temperature by about 0.3 degrees when the temperature is 30 degrees. The colder the temperature is, the more substantial the impact a change in wind speed will have on the wind chill temperature.

Cross Sections of a Surface

When an apple is sliced in half, the exposed surface is called a **cross section** of the apple. If the cross section (Figure 8.11a) is dipped in paint and pressed on a sheet of paper, the resultant image is a two-dimensional figure (see Figure 8.11b).

Figure 8.11a   Figure 8.11b

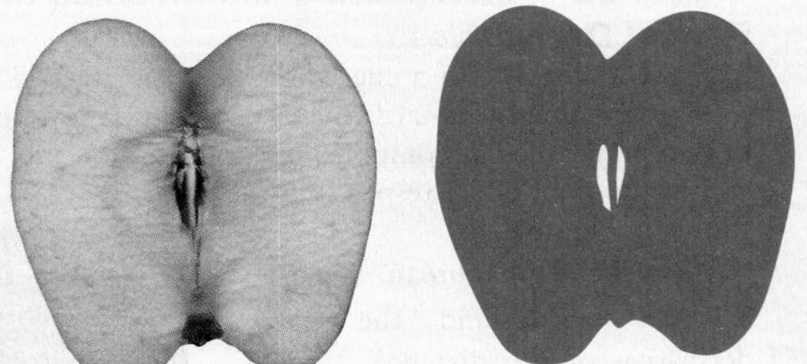

The shape of the cross section varies depending on the point at which the apple is sliced. Figure 8.12 shows three different cross sections of the same apple.

The equation for the border of the cross section can be derived from the original multivariable equation. (The shape of an apple may be modeled by a multivari-

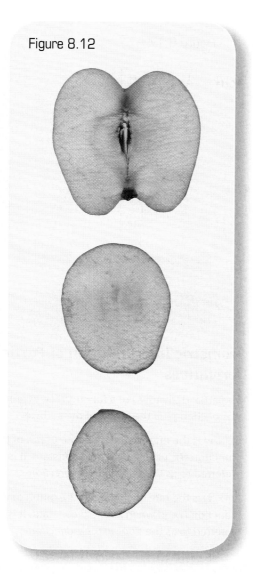

Figure 8.12

If we want to know what the cross section of the surface looks like at $y = 0$, we calculate $f(x, 0)$. The equation of the cross-section graph is given by

$$f(x, 0) = -x^3 + 4(0)^2$$
$$= -x^3$$

Slicing the surface at $y = 0$ yields the graph shown in Figure 8.14.

Figure 8.14

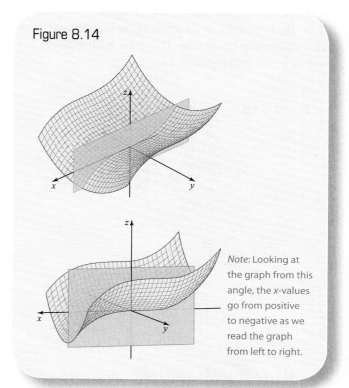

Note: Looking at the graph from this angle, the *x*-values go from positive to negative as we read the graph from left to right.

able piecewise equation. However, because of the complexity of the equation, we will focus on simpler models as we analyze cross sections.)

Consider the graph of the multivariable function $f(x, y) = -x^3 + 4y^2$, shown in Figure 8.13.

Graphing the cross section in two dimensions yields the curve $z = -x^3$ (Figure 8.15).

Figure 8.13

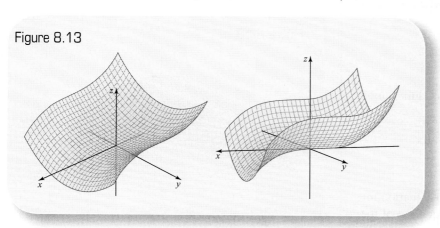

Figure 8.15

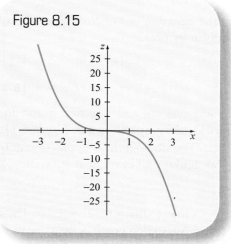

The slope of the cross-section graph is $\dfrac{dz}{dx} = -3x^2$.
What is the relationship between the slope of the cross-section graph and the partial derivative? Let's find f_x and evaluate it at $(x, 0)$.

$$f_x = -3x^2$$

Observe that $f_x(x, 0) = -3x^2$. The slope equation for the cross-section graph is the same as the equation of the partial derivative of f with respect to x evaluated at $(x, 0)$. Let's look at another cross section and see if the relationship holds true. This time we'll slice the surface at $y = 2$ (see Figure 8.16).

Figure 8.16

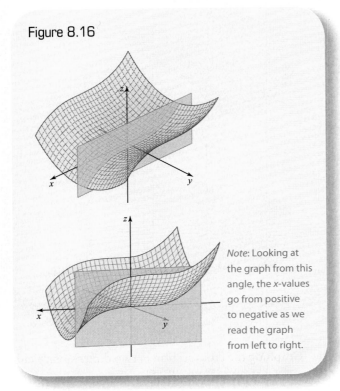

Note: Looking at the graph from this angle, the x-values go from positive to negative as we read the graph from left to right.

The equation of the cross-section graph when $y = 2$ is given by $f(x, 2)$.

$$f(x, 2) = -x^3 + 4(2)^2$$

$$= -x^3 + 16$$

Graphing the cross-section graph in two dimensions yields the graph shown in Figure 8.17.

The slope of the cross-section graph is $\dfrac{dz}{dx} = -3x^2$. We find f_x and evaluate it at $(x, 2)$.

$$f_x = -3x^2$$

Observe that $f_x(x, 2) = -3x^2$. The slope equation for the cross-section graph is the same as the equation of the partial derivative of f with respect to x evaluated at $(x, 2)$.

Figure 8.17

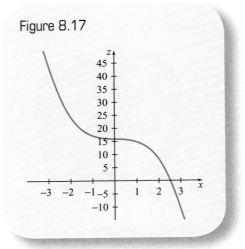

Geometric Interpretation of Partial Derivatives

The partial derivatives of a function $f(x, y)$ represent rates of change of the graph of the surface.

- $f_x(x, y)$ is the rate of change of the graph of the function $f(x, y)$ when y is held constant. It is referred to as the *slope in the x-direction*.

- $f_y(x, y)$ is the rate of change of the graph of the function $f(x, y)$ when x is held constant. It is referred to as the *slope in the y-direction*.

Consider the graph of the multivariable function $f(x, y) = -x^2 + 4x - y^2 + y + xy + 100$ (see Figure 8.18). The point $(3, 1, 106)$ lies on the graph.

Figure 8.18

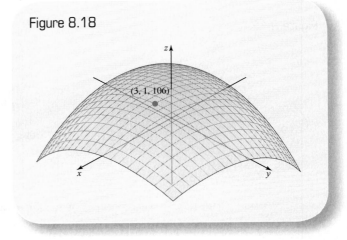

Let's calculate the partial derivatives.

$$f_x(x, y) = -2x + 4 + y$$

$$f_y(x, y) = -2y + 1 + x$$

Recall that $f_x(x, y)$ is the slope in the x-direction. Thus, $f_x(3, 1)$ is the slope of the graph in the x-direction at the point $(3, 1, 106)$.

$$f_x(x, y) = -2x + 4 + y$$

$$f_x(3, 1) = -2(3) + 4 + 1$$

$$= -1$$

There is a line tangent to the graph at $(3, 1, 106)$ that has slope -1. The line is of the form $z = -1x + b$, since the line is in the x-direction. Substituting in $x = 3$ and $z = 106$, we can find the exact equation of the line.

$$z = -1x + b$$

$$106 = -3 + b$$

$$b = 109$$

$$\text{So } z = -x + 109$$

The equation of the tangent line to the graph of f in the x-direction at the point $(3, 1, 106)$ is $z = -x + 109$. We draw the line on the graph of f (see Figure 8.19).

Figure 8.19

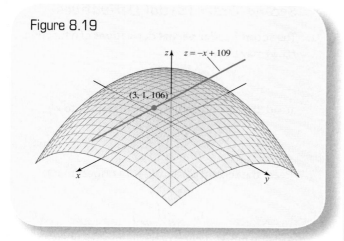

Recall that $f_y(x, y)$ is the slope in the y-direction. Thus, $f_y(3, 1)$ is the slope of the graph in the y-direction at the point $(3, 1, 106)$.

$$f_y(x, y) = -2y + 1 + x$$

$$f_y(3, 1) = -2(1) + 1 + 3$$

$$= 2$$

There is a line tangent to the graph at $(3, 1, 106)$ that has slope 2. The line is of the form $z = 2y + b$, since the line

is in the y-direction. Substituting in $y = 1$ and $z = 106$, we can find the exact equation of the line.

$$z = 2y + b$$

$$106 = 2(1) + b$$

$$106 = 2 + b$$

$$b = 104$$

$$\text{So } z = 2y + 104$$

So the equation of the tangent line to the graph of f in the y-direction at the point $(3, 1, 106)$ is $z = 2y + 104$. We draw the line on the graph of f (see Figure 8.20).

Figure 8.20

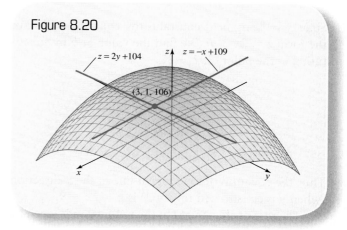

It is important to note that in three dimensions, the graph of the equation $z = 2y + 104$ is really the *plane* containing the green line that we have drawn, not the line itself. However, by requiring $x = 3$, we constrain the plane to the green line. Similarly, the graph of $z = -x + 109$ is the graph of the *plane* containing the blue line we have drawn, not the line itself. However, by requiring $y = 1$, we constrain the plane to the blue line. Without these additional constraints on x and y, respectively, the graph would look like Figure 8.21.

Figure 8.21

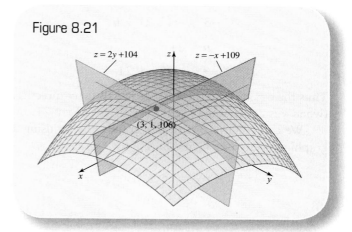

EXAMPLE 6 Finding Tangent Lines of a Multivariable Function

➡️ Given the function $f(x, y) = x^2 - xy^2$, determine the equations of the tangent lines in the x-direction and in the y-direction at the point $(3, 2, -3)$.

SOLUTION

The rate of change of f in the x-direction is given by

$$f_x(x, y) = 2x - y^2$$

Thus the slope of the tangent line in the x-direction at $(3, 2, -3)$ is

$$f_x(3, 2) = 2(3) - (2)^2$$
$$= 2$$

Since y is being held constant, the equation will have the form $z = 2x + b$. We find the value of b by substituting in the point $(3, 2, -3)$.

$$z = 2x + b$$
$$-3 = 2(3) + b$$
$$b = -9$$
$$\text{So } z = 2x - 9$$

Thus the equation of the tangent line in the x-direction (when y is constrained to $y = 2$) is $z = 2x - 9$.

The rate of change of f in the y-direction is given by

$$f_y(x, y) = -2xy$$

Thus the slope of the tangent line in the y-direction at $(3, 2, -3)$ is

$$f_y(3, 2) = -2(3)(2)$$
$$= -12$$

Since x is being held constant, the equation will have the form $z = -12y + b$. We find the value of b by substituting in the point $(3, 2, -3)$.

$$z = -12y + b$$
$$-3 = -12(2) + b$$
$$b = 21$$
$$\text{So } z = -12y + 21$$

Thus the equation of the tangent line in the y-direction (when x is constrained to $x = 3$) is $z = -12y + 21$.

We can confirm our results graphically by using a graphing utility (see Figure 8.22).

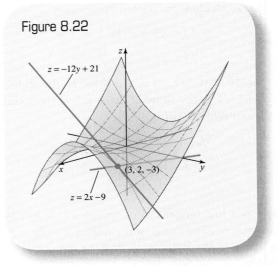

Figure 8.22

Second-Order Partial Derivatives

Recall that the derivative of a derivative of a function is called the second derivative of the function. Likewise, a partial derivative of a partial derivative of a function is called a **second-order partial derivative**.

Second-Order Partial Derivatives

The **second-order partial derivatives** of a function $f(x, y)$ may be written as

$$f_{xx} = \frac{\partial}{\partial x}\left(\frac{\partial f}{\partial x}\right)$$

(read "the second partial of f with respect to x")

$$f_{yy} = \frac{\partial}{\partial y}\left(\frac{\partial f}{\partial y}\right)$$

(read "the second partial of f with respect to y")

$$f_{xy} = \frac{\partial}{\partial y}\left(\frac{\partial f}{\partial x}\right)$$

(read "the mixed second partial derivative of f with respect to x then y")

$$f_{yx} = \frac{\partial}{\partial x}\left(\frac{\partial f}{\partial y}\right)$$

(read "the mixed second partial derivative of f with respect to y then x")

For all functions with continuous first and second partial derivatives, $f_{xy} = f_{yx}$. For all multivariable functions in this text, $f_{xy} = f_{yx}$.

EXAMPLE 7 Finding First and Second Partial Derivatives

The graph of the function $f(x, y) = x^2 + y^2 + 1$ is shown in Figure 8.23. Calculate the value of the first and second partial derivatives of the function at the point $(0, 0, 1)$.

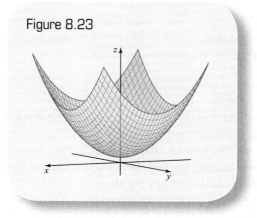

Figure 8.23

SOLUTION
We have

$$f_x = 2x$$
$$f_y = 2y$$
$$f_{xx} = 2$$
$$f_{yy} = 2$$
$$f_{xy} = 0$$
$$f_{yx} = 0$$

Evaluating each of the partial derivatives at $(0, 0, 1)$, we get

$$f_x(0, 0) = 2(0)$$
$$= 0$$
$$f_y(0, 0) = 2(0)$$
$$= 0$$

The second partial derivatives remain constant for all values of x and y. Thus

$$f_{xx}(0, 0) = 2$$
$$f_{yy}(0, 0) = 2$$

$$f_{xy}(0, 0) = 0$$
$$f_{yx}(0, 0) = 0$$

EXAMPLE 8 Finding First and Second Partial Derivatives

The graph of the function $f(x, y) = x^3 - xy^2$ is shown in Figure 8.24. Calculate the first and second partial derivatives of the function.

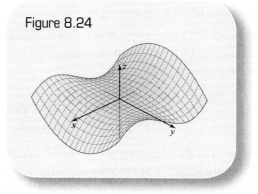

Figure 8.24

SOLUTION

$$f_x = 3x^2 - y^2$$
$$f_y = -2xy$$
$$f_{xx} = 6x$$
$$f_{yy} = -2x$$
$$f_{xy} = -2y$$
$$f_{yx} = -2y$$

In the next section, we will demonstrate the key role that second-order partial derivatives play in identifying the location of the maxima and minima of a surface.

8.2 Exercises

In Exercises 1–5, calculate the first and second partial derivatives of each function.

1. $f(x, y) = 3xy$

2. $f(x, y) = 4xy - y^2$

3. $W = t\sqrt{v}$

4. $V = \pi r^2 h$

5. $z = x \ln(xy)$

In Exercises 6–12, calculate and interpret the practical meaning of each partial derivative.

6. *Volume of a Cylinder* The volume of a cylinder is given by $V(h, r) = \pi r^2 h$, with h and r measured in inches. What is the practical meaning of $V_r(3, 2)$ and $V_h(3, 2)$?

7. *Volume of a Box* The volume of a rectangular box with a square base is given by $V(l, w) = lw^2$, with l and w measured in inches. What is the practical meaning of $V_l(4, 5)$ and $V_w(4, 5)$?

8. *Wind Chill Temperature* Wind chill temperature is a function of wind speed and air temperature and is modeled by

$$W(T, V) = 35.74 + 0.6215T$$
$$-35.75V^{0.16} + 0.4275TV^{0.16}$$
$$\text{degrees Fahrenheit}$$

where T is the actual air temperature (in degrees Fahrenheit) and V is the velocity of the wind (in miles per hour). (**Source:** National Climatic Data Center) What is the practical meaning of $W_T(27, 5)$ and $W_V(27, 5)$?

9. *Wind Chill Temperature* Wind chill temperature is a function of wind speed and air temperature and is modeled by

$$W(T, V) = 35.74 + 0.6215T$$
$$-35.75V^{0.16} + 0.4275TV^{0.16}$$
$$\text{degrees Fahrenheit}$$

where T is the actual air temperature (in degrees Fahrenheit) and V is the velocity of the wind (in miles per hour). (**Source:** National Climatic Data Center) What is the practical meaning of $W_T(-30, 15)$ and $W_V(-30, 15)$?

10. *Body Mass Index* The following formula is used to calculate body mass index:

$$B(H, W) = \frac{703 \cdot W}{H^2}$$

where W is weight in pounds and H is height in inches. What is the practical meaning of $B_H(67, 155)$ and $B_W(67, 155)$?

11. *Body Mass Index* The following formula is used to calculate body mass index:

$$B(H, W) = \frac{703 \cdot W}{H^2}$$

where W is weight in pounds and H is height in inches. What is the practical meaning of $B_H(67, 100)$ and $B_W(67, 100)$?

12. *Price-to-Earnings Ratio* The price-to-earnings ratio (P/E) is the most common measure of how expensive a stock is, and it is given by

$$R(E, P) = \frac{P}{E}$$

where P is the price of a share of the stock and E is the earnings per share. What is the practical meaning of $R_E(4, 32)$ and $R_P(4, 32)$?

In Exercises 13–15, use partial derivatives to determine the answer to each question.

13. *Price-to-Earnings Ratio* The price-to-earnings ratio (P/E) is the most common measure of how expensive a stock is, and it is given by

$$R(E, P) = \frac{P}{E}$$

where P is the price of a share of the stock and E is the earnings per share.

On July 13, 2010, PepsiCo, Inc., stock was selling for $63.45 per share and had a published earnings per share of $3.94. (**Source:** www.moneycentral.com) If the stock price increased by $1 or the earnings per share increased by $1, what effect would this have on the P/E ratio?

14. *Company Revenue* Based on data from 2006 to 2009, the revenue of PepsiCo may be modeled by

$$R(d, a) = 1.060d + 0.3055a + 23.70 \text{ billion dollars}$$

where a is the assets of the company (billions of dollars) and d is the amount of long-term debt (billions of dollars). In 2009, the company had $7.4 billion worth of long-term debt and $39.8 billion worth of assets. (**Source:** Modeled from PepsiCo Company 2009 Annual Report)

According to the model, what effect would adding $1 billion of assets in 2009 have on revenue? What effect would adding $1 billion of long-term debt have on revenue?

15. *Cobb-Douglas Model* Based on data from 1999 and 2000, a Cobb-Douglas model for Ford Motor Company is given by

$$P(L, C) = 0.5319L^{0.4558}C^{0.5442} \text{ thousand vehicles}$$

where L is the total labor cost and C is the capital expenditure cost, both in millions of dollars. In 2000, Ford Motor Company spent $25,783 million on labor and produced 7,424 thousand vehicles. (**Source:** Modeled from Ford Motor Company 2001 Annual Report)

According to the model, if Ford increased labor spending by $1 million or increased capital expenditures by $1 million in 2001, what effect would each change have on the production of vehicles?

In Exercises 16–20, a function graph with a cross section drawn at $x = 0$ and $y = -1$ is shown. Determine the equations of the tangent lines of the function in the x-direction and the y-direction at the point $(0, -1, f(0, -1))$.

16. $f(x, y) = x^4 - 4y^2$

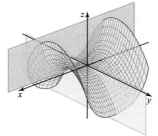

17. $f(x, y) = x^2 - xy + y^2$

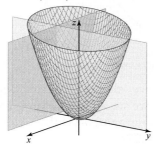

18. $f(x, y) = e^{-xy}$

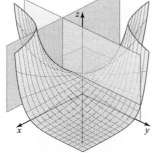

19. $f(x, y) = \dfrac{x - y}{x^2 + y^2 + 1}$

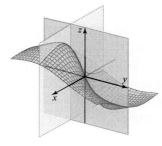

20. $f(x, y) = e^{x^2 + y^2}$

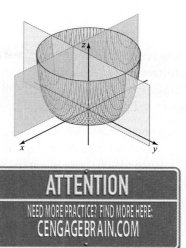

8.3 Multivariable Maxima and Minima

An investor has $3,000 to invest. She wants to get a $10,000 return on her investment without an excessive amount of risk. To achieve her goal, she has decided to invest a portion of her money in a certificate of deposit paying 6% interest compounded annually and the rest of her money in an aggressive growth fund that is expected to earn a maximum of 16% annually. How much money should she invest in each account if she wants to assume a moderate amount of risk? Questions such as these can be investigated using the techniques associated with finding *multivariable extrema* and *saddle points*.

In this section, we will demonstrate the techniques used to find relative extrema and saddle points of two-variable functions. The techniques used will be similar to those covered in our discussion of single-variable optimization in Chapter 4.

Let's first define relative extrema of multivariable functions.

> **Relative Extrema of Multivariable Functions**
> Consider the graph of a surface with equation $z = f(x, y)$. The value $f(a, b)$ is
>
> - a relative minimum if $f(a, b) \leq f(x, y)$ for all (x, y) in some region surrounding (a, b).
>
> - a relative maximum if $f(a, b) \geq f(x, y)$ for all (x, y) in some region surrounding (a, b).

Recall that relative extrema of single-variable functions cannot occur at endpoints. Similarly, relative extrema of two-variable functions cannot occur on the border of the domain region.

To help you visualize what relative extrema in three dimensions look like, we will label the relative extrema on the graphs of several multivariable functions (see Figure 8.25).

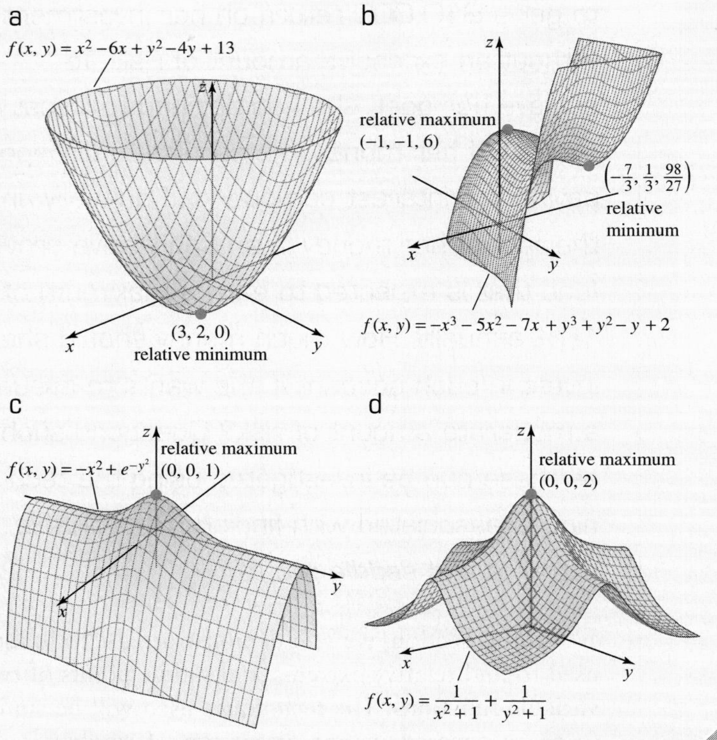

As was the case in two dimensions, relative extrema have horizontal tangent lines provided that the extrema don't occur at a sharp point on the graph.

In addition to relative extrema, a three-dimensional surface may have a *saddle point*. A **saddle point** is a point (a, b, c) that is a relative maximum of one cross-section graph and a relative minimum of another cross-section graph. The portion of the graph surrounding the saddle point looks like a saddle. Consider the graph of $f(x, y) = -x^2 + y^2 + 2$, shown in Figure 8.26.

Figure 8.26

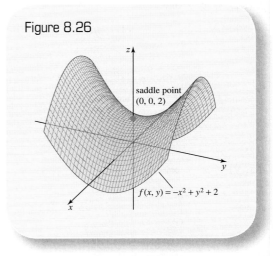

If we slice the graph with the plane $x = 0$, we see a cross section that looks like a concave-up parabola with a vertex at $(0, 0, 2)$ (see Figure 8.27).

Figure 8.27

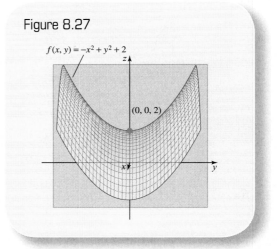

The smallest possible value of z (in the cross section) is reached at the vertex. Similarly, if we slice the graph with the plane $y = 0$, we see a cross section that looks like a concave-down parabola with a vertex at $(0, 0, 2)$ (see Figure 8.28).

Figure 8.25

a

$f(x, y) = x^2 - 6x + y^2 - 4y + 13$

$(3, 2, 0)$
relative minimum

b

relative maximum
$(-1, -1, 6)$

$\left(-\frac{7}{3}, \frac{1}{3}, \frac{98}{27}\right)$
relative minimum

$f(x, y) = -x^3 - 5x^2 - 7x + y^3 + y^2 - y + 2$

c

$f(x, y) = -x^2 + e^{-y^2}$ relative maximum $(0, 0, 1)$

d

relative maximum $(0, 0, 2)$

$f(x, y) = \dfrac{1}{x^2 + 1} + \dfrac{1}{y^2 + 1}$

Figure 8.28

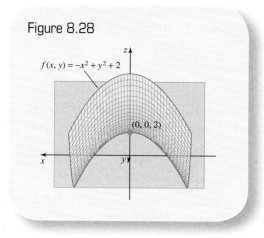

$f(x, y) = -x^2 + y^2 + 2$

$(0, 0, 2)$

The largest possible value of z (in the cross section) is reached at the vertex. Since the same point is a maximum value of one cross section and a minimum value of a different cross section, the point $(0, 0, 2)$ is a saddle point. Later in the section, we will discuss how to find saddle points using partial derivatives.

Recall from our discussion of single-variable functions that a value c in the domain of f was called a critical value of f if $f'(c) = 0$ or $f'(c)$ was undefined. All relative extrema occurred at critical values.

A similar definition is used for two-variable functions.

Critical Point of a Two-Variable Function

An ordered pair (c, d) in the domain of f is a **critical point** of f if

$$f_x(c, d) = 0 \text{ and } f_y(c, d) = 0$$

(Both statements must be true in order for (c, d) to be a critical point.)

When we refer to the critical point (c, d), we mean that (c, d) is the pair of input values whose output is $f(c, d)$. Let's find the critical point of $f(x, y) = x^2 + y^2 + 1$.

$$f_x = 2x$$
$$f_y = 2y$$

Setting each partial derivative equal to zero and solving, we find that

$f_x = 2x$	$f_y = 2y$
$0 = 2x$	$0 = 2y$
$x = 0$	$y = 0$

Therefore, the ordered pair $(0, 0)$ is a critical point of f. The domain region of the function is defined by $-\infty < x < \infty$ and $-\infty < y < \infty$. Since $(0, 0)$ is in the interior of the region, it is a relative extremum candidate.

To determine whether a maximum or minimum occurs at a critical point, we will apply the Second Derivative Test for Two-Variable Functions.

Second Derivative Test for Two-Variable Functions (D-Test)

To determine if a relative extremum of a function f occurs at a critical point (c, d), first calculate

$$D(c, d) = f_{xx}(c, d) \cdot f_{yy}(c, d) - (f_{xy}(c, d))^2$$

Then

- f has a **relative maximum** at (c, d) if $D(c, d) > 0$ and $f_{xx}(c, d) < 0$.

- f has a **relative minimum** at (c, d) if $D(c, d) > 0$ and $f_{xx}(c, d) > 0$.

- f has a **saddle point** at (c, d) if $D(c, d) < 0$.

- The test is inconclusive if $D(c, d) = 0$.

To apply the D-Test, we must first find the second partial derivatives. For the function $f(x, y) = x^2 + y^2 + 1$, we have the first partial derivatives $f_x = 2x$ and $f_y = 2y$. Therefore,

$$f_{xx} = 2$$
$$f_{yy} = 2$$
$$f_{xy} = 0$$

We evaluate each of the second partial derivatives at the critical point $(0, 0)$. In this case, the second partials are constant for all values of x and y. Thus

$$f_{xx}(0, 0) = 2$$
$$f_{yy}(0, 0) = 2$$
$$f_{xy}(0, 0) = 0$$

We now calculate D.

$$D(0, 0) = 2 \cdot 2 - 0^2$$
$$= 4$$

Since $D > 0$, either a maximum or a minimum occurs at $(0, 0)$. Since $f_{xx}(0, 0) > 0$, a relative minimum occurs at $(0, 0)$. Evaluating f at $(0, 0)$, we determine that

$$f(0, 0) = 0^2 + 0^2 + 1$$
$$= 1$$

The relative minimum occurs at $(0, 0, 1)$.

EXAMPLE 1 Locating Relative Extrema and Saddle Points

➡️ Find the location of the relative extrema and saddle points of $f(x, y) = x^2 + y^2 + xy - 2$. The graph of the surface is shown in Figure 8.29.

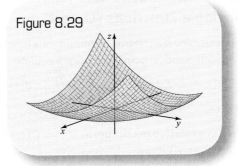

Figure 8.29

SOLUTION

The partial derivatives of f are

$$f_x = 2x + y$$
$$f_y = 2y + x$$

We set each partial derivative equal to zero and solve.

$$0 = 2x + y \qquad\qquad 0 = 2y + x$$
$$y = -2x \qquad\qquad\quad y = -0.5x$$

Since both equalities must be true at a critical point, we set the functions equal to each other and solve.

$$-2x = -0.5x$$
$$1.5x = 0$$
$$x = 0$$

Back substituting into the equation $y = -2x$, we determine that

$$y = -2(0)$$
$$= 0$$

The critical point of the function is $(0, 0)$. We now find the second partial derivatives.

$$f_{xx} = 2$$
$$f_{yy} = 2$$
$$f_{xy} = 1$$

Since the second partials are constant, evaluating each second partial at the critical point will result in the same values.

$$D(0, 0) = 2 \cdot 2 - 1 \cdot 1$$
$$= 3$$

Since $D(0, 0) > 0$ and $f_{xx} > 0$, a relative minimum occurs at $(0, 0)$. The value of the function at $(0, 0)$ is $f(0, 0)$.

$$f(0, 0) = (0)^2 + (0)^2 + (0)(0) - 2$$
$$= -2$$

The relative minimum occurs at $(0, 0, -2)$. There are no relative maxima or saddle points.

EXAMPLE 2 Locating Relative Extrema and Saddle Points

➡️ Find the location of the relative extrema and saddle points of $f(x, y) = e^{-xy}$. The graph of the surface is shown from two different viewpoints in Figure 8.30.

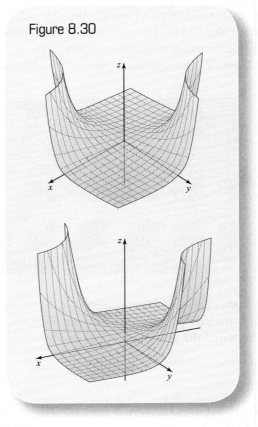

Figure 8.30

SOLUTION

We'll first find the partial derivatives of f.

$$f_x = -ye^{-xy}$$
$$f_y = -xe^{-xy}$$

To find the critical points, we'll set each partial equal to zero and solve.

$$0 = -ye^{-xy}$$

We solve the equation by setting each factor to zero.

$$-y = 0 \qquad e^{-xy} = 0$$

$$y = 0$$

Since $e^{-xy} \neq 0$ for all values of x and y, the only value that makes the equation zero is $y = 0$. We'll now set f_y equal to zero.

$$0 = -xe^{-xy}$$

We solve the equation by setting each factor to zero.

$$-x = 0 \qquad e^{-xy} = 0$$

$$x = 0$$

Since $e^{-xy} \neq 0$ for all values of x and y, the only value that makes the equation zero is $x = 0$. Thus, the critical point of the function is $(0, 0)$.

We'll now find the second partials and calculate the value of D at $(0, 0)$.

$$f_{xx} = y^2 e^{-xy}$$
$$f_{yy} = x^2 e^{-xy}$$
$$f_{xy} = -1(e^{-xy}) + (-xe^{-xy})(-y)$$
$$= -e^{-xy}(1 - xy)$$
$$f_{xx}(0, 0) = 0$$
$$f_{yy}(0, 0) = 0$$
$$f_{xy}(0, 0) = -e^{(0)(0)}(1 - (0)(0))$$
$$= -1$$
$$D(0, 0) = 0 \cdot 0 - (-1)^2$$
$$= -1$$

Since $D < 0$, a saddle point occurs at the critical point $(0, 0)$. The coordinates of the saddle point are $(0, 0, 1)$. There are no relative extrema.

EXAMPLE 3 Finding Relative Extrema and Saddle Points

Find the location of the relative extrema and saddle points of $f(x, y) = 0.1x^2y^2$. Two different views of the graph of the surface are shown in Figure 8.31.

SOLUTION

The partial derivatives of f are given by

Figure 8.31

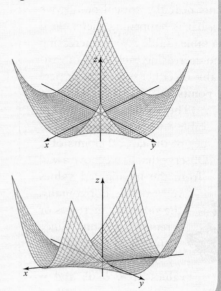

$$f_x = 0.2xy^2$$
$$f_y = 0.2x^2y$$

Setting the partial derivatives equal to zero and solving yields

$$0 = 0.2xy^2 \qquad 0 = 0.2x^2y$$
$$0 = xy^2 \qquad 0 = x^2y$$
$$x = 0 \text{ or } y = 0 \qquad x = 0 \text{ or } y = 0$$

Any point of the form $(a, 0)$ or $(0, b)$ is a critical point of the function (a and b are real numbers). The function f has an infinite number of critical points.

The second partial derivatives are given by

$$f_{xx} = 0.2y^2$$
$$f_{yy} = 0.2x^2$$
$$f_{xy} = 0.4xy$$
$$f_{yx} = 0.4xy$$
$$D = (0.2y^2)(0.2x^2) - (0.4xy)^2$$
$$= 0.04x^2y^2 - 0.16x^2y^2$$
$$= -0.12x^2y^2$$

Observe that $D(a, 0) = 0$ and $D(0, b) = 0$ for all values of a and b.

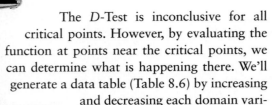

The *D*-Test is inconclusive for all critical points. However, by evaluating the function at points near the critical points, we can determine what is happening there. We'll generate a data table (Table 8.6) by increasing and decreasing each domain variable on either side of the critical points.

The values highlighted in Table 8.6 are the function values of the critical points of *f*. Observe that as we move away from the highlighted values vertically or horizontally, the value of *f* increases or remains constant. Since $f(a, 0) \le f(x, y)$ and $f(0, b) \le f(x, y)$ for all values of *a, b, x,* and *y,* relative minima occur at all of the critical points. There are no relative maxima or saddle points.

Table 8.6

x ↓ y→	−2	−1	0	1	2
−2	1.6	0.4	0	0.4	1.6
−1	0.4	0.1	0	0.1	0.4
0	0	0	0	0	0
1	0.4	0.1	0	0.1	0.4
2	1.6	0.4	0	0.4	1.6

EXAMPLE 4 Using Multivariable Functions to Forecast Investment Growth

An investor has $3,000 to invest in two separate accounts. The first account is a certificate of deposit guaranteed to earn 6% annually. The second account is an aggressive growth fund expected to earn between −4% and 16% annually. The combined dollar value of the return on the investment accounts after 20 years is given by

$$V = x(1.06)^{20} + (3000 - x)(1 + y)^{20} - 3000$$

where *x* is the amount of money invested in the certificate of deposit and *y* is the growth rate (as a decimal) of the aggressive growth fund. (If the growth rate is 8%, *y* = 0.08.) A graph of the function is shown in Figure

8.32. The gray plane is a graph of the function *z* = 0 and is drawn to provide a point of reference. When the graph of *V* is above the gray plane, the investor makes money. When the graph of *V* is below the gray plane, the investor loses money.

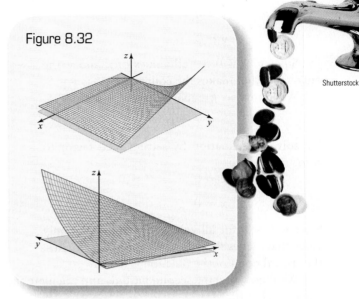

Figure 8.32

Find all critical points of *V* and determine the location of relative extrema and saddle points.

SOLUTION

$$V_x = (1.06)^{20} - (1 + y)^{20}$$

$$V_y = 20(3000 - x)(1 + y)^{19}$$

We set each partial derivative equal to zero.

$$0 = (1.06)^{20} - (1 + y)^{20}$$

$$(1 + y)^{20} = (1.06)^{20}$$

$$1 + y = 1.06$$

$$y = 0.06$$

$$0 = 20(3000 - x)(1 + y)^{19}$$

$$3000 - x = 0 \qquad (1 + y)^{19} = 0$$

$$x = 3000 \qquad y = -1$$

Recall that in order to be a critical point, the ordered pair must satisfy both equations. Since −1 ≠ 0.06, we throw out the value *y* = −1. Thus the only critical point is (3000, 0.06).

We'll now find the second partials.

$$V_{xx} = 0$$

$$V_{yy} = 380(3000 - x)(1 + y)^{18}$$

$$V_{xy} = -20(1 + y)^{19}$$

$$D = (0)(380(3000 - x)(1 + y)^{18})$$
$$- (-20(1 + y)^{19})^2$$
$$= -400(1 + y)^{38}$$

$$D(3000, 0.06) = -400(1 + 0.06)^{38}$$

$$D(3000, 0.06) < 0 \qquad \text{We only need to know the sign of } D$$

Since $D < 0$, there is a saddle point that occurs at $(3000, 0.06, 6621.41)$. However, from the context of the problem, we know that $0 \leq x \leq 3000$. Therefore, this point sits on the border of the domain region. As with relative extrema, a saddle point may not sit on the border of the domain region. We conclude that the function does not have any relative extrema or saddle points on the domain region defined by $0 \leq x \leq 3000$ *and* $-0.04 \leq y \leq 0.16$.

Let's look at the table of data for the function in the region surrounding the critical point given in Example 4 (Table 8.7).

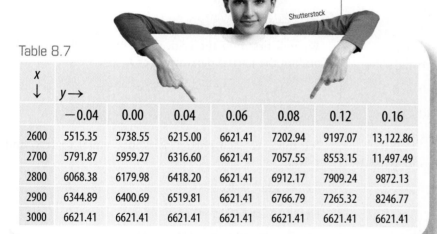

Shutterstock

Table 8.7

X ↓	y →						
	−0.04	0.00	0.04	0.06	0.08	0.12	0.16
2600	5515.35	5738.55	6215.00	6621.41	7202.94	9197.07	13,122.86
2700	5791.87	5959.27	6316.60	6621.41	7057.55	8553.15	11,497.49
2800	6068.38	6179.98	6418.20	6621.41	6912.17	7909.24	9872.13
2900	6344.89	6400.69	6519.81	6621.41	6766.79	7265.32	8246.77
3000	6621.41	6621.41	6621.41	6621.41	6621.41	6621.41	6621.41

From the table, we see that investing all of the money in the certificate of deposit or earning exactly 6% on the amount of money invested in the aggressive growth account will yield a combined return of $6,621.41 on the $3,000 investment. As the amount of money invested in the certificate of deposit (x) decreases, the amount of the return becomes more volatile. If $2,600 is invested in the certificate of deposit and $400 is invested in the aggressive growth account, the combined return could be as low as

$5,515.35 or as high as $13,122.86. Many investors are willing to risk losing $1,106.06 ($6,621.41 − $5,515.35) for the potential to earn an additional $6,501.45 ($13,122.86 − $6,621.41). Investors who are willing to risk all $3,000 on the aggressive growth account could lose as much as $1,673.99 of their initial investment or earn as much as $55,382.28 on top of their initial investment. Most financial advisers encourage investors to diversify their investments among a variety of investment accounts with varying levels of risk in order to achieve a return that meets the investor's goals.

EXAMPLE 5 Using Multivariable Functions to Forecast Product Sales

A software company produces two versions of its lead product: one for the Windows operating system and the other for the Macintosh operating system. Through market research, the company determines that at a price of p dollars, approximately q copies of the Windows version will be sold, and at a price of s dollars, approximately v copies of the Macintosh version will be sold. The relationship between the price of each item and sales may be modeled by

$$q = \frac{100,000}{p} + 100 - 0.2p$$

and

$$v = \frac{20,000}{s} + 50 - 0.1s$$

Determine the price the company should charge for each version in order to maximize the combined revenue from sales.

SOLUTION

The revenue from the sale of q copies of the Windows version is given by

$$pq = p\left(\frac{100,000}{p} + 100 - 0.2p\right)$$
$$= 100,000 + 100p - 0.2p^2$$

Revenue from the sale of v copies of the Macintosh version is given by

$$sv = s\left(\frac{20,000}{s} + 50 - 0.1s\right)$$
$$= 20,000 + 50s - 0.1s^2$$

The combined revenue from the sales of the two products is given by

$$R(p, s) = 100,000 + 100p - 0.2p^2 + 20,000 + 50s - 0.1s^2$$
$$= 100p - 0.2p^2 + 50s - 0.1s^2 + 120,000$$

From the graph of R on the region defined by $0 \le p \le 500$ and $0 \le s \le 500$ (see Figure 8.33), we can see that the revenue function has a relative maximum.

Figure 8.33

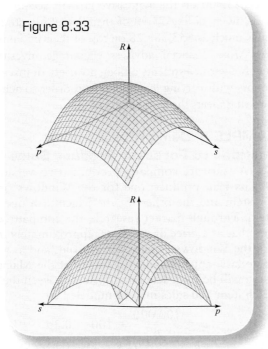

To find the location of the relative maximum, we first find the partial derivatives.

$$R_p = 100 - 0.4p$$

$$R_s = 50 - 0.2s$$

We then set each partial derivative equal to zero and solve.

$$0 = 100 - 0.4p \qquad\qquad 0 = 50 - 0.2s$$

$$0.4p = 100 \qquad\qquad 0.2s = 50$$

$$p = 250 \qquad\qquad s = 250$$

The critical point is $(250, 250)$. The second partial derivatives are

$$R_{pp} = -0.4$$

$$R_{ss} = -0.2$$

$$R_{ps} = 0$$

$$D(250, 250) = (-0.4)(-0.2) - 0$$

$$= 0.08$$

Since $D > 0$ and $R_{pp} < 0$, a relative maximum occurs at $(250, 250)$. To maximize revenue, the company should charge \$250 for each version of the software. The maximum combined revenue is \$138,750, with \$112,500 coming from the Windows version and \$26,250 coming from the Macintosh version.

EXAMPLE 6 Finding Relative Extrema and Saddle Points

Find the relative extrema and saddle points of $f(x, y) = x^4 - 4x^3 + 4x^2 - y^2$. Two different views of the graph are shown in Figure 8.34.

Figure 8.34

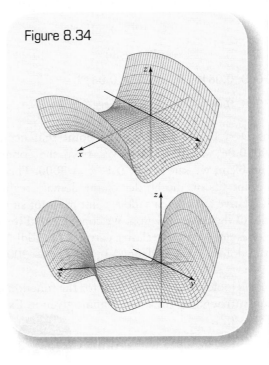

SOLUTION

We will first find the critical points of f. We begin by finding the partial derivatives of f.

$$f_x = 4x^3 - 12x^2 + 8x$$

$$f_y = -2y$$

We then set each partial derivative equal to zero and solve.

$$4x^3 - 12x^2 + 8x = 0 \qquad\qquad -2y = 0$$

$$4x(x^2 - 3x + 2) = 0 \qquad\qquad y = 0$$

$$4x(x - 1)(x - 2) = 0$$

$$x = 0, 1, 2$$

We must pair up each x-value with each y-value to form the critical points. The critical points are $(0, 0)$, $(1, 0)$, and $(2, 0)$.

Next we will find the second partial derivatives and perform the D-Test on each critical point.

$$f_{xx} = 12x^2 - 24x + 8$$

$$= 4(3x^2 - 6x + 2)$$

Table 8.8

Critical Point	$f_{xx} = 4(3x^2 - 6x + 2)$	$D = -8(3x^2 - 6x + 2)$	Graphical Interpretation
$(0, 0)$	8	-16	Saddle point
$(1, 0)$	-4	8	Relative maximum
$(2, 0)$	8	-16	Saddle point

$$f_{yy} = -2$$
$$f_{xy} = 0$$
$$D = (12x^2 - 24x + 8)(-2) - (0)(0)$$
$$= -8(3x^2 - 6x + 2)$$

When there are many critical points, it is often helpful to construct a table of data (see Table 8.8).

In order to understand better what a saddle point looks like, we'll zoom in on the region surrounding the critical point at $(0, 0)$ (see Figure 8.35).

Figure 8.35

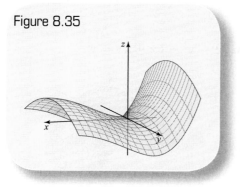

Looking at the graph from the xz plane (Figure 8.36), it looks as if a relative minimum occurs at the critical point.

Figure 8.36

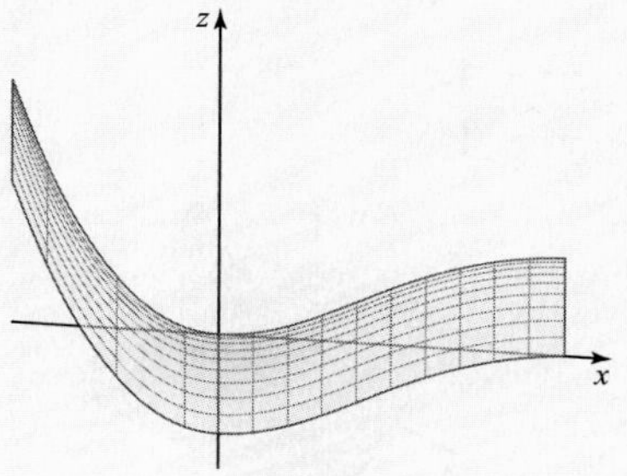

However, looking at the graph from the yz plane (Figure 8.37), it looks as if a relative maximum occurs at the critical point.

Figure 8.37

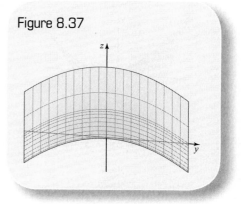

When a graph has a saddle point, the point will look like a minimum from one plane and a maximum from the other plane. (The planes will not necessarily be the xz plane and yz plane.)

To determine the actual coordinates of each of the extrema and saddle points, we evaluate the function at the critical points (see Table 8.9).

Table 8.9

x $\downarrow$	y
0	0
1	1
2	0

From the table, we see the relative maximum occurs at $(1, 0, 1)$. Saddle points occur at $(0, 0, 0)$ and $(2, 0, 0)$. These results are consistent with our earlier conclusions.

8.3 Exercises

In Exercises 1–20, find the relative extrema and saddle points of the function.

1. $f(x, y) = x^2 + 4y^2 - 10$

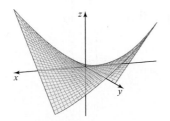

2. $f(x, y) = x^2 - 4xy$

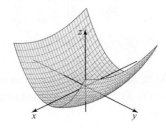

3. $f(x, y) = -x^2 - y^3 + 3y - x$

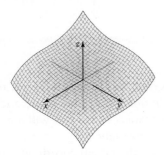

4. $f(x, y) = x^4 - 4x^2 + y^4 - 9y^2 + 1$

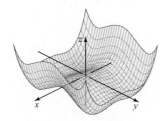

5. $f(x, y) = x^2y - y^2 + x$

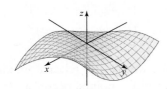

6. $f(x, y) = x^4 - 4x + y^4 - 9y^2 + 1$

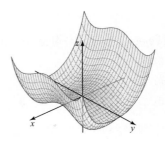

7. $f(x, y) = x^4 - 4x + y^3 - y^2$

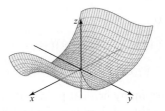

8. $f(x, y) = x^3 - 2x + y^3 - 2y$

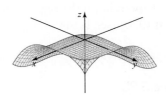

9. $f(x, y) = x^3 - 6x + xy - y^2$

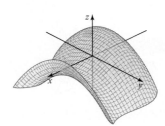

10. $f(x, y) = -\dfrac{x^2 - 4}{y^2 + 4}$

11. $f(x, y) = \dfrac{x^2 + 1}{y^2 + 1}$

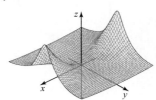

12. $f(x, y) = e^{-x^2 - y^2}$

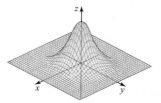

13. $f(x, y) = \dfrac{-1}{x^2 + 1} + y^4 - 4y^2$

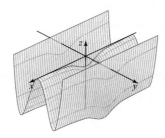

14. $f(x, y) = e^{xy - 4}$

15. $f(x, y) = x^3 - 3x + y^3 - 12y$

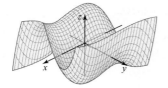

16. $f(x, y) = x^2 - 4x + y^2 - y - xy$

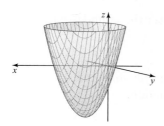

17. $f(x, y) = x^2 - x - 3y^2 + 3y$

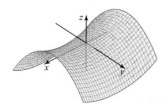

18. $f(x, y) = \ln(x^2 + 1) + \ln(y^2 + 1)$

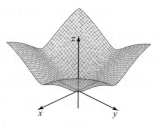

19. $f(x, y) = \dfrac{x^2 + y^2 + 1}{e^{x^2 + y^2}}$

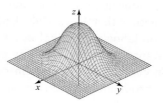

20. $f(x, y) = \dfrac{(x - 1)^2 + (y - 2)^2}{e^{(x - 1)^2 + (y - 2)^2}}$

8.4 Constrained Maxima and Minima and Applications

Based on data from 2008 and 2009, a Cobb-Douglas production model for the Coca-Cola Company is given by

$$P(L, C) = 0.01488L^{0.1135} C^{0.8865}$$

billion unit cases of liquid beverage

where L is the labor cost (including sales, payroll, and other taxes) in millions of dollars and C is the cost of capital expenditures (property, plant, and equipment) in millions of dollars. One unit case of beverage equals 24 eight-ounce servings. (**Source:** Modeled from Coca-Cola Company 2009 Annual Report)

In 2009, Coca-Cola Company spent $357 million on labor and produced 24.4 billion unit cases of liquid beverage. Assuming that Coca-Cola had $2,350 million to spend on labor and capital in 2009, how much should it have spent on labor and how much should it have spent on capital expenditures in order to maximize unit case production? Questions such as this may be answered by solving a constrained optimization problem. (See Example 4.)

In this section, we will demonstrate how to use the Lagrange Multiplier Method to solve constrained multivariable optimization problems. This technique may be used in two-variable and multivariable optimization problems. We will first demonstrate the technique with an example and then formally define the process.

EXAMPLE 1 Minimizing a Multivariable Function with the Lagrange Multiplier Method

Minimize $f(x, y) = x^2 - 2y$ subject to the constraint $-2x + y = 4$ (see Figure 8.38).

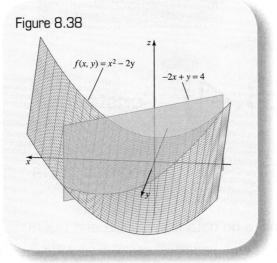

Figure 8.38

SOLUTION

We are to minimize $f(x, y) = x^2 - 2y$ subject to the constraint $-2x + y = 4$. As shown in Figure 8.38, the graph of the constraint is a plane that cuts through

the graph of $f(x, y) = x^2 - 2y$. We are to minimize f along the intersection of the graph of f and the graph of the constraint. We have

$$-2x + y = 4$$
$$-2x + y - 4 = 0$$

We will call the constraint function $g(x, y)$. That is,

$$g(x, y) = -2x + y - 4$$

Observe that $g(x, y) = 0$. We will now introduce a new function $F(x, y, \lambda)$. The third input variable of the function is called the **Lagrange multiplier** and is represented by the Greek letter λ (read "lambda"). The function $F(x, y, \lambda)$ is called the **Lagrange function** and is equal to the function to be maximized plus λ times the constraint function. That is,

$$F(x, y, \lambda) = f(x, y) + \lambda g(x, y)$$
$$= x^2 - 2y + \lambda(-2x + y - 4)$$
$$= x^2 - 2y - 2\lambda x + \lambda y - 4\lambda$$

We will next take the partial derivative of F with respect to each of the input variables.

$$F_x = 2x - 2\lambda$$
$$F_y = -2 + \lambda$$
$$F_\lambda = -2x + y - 4$$

By design, $F_\lambda = g(x, y) = 0$. We will now set the other two partial derivatives equal to zero and solve for λ.

$F_x = 2x - 2\lambda$	$F_y = -2 + \lambda$
$0 = 2x - 2\lambda$	$0 = -2 + \lambda$
$2\lambda = 2x$	$\lambda = 2$
$\lambda = x$	

Now we set the lambdas equal to each other and solve.

$$x = 2$$

So $x = 2$. We will substitute this result back into the constraint equation.

$$g(x, y) = 0$$
$$-2x + y - 4 = 0$$
$$-2x + y = 4$$
$$-2(2) + y = 4$$
$$-4 + y = 4$$
$$y = 8$$

The critical point of the *objective function* subject to the constraint is $(2, 8)$. (Recall that the objective function is the function that is to be maximized or minimized.) It is important to note that the D-Test works only for unconstrained optimization and cannot be used to solve a constrained problem. Although the Lagrange Multiplier Method finds the critical points of the constrained objective function, it does not tell us whether maxima or minima (or neither) occur at these points. Let's construct a table of values for f at points near the critical point that also satisfy the constraint equation $-2x + y = 4$ (see Table 8.10). Points satisfying the constraint equation nearby (2, 8) include (0, 4), (1, 6), (3, 10), and (4, 12).

Table 8.10

x	y	$f(x, y)$
0	4	−8
1	6	−11
2	8	−12
3	10	−11
4	12	−8

The constrained minimum of f is (2, 8, −12).

The approach used in Example 1 is referred to as the Lagrange Multiplier Method. The steps of the method are detailed in the following box.

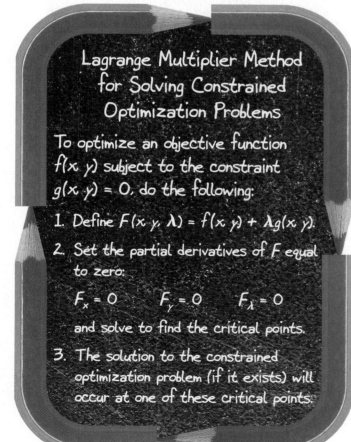

Lagrange Multiplier Method for Solving Constrained Optimization Problems

To optimize an objective function $f(x, y)$ subject to the constraint $g(x, y) = 0$, do the following:

1. Define $F(x, y, \lambda) = f(x, y) + \lambda g(x, y)$.

2. Set the partial derivatives of F equal to zero:

$$F_x = 0 \qquad F_y = 0 \qquad F_\lambda = 0$$

and solve to find the critical points.

3. The solution to the constrained optimization problem (if it exists) will occur at one of these critical points.

EXAMPLE 2 Maximizing a Multivariable Function with the Lagrange Multiplier Method

The United States Postal Service classifies a package as Parcel Post if its length plus its girth is less than or equal to 130 inches. (**Source:** U.S. Postal Service) (The girth is the distance around the package at its thickest point.) What are the dimensions of the rectangular packing box with a square base that has the greatest volume and meets the requirements for Parcel Post?

SOLUTION

We are looking for the dimensions of the rectangular box with maximum volume (see Figure 8.39).

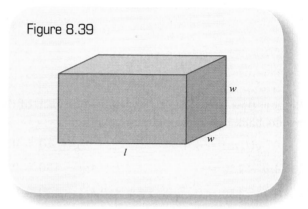

Figure 8.39

The volume of the box is a multivariable function given by

$$V(l, w) = lw^2$$

We have the constraint that the length plus the girth must not exceed 130 inches. A package with maximum volume will use all of the 130 inches allowed. Since the length of the package is l and the girth of the package is $4w$, we have

$$130 = l + 4w$$

$$0 = l + 4w - 130$$

We will call the constraint function $g(l, w)$. That is,

$$g(l, w) = l + 4w - 130$$

Observe that $g(l, w) = 0$.

We will now find the Lagrange function $F(l, w, \lambda)$. Recall that it is equal to the function to be maximized plus λ times the constraint function. That is,

$$F(l, w, \lambda) = V(l, w) + \lambda g(l, w)$$

$$= lw^2 + \lambda(l + 4w - 130)$$

$$= lw^2 + \lambda l + 4\lambda w - 130\lambda$$

We will next take the partial derivative of F with respect to each of the input variables.

$$F_l = w^2 + \lambda$$

$$F_w = 2lw + 4\lambda$$

$$F_\lambda = l + 4w - 130$$

By design, $F_\lambda = g(l, w) = 0$. We will now set the other two partial derivatives equal to zero and solve for λ.

$$0 = w^2 + \lambda \qquad\qquad 0 = 2lw + 4\lambda$$

$$\lambda = -w^2 \qquad\qquad -2lw = 4\lambda$$

$$\lambda = -0.5lw$$

Now we set the lambdas equal to each other and solve.

$$-w^2 = -0.5lw$$

$$w^2 - 0.5lw = 0$$

$$w(w - 0.5l) = 0$$

So $w = 0$ or $w = 0.5l$. We will substitute each of these results back into the constraint equation.

$$g(l, w) = 0 \qquad\qquad g(l, w) = 0$$

$$l + 4w - 130 = 0 \qquad\qquad l + 4w - 130 = 0$$

$$l + 4(0) - 130 = 0 \quad \text{or} \quad l + 4(0.5l) - 130 = 0$$

$$l = 130 \qquad\qquad 3l = 130$$

$$l \approx 43.33$$

We now back substitute these results to solve for w.

$$g(l, w) = 0 \qquad\qquad g(l, w) = 0$$

$$l + 4w - 130 = 0 \qquad\qquad l + 4w - 130 = 0$$

$$130 + 4w - 130 = 0 \quad \text{or} \quad 43.33 + 4w - 130 = 0$$

$$4w = 0 \qquad\qquad 4w = 86.67$$

$$w = 0 \qquad\qquad w \approx 21.67$$

Since it doesn't make sense to have a box with width 0 inches, we throw out the extraneous solution (130, 0). The box with maximum volume is 43.33 inches long, 21.67 inches high, and 21.67 inches wide.

As stated before, the Lagrange Multiplier Method finds the critical points of the constrained objective function, but it does not tell us whether maxima or minima (or neither) occur at these points. However, in most applied problems, we can figure out the status of the points from the context of the problem. In Example 2, we threw out the critical point (130, 0) because it didn't make sense to talk about the volume of a box with width 0. Since conceptually we knew that there had to be a box with maximum volume, we concluded that the other critical point would yield the optimal solution.

Locating critical points is sometimes a challenge. However, approaching the task in a methodical way makes it easier. Consider the following technique for finding critical points.

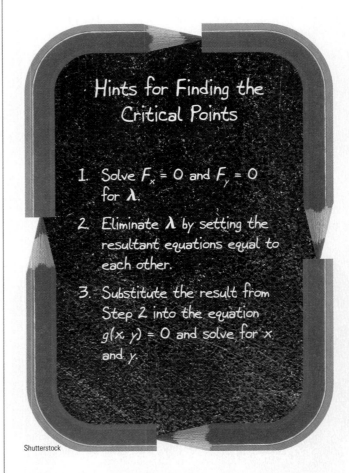

Hints for Finding the Critical Points

1. Solve $F_x = 0$ and $F_y = 0$ for λ.

2. Eliminate λ by setting the resultant equations equal to each other.

3. Substitute the result from Step 2 into the equation $g(x, y) = 0$ and solve for x and y.

EXAMPLE 3 Maximizing Volume with the Lagrange Multiplier Method

A box designer wants to construct a closeable box with a square base, as shown in Figure 8.40. The box will be used to store products in a warehouse.

Figure 8.40

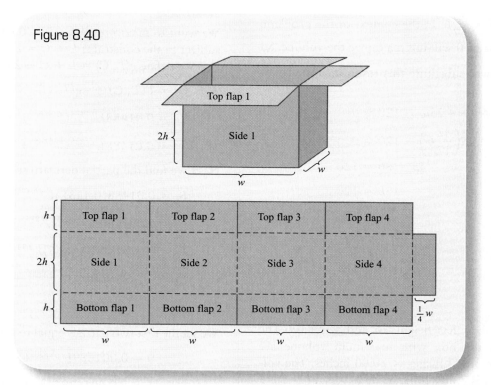

The designer requires that the sum of the height and width of the piece of cardboard used to make the box be at most 120 inches. What are the dimensions of the box with the maximum volume?

SOLUTION

We are asked to maximize the volume of the box subject to the constraint. The volume of the box is

$$V(h, w) = (2h)(w)(w)$$

$$= 2hw^2$$

The constraint is given by

$$(h + 2h + h) + (w + w + w + w + 0.25w) = 120$$

$$4h + 4.25w = 120$$

$$4h + 4.25w - 120 = 0$$

We'll call this function $g(h, w)$.

$$g(h, w) = 4h + 4.25w - 120$$

The Lagrange function $F(h, w, \lambda)$ is given by

$$F(h, w, \lambda) = V(h, w) + \lambda g(h, w)$$

$$= 2hw^2 + \lambda(4h + 4.25w - 120)$$

$$= 2hw^2 + 4h\lambda + 4.25w\lambda - 120\lambda$$

We'll now find the partial derivatives of F.

$$F_h = 2w^2 + 4\lambda$$

$$F_w = 4hw + 4.25\lambda$$

$$F_\lambda = 4h + 4.25w - 120$$

Observe that $F_\lambda = g(h, w) = 0$. We will now set the other two partial derivatives equal to zero and solve for λ.

$$0 = 2w^2 + 4\lambda \qquad\qquad 0 = 4hw + 4.25\lambda$$

$$-4\lambda = 2w^2 \qquad\qquad 4.25\lambda = -4hw$$

$$\lambda = -0.5w^2 \qquad\qquad \lambda = -\frac{4}{4.25}hw$$

$$= -\frac{1}{2}w^2 \qquad\qquad = -\frac{16}{17}hw$$

We now set the equations equal to each other and solve.

$$-\frac{1}{2}w^2 = -\frac{16}{17}hw$$

$$w^2 = \frac{32}{17}hw$$

$$w^2 - \frac{32}{17}hw = 0$$

$$w\left(w - \frac{32}{17}h\right) = 0$$

So $w = 0$ or $w = \frac{32}{17}h$. In the context of the problem, we know that $w = 0$ will not maximize the volume. So $w = \frac{32}{17}h$. We will substitute this result into the equation of $g(h, w)$.

$$4h + 4.25w - 120 = 0$$

$$4h + 4.25\left(\frac{32}{17}h\right) - 120 = 0$$

$$4h + 8h = 120$$

$$12h = 120$$

$$h = 10$$

$$w = \frac{32}{17}h$$

$$= \frac{32}{17}(10)$$

$$\approx 18.82$$

Note that the box has height $2h$ and width w. The dimensions of the box with maximum volume are 18.82 inches $\times$ 18.82 inches $\times$ 20.00 inches. The volume of this box is 7,084 cubic inches.

Although the problem in Example 3 could have been solved using optimization methods from Chapter 4, there is value in the Lagrange Multiplier Method, especially when working with functions of more than two variables.

EXAMPLE 4 Using the Lagrange Multiplier Method in a Real-World Setting

Based on data from 2008 and 2009, a Cobb-Douglas production model for the Coca-Cola Company is given by

$$P(L, C) = 0.01488L^{0.1135}C^{0.8865}$$

billion unit cases of liquid beverage

where L is the labor cost (including sales, payroll, and other taxes) in millions of dollars and C is the cost of capital expenditures (property, plant, and equipment) in millions of dollars. One unit case of beverage equals 24 eight-ounce servings. (**Source:** Modeled from Coca-Cola Company 2009 Annual Report)

In 2009, Coca-Cola Company spent \$357 million on labor and produced 24.4 billion unit cases of liquid beverage. Assuming that Coca-Cola had \$2,350 million to spend on labor and capital in 2009, how much should it have spent on labor and how much should it have spent on capital expenditures in order to maximize unit case production?

SOLUTION

We want to maximize $P(L, C) = 0.01488L^{0.1135}C^{0.8865}$ subject to the constraint $L + C = 2350$.

We define $g(L, C) = L + C - 2350$. Then

$$F(L, C) = P(L, C) + \lambda g(L, C)$$

$$= 0.01488L^{0.1135}C^{0.8865} + \lambda(L + C - 2350)$$

$$= 0.01488L^{0.1135}C^{0.8865} + L\lambda + C\lambda - 2350\lambda$$

Next we find the partial derivatives of F.

$$F_L = 0.01488(0.1135)L^{-0.8865}C^{0.8865} + \lambda$$

$$F_L = 0.001689L^{-0.8865}C^{0.8865} + \lambda$$

$$F_C = (0.01488)(0.8865)L^{0.1135}C^{-0.1135} + \lambda$$

$$F_C = 0.01319L^{0.1135}C^{-0.1135} + \lambda$$

$$F_\lambda = L + C - 2350$$

We set the first two partials equal to zero and solve for λ.

$$0 = 0.001689L^{-0.8865}C^{0.8865} + \lambda$$

$$\lambda = -0.001689L^{-0.8865}C^{0.8865}$$

$$0 = 0.01319L^{0.1135}C^{-0.1135} + \lambda$$

$$\lambda = -0.01319L^{0.1135}C^{-0.1135}$$

We then set the equations equal to each other and solve.

$$-0.001689L^{-0.8865}C^{0.8865} = -0.01319L^{0.1135}C^{-0.1135}$$

$$\frac{C^{0.8865}}{C^{-0.1135}} = \frac{-0.01319L^{0.1135}}{-0.001689L^{-0.8865}}$$

$$C = 7.809L$$

We substitute this result into the third equation, $F_\lambda = L + C - 2350$. Since $F_\lambda = g(L, C)$, $F_\lambda = 0$. Thus $L + C - 2350 = 0$.

$$L + 7.809L - 2350 = 0 \qquad \text{Since } C = 7.809L$$

$$8.809L = 2350$$

$$L \approx 266.8$$

We know that $L + C = 2350$; thus

$$266.8 + C = 2350$$

$$C = 2083.2$$

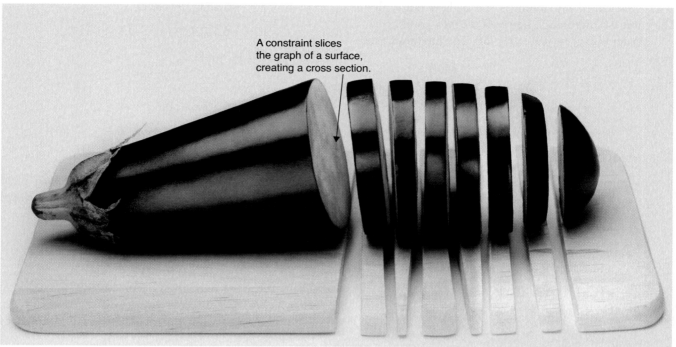

A constraint slices
the graph of a surface,
creating a cross section.

According to our model, when $266.8 million is invested in labor and $2,083.2 million is invested in capital expenditures, unit case production will be maximized. Since $P(266.8, 2083.2) = 24.55$, we conclude that a maximum of 24.55 billion unit cases could have been produced in 2009. This is more than the 24.4 billion unit cases that were produced.

Graphical Interpretation of Constrained Optimization Problems

Graphically speaking, a constraint "slices" the graph of the surface, creating a cross section. A constrained maximum is the largest value of the objective function when the domain is restricted to the cross section. Similarly, a constrained minimum is the smallest value of the objective function when the domain is restricted to the cross section. Using a graphing utility to graph the objective function and its constraint helps us identify the location of constrained extrema visually. However, we will check our work using the Lagrange Multiplier Method. It is easy to overlook extrema when relying on graphical methods alone, as will be illustrated in Example 5.

EXAMPLE 5 Maximizing a Multivariable Function with the Lagrange Multiplier Method

➜ Maximize $f(x, y) = -x^3 + 6x^2 - y^2 + 10$ subject to the constraint $10x - y = 50$.

SOLUTION

The graph of the function f is shown in Figure 8.41, together with the graph of the constraint equation.

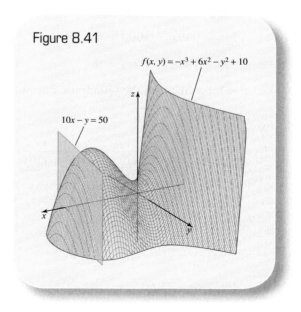

Figure 8.41

$f(x, y) = -x^3 + 6x^2 - y^2 + 10$

$10x - y = 50$

The constrained maximum is the highest point along the curve of intersection between the two graphs.

We let $g(x, y) = 10x - y - 50$. The Lagrange function is

$$F(x, y, \lambda) = f(x, y) + \lambda g(x, y)$$
$$= -x^3 + 6x^2 - y^2 + 10 + \lambda(10x - y - 50)$$
$$= -x^3 + 6x^2 - y^2 + 10 + 10x\lambda - y\lambda - 50\lambda$$

The partial derivatives are

$$F_x = -3x^2 + 12x + 10\lambda$$

$$F_y = -2y - \lambda$$

$$F_\lambda = 10x - y - 50$$

We set F_x and F_y equal to zero and solve for λ.

$$0 = -3x^2 + 12x + 10\lambda \qquad 0 = -2y - \lambda$$

$$-10\lambda = -3x^2 + 12x \qquad \lambda = -2y$$

$$\lambda = 0.3x^2 - 1.2x$$

We set the equations equal to each other and solve for y in terms of x.

$$0.3x^2 - 1.2x = -2y$$

$$y = -0.15x^2 + 0.6x$$

Since $g(x, y) = 0$,

$$10x - y - 50 = 0$$

$$10x - (-0.15x^2 + 0.6x) - 50 = 0$$

$$0.15x^2 + 9.4x - 50 = 0$$

$$15x^2 + 940x - 5000 = 0$$

$$3x^2 + 188x - 1000 = 0$$

We solve for x by using the Quadratic Formula.

$$x = \frac{-b \pm \sqrt{b^2 - 4ac}}{2a}$$

$$= \frac{-(188) \pm \sqrt{(188)^2 - 4(3)(-1000)}}{2(3)}$$

$$= \frac{-188 \pm \sqrt{47,344}}{6}$$

$$= \frac{-188 \pm 217.6}{6}$$

$$= 4.931 \text{ or } -67.60$$

$$y = -0.15x^2 + 0.6x$$

$$y = -0.15(4.931)^2 + 0.6(4.931)$$

$$y = -0.6886$$

$$y = -0.15x^2 + 0.6x$$

$$y = -0.15(-67.60)^2 + 0.6(-67.60)$$

$$y = -726.0$$

The critical points are $(4.931, -0.6886)$ and $(-67.60, -726.0)$. Evaluating f at each of the critical points, we get

$$f(4.931, -0.6886) = 35.52 \text{ and}$$

$$f(-67.60, -726.0) = -190,732$$

The constrained maximum occurs at $(4.931, -0.6886, 35.52)$. We also discovered a constrained minimum. (The minimum is not shown in the original graph of the function.)

8.4 Exercises

In Exercises 1–10, solve the constrained optimization problem using the Lagrange Multiplier Method.

1. Minimize $f(x, y) = x^2 - 2y$ subject to the constraint $-2x + y = 5$.

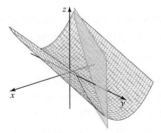

2. Maximize $f(x, y) = -x^2 - y^2 + 16$ subject to the constraint $x - 2y = 4$.

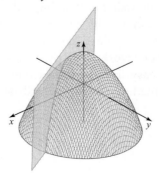

3. Maximize $f(x, y) = x^2 - y^2$ subject to the constraint $y = 3x - 1$.

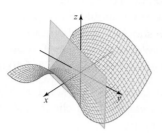

4. Minimize $f(x, y) = xy$ subject to the constraint $2x - y = 4$.

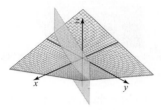

5. Maximize *and* minimize $f(x, y) = -x^3 + 4x^2 + 2y^2$ subject to the constraint $y = x$.

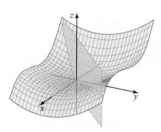

6. Minimize $f(x, y) = x^2 + 3y^2$ subject to the constraint $2x + y = -4$.

7. Minimize $f(x, y) = 4xy$ subject to the constraint $5x - y = 2$.

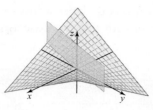

8. Maximize $f(x, y) = -4xy$ subject to the constraint $-2x + y = -4$.

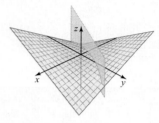

9. Maximize *and* minimize $f(x, y) = x^3 - 20x + 2y^3 + y^2$ subject to the constraint $y = 2x + 3$.

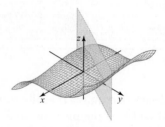

10. Maximize $f(x, y) = x^2y + 2y^2$ subject to the constraint $x - 4y = 24$.

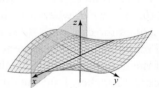

In Exercises 11–15, use the Lagrange Multiplier Method to find the answer to the question.

11. *Box Design* A box designer wants to construct a closeable box with a square base as shown in the figure.

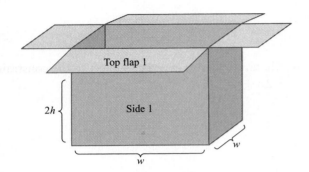

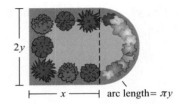

   The designer requires that the sum of the height and width of the piece of cardboard used to make the box be at most 180 inches. What are the dimensions of the box with the maximum volume?

12. *Can Volume* The volume of a cylindrical can is given by $V(h, r) = \pi r^2 h$, where r is the radius and h is the height of the can, both measured in inches. Because of manufacturing equipment limitations, the sum of the diameter and the height of the can cannot exceed 12 inches. Find the dimensions of the can with maximum volume.

13. *Cobb-Douglas Model* Based on data from 1999 and 2000, a Cobb-Douglas model for Ford Motor Company is given by

$$P(L, C) = 0.5319L^{0.4558}C^{0.5442} \text{ thousand vehicles}$$

   where L is the total labor cost and C is the capital expenditure cost, both in millions of dollars. In 2000, Ford Motor Company spent $25,783 million on labor and produced 7,424 thousand

vehicles. (**Source:** Modeled from Ford Motor Company 2001 Annual Report)

   Assuming that the model is valid for future years, how much should Ford spend on labor and how much should it spend on capital in order to maximize vehicle production if its annual budget for labor and capital expenditures is $40,000 million?

14. *Landscape Design* A landscape architect is designing a garden for a client as shown in the following figure.

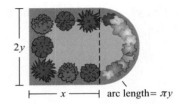

   She intends to line the flat edges of the garden with shrubs and the curved edge of the garden with flowers. The shrubs cost $25 a linear foot, and the flowers cost $20 a linear foot, including labor. The client is willing to pay up to $1,000 and wants as large a garden as possible. What are the dimensions of the garden? (*Hint:* The area of the garden is given by

$$A(x, y) = 2xy + \frac{1}{2}\pi y^2.)$$

15. *Cobb-Douglas Model* Based on data from 2000 to 2001, a Cobb-Douglas model for the Coca-Cola Company is given by

$$P(L, C) = 0.02506L^{0.0481}C^{0.9519} \text{ billion unit cases of beverage}$$

   where L is the amount of money spent on labor (including sales, payroll, and other taxes) and C is the amount of money spent on capital expenditures, both in millions of dollars. (One unit case of beverage equals 24 eight-ounce servings.) In 2001, Coca-Cola spent a combined total of $917 million on labor and capital expenditures and produced 17.8 billion unit cases of beverage. (**Source:** Coca-Cola 2001 Annual Report) According to the model, what is the maximum number of unit cases that can be produced with a labor and capital budget of $917 million?

Index

A

Absolute extrema, 98–106
  critical values, 101–102
  defined, 98–99
  exercises, 106–107
  First Derivative Test, 102–104
  graphs, 98–101
  juvenile arrest rates example, 104–105
  maximum revenue forecasts, 105–106
  value of derivative at, 99
  water consumption example, 105
Acceleration, 122–123
Accumulated change, 182–184
Advertising spending, 70
Algebraic method, of finding derivative function, 66–68
Alignment of data, 25, 41
Alternative fuel vehicles, quadratic regression to forecast number of, 30–31
Antiderivatives. *See* Indefinite integrals
Antidifferentiation. *See* Integrals and integration
Apple iPod sales, 61–62
Appreciation, 42
Area
  Fundamental Theorem of Calculus, 166–168
  landscaping cost minimization, 112–113
  resource use minimization, 110–112
  Riemann sums for estimation, 146–149
  summation notation, 156–159
  between two curves, 178–185
Automobiles
  accidents, 162
  alternative fuel vehicles, 30–31
  value forecasts, 42
  velocity, 134
Average rate of change, 46–51
  defined, 47
  from equation, 48
  exercises, 49–51
  graphs, 48–49
  introduction, 46–47
  from table, 48
Axis of symmetry, 24

B

Base, of exponential function, 36
Billiards, 213–214
Blu-ray Disc shipments, 196–197
Body weight, 71

B (continued)

Books, 53, 139–140
Bottled water consumption examples, 68, 88–89, 105
Boxes
  maximizing volume, 234–236
  minimizing use of, 110–112
  surface area, 208–209
Break-even point, 108

C

Calculators
  derivative values, 104
  exponential regression feature, 40
  intersect feature, 30
  maximum command, 122
  quadratic regression feature, 25
  Riemann sum programs, 154
Cars. *See* Automobiles
Cartesian coordinate system, 5
Chain Rule, 82–88
Change
  accumulated change, 182–184
  average rate of (*See* Average rate of change)
  instantaneous rate of (*See* Instantaneous rate of change)
Circles, 92–93
Cobb-Douglas production function, 202, 207–208, 231–232, 236–237
Coca-Cola Company
  Cobb-Douglas production model, 207–208, 231–232, 236–237
  marginal revenue of, 156
  revenue, 168–169
Composition of functions, 83–84
Compound interest, 188, 212–213
Concave up/concave down, 22–23, 117
Concavity, 117–119. *See also* Second derivative
Constant Multiple Rule, 77
Constant Multiple Rule for Integrals, 137
Constant Rule, 76
Constant Rule for Integrals, 137–138
Constrained optimization problems, 231–240
Continuity, 96–97
Continuous functions, 96–97
Cooling rates, 190
Costs, 108, 139–140
Critical point
  defined, 101
  finding, 101–102
  of multivariable functions, 234
  of two-variable functions, 223

Critical values, 101–102
Cross sections, of surface, 214–218
Cumulative revenue, 177
Curve sketching, 123–127

Decay factor, 36, 42
Definite integrals
  defined, 159
  exercises, 162–164
  Fundamental Theorem of Calculus used
     to calculate, 165–166
  graphical meaning, 160
  and indefinite integrals, 159–160
  integrand, 159
  integration by substitution, 169–170
  lower limit of integration, 159
  properties, 161–162
  upper limit of integration, 159
Degree of polynomial, 22
Demand equation, 107
Dependent variables, 4, 204
Depreciation, 42
Derivative functions, 66–68
Derivative notation, 74–75, 86–88
Derivatives, 46–73
  average rate of change, 46–51
  Chain Rule, 82–88
  Constant Multiple Rule, 77
  Constant Rule, 76
  curve sketching, 123–127
  defined, 55
  estimating, 68
  Exponential Rule, 88–90
  of function algebraically, 66–68
  of function at multiple points, 66–67
  of function at point, 55–56
  graphs, 58–62
  implicit differentiation, 92–95
  instantaneous rate of change, 51–57
  interpreting meaning of, 70–73
  Logarithmic Rule, 90–91
  maxima and minima (*See* Maxima and minima)
  numerical estimates, 62–64
  partial derivatives (*See* Partial derivatives)
  Power Rule, 76–77
  Product Rule, 79–81
  Quotient Rule, 81
  sign charts, 103
  Sum and Difference Rule, 77–78
Dietary fiber intake, 14–15, 19
Difference quotient
  average rate of change, 48–49
  defined, 47

instantaneous rate of change, 52
  limits, 53–56
Differentiable functions, defined, 75
Differential equations, 185–192
  exercises, 191–192, 197–201
  first-order differential equations, 186
  general solution for, 187–188
  limited growth models, 192–195
  logistic growth model, 195–197
  Newton's Law of Heating and Cooling,
     185–186, 188–190
  particular solution for, 186–187, 188
  separable differential equations, 186–188
Differentials, 141
Differentiate, defined, 75
Differentiation, 74–95
  basic rules, 74–78
  Chain Rule, 82–88
  Constant Multiple Rule, 77, 137
  Constant Rule, 76, 137–138
  definitions, 75
  Exponential Rule, 88–90, 138
  implicit differentiation, 92–95
  Logarithmic Rule, 90–91
  Power Rule, 76–77, 137
  Product Rule, 79–81, 172–174
  Quotient Rule, 81
  Sum and Difference Rule, 77–78, 138
Direct mail advertising spending, 70
Discontinuous functions, 97
Distance, 153–154
Domain, 8–9, 204
D-Test, 223–224

Earnings. *See* Wages and earnings
E-commerce sales, 74, 77–78
Electronic Arts, Inc., 178–180, 182–184
Event ticket online sales, 164, 165–166, 177
Exponential functions, 35–45
  aligned data sets, 40–41
  base, 36
  car's value forecasts, 42
  decay factor, 36, 42
  defined, 36
  exercises, 43–45
  graphs, 37–38
  growth factor, 36
  horizontal asymptote, 37
  inflation example, 42–43
  initial value, 36
  introduction, 35–37
  population growth example, 35, 40–41
  from table, 38–40

Exponential Rule, 88–90
Exponential Rule for Integrals, 138
Exponents, properties of, 38
Extrapolation, 25
Extrema, defined, 97. *See also* Absolute extrema; Relative
extrema

F

First Derivative Test, 102–103
First-order differential equations, 186
Fish size, forecasting with Von Bertalanffy Limited Growth
Model, 194–195
Fixed costs, 108
Fuel costs, 3, 4
Functions
    composition of, 83–84
    curve sketching, 123–127
    defined, 2
    derivatives (*See* Derivatives)
    domain and range, 8–9
    employee earnings example, 4–5
    exercises, 9–11
    exponential functions (*See* Exponential functions)
    graphs, 5–8
    inputs and outputs, 3
    introduction, 2–3
    linear functions (*See* Linear functions)
    multivariable functions (*See* Multivariable functions)
    quadratic functions (*See* Quadratic functions)
    symbolic notation, 4–5
    tables representing, 3–4
    of two variables, 203
Fundamental Theorem of Calculus, 164–171
    accumulated change in revenue calculation, 168–169
    area calculation, 166–168
    defined, 165
    definite integral calculation, 165–166
    exercises, 170–171
    introduction, 164–165

G

Gas costs, 3, 4
Generalized Exponential Rule, 90
Generalized Power Rule, 85
Graphs and graphing
    average rate of change, 48–49
    concavity, 117–120
    constrained optimization problems, 237–238
    cross sections of surface, 214–218
    curve sketching process, 123–127
    definite integrals, 160
    derivatives, 58–62

difference quotient, 48–49
exponential functions, 37–38
functions, 5–8
inflection points, 118
linear functions, 11, 17–18
multivariable functions, 204–206
quadratic functions, 22–25
relative and absolute extrema, 98–101
Growth, in population, 35, 40–41, 56
Growth factor, 36
Growth models
    limited growth model, 192–195
    logistic growth model, 195–197

H

Heat index, 203
Home sale prices, 63–64
Homicides, 66, 67, 116, 119–120
Horizontal asymptote, 37

I

Ice cream consumption, 29
Implicit differentiation, 92–95
Indefinite integrals, 134–140
    applications, 139–140
    basic integration rules, 137–139
    car's velocity example, 134
    defined, 136
    and definite integrals, 159–160
    exercises, 140
    finding, 136
    introduction, 134–136
Independent variables, 4, 204
Inflation, 42–43
Inflection points
    curve sketching, 125
    defined, 118
Information, differential equations to forecast
    spread of, 192–194
Initial value, 36
Inputs, 3
Instantaneous rate of change
    calculation, 51–55
    consumer spending on books example, 53
    derivative rules, 77–78
    exercises, 56–57
    introduction, 51–52
    limits, 53–56
    at multiple points, 67, 68
    population example, 56
    running speed estimates, 51–52
Integrable functions, 159

Integral Rule for $1/x$, 138–139
Integrals and integration, 134–171
  area between two curves, 178–185
  basic rules, 137–139
  definite integrals (*See* Definite integrals)
  definitions, 136
  differential equations (*See* Differential equations)
  Fundamental Theorem of Calculus, 164–171
  indefinite integrals (*See* Indefinite integrals)
  lower limit of integration, 159
  natural logarithm function, 177
  by parts method, 172–178
  by substitution method, 141–145, 169–170
  sums to approximate area (*See* Riemann sums)
  upper limit of integration, 159
Integral sign, defined, 136
Integrand, defined, 136, 159
Integration by parts, 172–178
Integration by substitution, 141–145, 169–170
Interest, 187–188, 212–213
Internet use, 91
Investments
  multivariable functions to forecast value, 207,
    226–227
  rate of change of account earning compound
    interest, 187–188, 212–213

J

Juvenile arrest rates, 96, 104–105

L

Lagrange function, 232
Lagrange multiplier, 232
Lagrange Multiplier Method, 232–240
  Coca-Cola Cobb-Douglas production
    model, 231–232, 236–237
  defined, 233
  exercises, 238–240
  multivariable function maximization, 233–234
  multivariable function minimization, 232–233
  volume maximization, 234–236
Landscaping costs, 112–113
Left-hand limits, 53
Left-hand sums, 149–151, 157
Leibniz notation, 75, 86–88
Limited growth model, 192–195
Limits, 53–56
Linear equations, 14–17
Linear functions, 11–19
  characteristics, 11
  dietary fiber intake example, 14–15, 17
  exercises, 18–19

  graphs, 11, 17–18
  intercepts, 13–14
  introduction, 11–12
  linear equations, 14–17
  sales tax as function of sales price example,
    15–16
  slope, 11–13
Lines
  equation of, 16
  point-slope form, 17
  slope-intercept form, 14, 16
  slope of, 12–13
  standard form, 17
Logarithmic Rule, 90–91
Logistic growth model, 195–197
Lower limit of integration, 159
Lumber and wood products industry,
    employee earnings, 82–83, 87–88

M

Magnitude, of slope, 13
Marginal cost, 108
Marginal profit, 108
Marginal revenue, 108
Maxima and minima, 96–116
  absolute extrema (*See* Absolute extrema)
  area and volume application, 110–113
  continuity, 96–97
  curve sketching, 123–127
  demand equation, 107
  exercises, 113–116
  introduction, 96
  multivariable maxima and minima
    (*See* Multivariable maxima and minima)
  point of diminishing returns, 121–122
  position, velocity, and acceleration, 122–123
  related rates, 129–133
  relative extrema (*See* Relative extrema)
  revenue, cost, and profit application, 108–110
  second derivative, 116–121
Maximum profit, 108–109
Maximum revenue, 105–106, 109–110
McDonald's, 185–186
Medicare costs, 28
Milk consumption, 17
Minima. *See* Maxima and minima
Mining industry jobs, 160–161
Moving costs, 110–112
Multivariable functions, 202–211
  Cobb-Douglas production function, 202,
    207–208, 232–233
  defined, 203
  exercises, 209–211
  graphs, 204–206

introduction, 202–205
investment value forecasts, 207, 226–227
product sales forecasts, 227–228
surface area of box, 208–209
tangent lines of, 218
Multivariable maxima and minima, 221–240
constrained optimization problems, 231–240
critical point, 223
investment growth forecasts, 221
relative extrema, 222–226, 228–229
Second Derivative Test, 223–224

Natural logarithm function, integration of, 177
Net sales, 28–29
Newton's Law of Heating and Cooling, 188–190
Nintendo DS game console sales, 121–122
Numerical derivatives, 62–64

Oil spill in Gulf of Mexico, 129, 134
Origin, 5
Outputs, 3

Parabolas, 22–25
Partial derivatives, 211–221
of compound interest function, 212–213
cross sections of surface, 214–218
defined, 211
exercises, 219–221
finding, 211–212
geometric interpretation, 216
meaning of, 213–214
second-order partial derivatives, 218–219
wind chill temperature changes, 211, 214
Percentages, 41–42
Point of diminishing returns, 121–122
Point-slope form, of linear equations, 16–17
Polynomials
curve sketching, 123
defined, 22
Pool game, 213–214
Population growth, 35, 40–41, 56
Position, 122–123, 139
Power Rule, 76–77
Power Rule for Integrals, 137
Prescription drug sales and spending,
26–27, 58–60
Product Rule, 79–81, 172–174

Product sales forecasts
multivariable functions, 227–228
tangent-line approximations, 61–62
Profit, 108–109

Quadratic functions, 20–34
alternative fuel vehicle forecasts, 30–31
defined, 22
exercises, 31–34
graphs, 22–25
Medicare cost forecasts, 28
net sales model, 28
prescription drug sale forecasts, 26–27
Walmart Supercenter openings, 20–21, 25
when not to use, 29
Quotient Rule, 81

Radicals, integration by substitution of
function containing, 143–144
Range, 9, 204
Related rates, 129–133
Relative extrema, 98–106
critical values, 101–102
curve sketching, 124–125
defined, 98–99
exercises, 106–107
finding, 100–101, 224–226, 228–229
First Derivative Test, 102–104
graphs, 98–101
juvenile arrest rates example, 104–105
maximum revenue forecasts, 105–106
of multivariate functions, 222–226, 228–229
Second Derivative Test, 120–121, 223
value of derivative at, 99
water consumption example, 105
Restaurant sales, 86–87
Retail and food services spending, 79, 80–81
Revenue
accumulated change in revenue, 168–169
cumulative revenue, 177
defined, 108
marginal revenue, 108
maximum revenue forecasts, 105–106, 109–110
rate of change in annual net revenue, 178–180
Riemann sums, 145–156
defined, 149
distance traveled estimates, 153–154
to estimate area, 146–149
exercises, 155–156
flow rate of water, 145

Riemann sums, *continued*
  introduction, 145–146
  left-hand and right-hand sums, 149–152
  summation notation, 157–159
  water volume, 154–155
Right-hand limits, 54
Right-hand sums, 151–152, 158

S

Saddle points
  defined, 222–223
  finding, 224–226, 228–229
Sales tax, 15–16
Scatter plots, 29–30, 40–41
Secant lines, 48–49
Second derivative, 116–121
Second Derivative Test
  defined, 120–121
  for two-variable functions, 223–224
Second-order partial derivatives, 218–219
September 11, 2001, 192–194
Shipping costs, 110, 233–234
Singular point, 101
Singular value, 101
Slope
  curves, 60
  linear functions, 11–13
  secant lines, 49
  tangent lines, 60
Slope-intercept form, 14, 16
Social Security benefits, 70–71
Software, Riemann sum programs, 154
Speed, instantaneous rate of change, 52
Standard form, of linear equations, 16–17
Stationary point, 101
Stationary value, 101
Substitution, integration by, 141–145, 169–170
Sum and Difference Rule, 77–78
Sum and Difference Rule for Integrals, 138
Summation notation, 156–159
Sum of the signed areas, 160. *See also* Fundamental
  Theorem of Calculus
Sums, using to approximate area. *See* Riemann sums

Surface area, 208–209
Surfaces
  cross sections of, 214–218
  defined, 203
Swimming pools, amount of time to fill, 131
Symbolic notation, 4–5

T

Tables
  finding derivatives using, 62–64
  finding exponential functions using, 38–40
  linear model from, 15–16
  representing functions, 3–4
Tangent lines
  approximations, 61–62, 71
  equation of, 60–61
  of multivariable function, 218
Television sets, number in U.S., 71
Temperature, Newton's Law of Heating and Cooling,
  188–190
Tuition, average rate of change in, 46–48

U

United Arab Emirates (UAE), population growth, 35
Units of rate of change, 48
Upper limit of integration, 159

V

Variable costs, 108
Velocity, 122–123
Vertical Line Test, 6–7
Volume
  Lagrange Multiplier Method maximization,
    234–236
  landscaping cost minimization, 112–113
  occupied by liquid in cylindrical can, 130
  resource use minimization, 110–112
  Riemann sums to estimate water volume, 154–155
Von Bertalanffy Limited Growth Model, 194–195

W

Wages and earnings
  equation as function of hours worked, 4–5
  growth in industry, 82–83, 87–88
Walmart, 20–21, 25
Water consumption, 68, 88–89, 105
Water temperature, 189–190
Water volume, Riemann sums to estimate, 154–155
Wind chill temperature, 131–132, 211, 214

X

x-intercept, 13–14, 124

Y

y-intercept, 13–14, 124

TURN TO A TRUSTED RESOURCE

With the Student Solutions Manual, now available online, you will have all the learning resources you need for your course in one convenient place, at no additional charge!

Solutions to all odd-numbered exercises, organized by chapter, are easily found online.

Simply visit **login.cengagebrain.com** to access this trusted resource.

STUDY
YOUR WAY

At no additional cost, you have access to online learning resources that include **tutorial videos, printable flashcards, quizzes,** and more!

Watch videos that offer step-by-step conceptual explanations and guidance for each chapter in the text.

Along with the printable flashcards and other online resources, you will have a multitude of ways to check your comprehension of key mathematical concepts.

You can find these resources at **login.cengagebrain.com.**

LEARN YOUR WAY

With **Applied CALC**, you have a multitude of study aids at your fingertips.

The Student Solutions Manual contains worked-out solutions to every odd-numbered exercise, to further reinfore your understanding of mathematical concepts in **Applied CALC**.

In addition to the Student Solutions Manual, Cengage Learning's **CourseMate** offers exercises and questions that correspond to every section and chapter in the text for extra practice.

For access to these study aids, sign in at **login.cengagebrain.com**.

Learning Objectives

1.1 Functions

Objectives

- Distinguish between functions and nonfunctions in tables, graphs, and words
- Use function notation
- Graph functions using technology
- Determine the domain of a function

In this section, you learned to distinguish between functions and nonfunctions, and you practiced using function notation. You learned how to use technology to graph a function, and you discovered how to find the domain of a function. Mastering each of these techniques will help you understand the subsequent concepts covered in this chapter.

1.2 Linear Functions

Objectives

- Calculate and interpret the meaning of the slope of a linear function
- Interpret the physical and graphical meaning of x- and y-intercepts
- Formulate the equation of a line given two points
- Recognize the slope-intercept, point-slope, and standard forms of a line

In this section, you learned how to calculate the slope of a linear function and how to interpret the physical and graphical meaning of slope, x-intercept, and y-intercept. You also discovered how to calculate the equation of a linear function from a set of data and learned three different ways to write a linear equation.

Keywords

Function	A rule that associates each input with *exactly one* output.
Domain	The set of all possible values of the independent variable of a function.
Range	The set of all possible values of the dependent variable of a function.
Graph of a linear function	The line passing through any two points (x_1, y_1) and (x_2, y_2) with $x_1 \neq x_2$.

Slope	The change in the output that occurs when the input is increased by one unit.
y-intercept	The point on the graph where the function intersects the y-axis. It occurs when the value of the independent variable is 0. It is formally written as an ordered pair $(0, b)$, but b itself is often called the y-intercept.
x-intercept	The point on the graph where the function intersects the x-axis. It occurs when the value of the dependent variable is 0. It is formally written as an ordered pair $(a, 0)$, but a itself is often called the x-intercept.
Slope-intercept form of a line	A linear function with slope m and y-intercept $(0, b)$ has the equation $y = mx + b$.
Standard form of a line	A linear equation written as $$Ax + By = C$$ where A, B, and C are real numbers. If $A = 0$, the graph of the equation is a horizontal line. If $B = 0$, the graph of the equation is a vertical line. In the equation, A and B cannot both be zero.
Point-slope form of a line	A linear function written as $$y - y_1 = m(x - x_1)$$ with slope m that passes through the point (x_1, y_1).
Independent variable	Input variable of a function.
Dependent variable	Output variable of a function.
Origin	The point $(0, 0)$.
Vertical Line Test	The test that determines if a graph is a function.
Magnitude	The absolute value of the slope.

Go to CourseMate for Applied CALC for additional practice, printable flashcards, and more! Access at **login.cengagebrain.com**.

2.1 Quadratic Function Models

Objectives
- Use nonlinear functions to model real-life phenomena
- Interpret the meaning of mathematical models in their real-world context
- Model real-life data with quadratic functions

In this section, you learned how to use quadratic regression to model a data set. You also discovered the importance of analyzing a mathematical model before using it to calculate unknown values.

2.2 Exponential Function Models

Objectives
- Use exponential functions to model real-life phenomena
- Interpret the meaning of mathematical models in their real-world context
- Graph exponential functions
- Find the equation of an exponential function from a table
- Model real-life data with exponential functions

In this section, you learned how exponential functions can be used to model rapidly increasing (or decreasing) data sets. You graphed exponential functions and developed exponential models from tables of data and verbal descriptions.

Keywords

Quadratic function	A polynomial function of the form $f(x) = ax^2 + bx + c$ with $a \neq 0$. The graph of a quadratic function is a **parabola**.
Exponential function	If a and b are real numbers with $a \neq 0$, $b > 0$, and $b \neq 1$, then the function $$y = ab^x$$ is called an **exponential function**.

Properties of exponents	If b, m, and n are real numbers with $b > 0$, the following properties hold.

Property
1. $b^{-n} = \dfrac{1}{b^n}$
2. $b^m \cdot b^n = b^{m+n}$
3. $\dfrac{b^m}{b^n} = b^{m-n}$
4. $b^{mn} = (b^m)^n = (b^n)^m$

Annual growth rate	The percentage growth of a quantity over a one-year period.
Degree of a polynomial	The exponent of a polynomial function that has the greatest value.
Axis of symmetry	The vertical line that passes through the vertex of a parabola.
Extrapolation	Predicting unknown function values for inputs that are larger than the maximum input value of the raw data or smaller than the minimum input value of the raw data.
Initial value	The value of a function when the domain value is 0. For an exponential function, $y = ab^x$, the initial value is a.
Base	The value b of the exponential function $y = ab^x$.
Growth factor	A term for the base b when $b > 1$.
Decay factor	A term for the base b when $0 < b < 1$.
Horizontal asymptote	A horizontal line that the graph of a function may approach as the input values approach positive or negative infinity.
Aligned data set	A data set whose input or output values have been adjusted by a constant value.
Depreciation	A negative percentage rate of growth.
Appreciation	A positive percentage rate of growth.

Go to CourseMate for Applied CALC for additional practice, printable flashcards, and more! Access at **login.cengagebrain.com**.

3.1 Average Rate of Change

Objectives
- Calculate the average rate of change of a function over an interval

In this section, you learned how to calculate the average rate of change of a function over a specified interval. You also learned that the units of the average rate of change are the units of the *output* divided by the units of the *input*. You saw that, graphically speaking, the difference quotient is the slope of a secant line.

3.2 Limits and Instantaneous Rates of Change

Objectives
- Estimate the instantaneous rate of change of a function at a point
- Use derivative notation and terminology to describe instantaneous rates of change

In this section, you learned how to use the difference quotient to estimate an instantaneous rate of change. You also discovered that the derivative may be used to calculate the exact value of an instantaneous rate of change of a function at a point.

3.3 The Derivative as a Slope: Graphical Methods

Objectives
- Find the equation of the tangent line of a curve at a given point
- Numerically approximate derivatives from a table of data
- Graphically interpret average and instantaneous rates of change
- Use derivative notation and terminology to describe instantaneous rates of change

In this section, you learned that the derivative at a point is the slope of the tangent line to the graph at that point. You discovered that the tangent line may be used to approximate the value of a function. You also learned how to estimate a derivative from a table of data.

3.4 The Derivative as a Function: Algebraic Methods

Objectives
- Use the limit definition of the derivative to find the derivative of a function
- Use derivative notation and terminology to describe instantaneous rates of change

In this section, you learned how to find the derivative function algebraically. You discovered that it is often easier to expand and simplify $f(x + h)$ before substituting it into the derivative formula. Additionally, you found that when the h in the denominator cannot be eliminated algebraically, you can estimate the derivative numerically.

3.5 Interpreting the Derivative

Objectives
- Interpret the meaning of the derivative in the context of a word problem

In this section, you learned how to interpret the meaning of the derivative verbally in its real-world context. You also discovered two different ways of representing the concept.

Keywords

Average rate of change	The average increase in the output resulting from a one-unit increase in the input.
Secant line	A line connecting any two points of a graph.
Instantaneous rate of change	The rate at which a quantity is changing at a given point.
Left-hand limit	The output value that a function approaches as the input variable approaches a constant value, c, through values less than c.
Right-hand limit	The output value that a function approaches as the input variable approaches a constant value, c, through values greater than c.
Derivative	The slope of the tangent line.
Slope of the secant line	The graphical interpretation of an average rate of change.
Tangent line	The line passing through a point $(c, f(c))$ with slope $f'(c)$.

Key Concepts

The Difference Quotient: An Average Rate of Change

The average rate of change of a function $y = f(x)$ over an interval $[a, b]$ is

$$\frac{f(b) - f(a)}{b - a}$$

The Difference Quotient as an Estimate of an Instantaneous Rate of Change

The *instantaneous rate of change* of a function $y = f(x)$ at a point $(a, f(a))$ may be *estimated* by calculating the **difference quotient of f at a**,

$$\frac{f(a + h) - f(a)}{h}$$

using an h arbitrarily close to 0. (If $h = 0$, the difference quotient is undefined.)

The Limit of a Function

If $f(x)$ is defined for all values of x near c, then

$$\lim_{x \to c} f(x) = L$$

means that as x approaches c, $f(x)$ approaches L.

We say that the limit exists if:

1. L is a finite number and
2. approaching c from the left or right yields the same value of L.

The Derivative of a Function at a Point

The derivative of a function $y = f(x)$ at a point $(a, f(a))$ is

$$f'(a) = \lim_{h \to 0} \frac{f(a + h) - f(a)}{h}$$

$f'(a)$ is read "f prime of a" and is the *instantaneous rate of change* of the function f at the point $(a, f(a))$.

The Graphical Meaning of the Derivative

The derivative of a function f at a point $(a, f(a))$ is the **slope of the tangent line** to the graph of f at that point. The slope of the tangent line at a point is also referred to as the **slope of the curve** at that point.

Numerical Estimate of the Derivative

The derivative of a function f at a point $(a, f(a))$ may be approximated from a table by

$$f'(a) \approx \frac{f(a + h) - f(a - h)}{(a + h) - (a - h)}$$

$$\approx \frac{f(a + h) - f(a - h)}{2h}$$

where h is the horizontal distance between a and $a + h$.

The Derivative Function

The **derivative function** of a function f is given by

$$f'(x) = \lim_{h \to 0} \frac{f(x + h) - f(x)}{h}$$

if the limit exists.

Go to CourseMate for Applied CALC for additional practice, printable flashcards, and more! Access at **login.cengagebrain.com**.

Learning Objectives

4.1 Basic Derivative Rules

Objectives
- Use various forms of derivative notation
- Use the Constant Rule, Power Rule, Constant Multiple Rule, and Sum and Difference Rule to find the derivative of a function

Calculating the difference quotient for a complex function is cumbersome and prone to error. Various alternative forms of derivative notation simplify the process. Several shortcuts can greatly enhance our efficiency in calculating derivatives. In this section you learned how to use the Constant Rule, the Power Rule, the Constant Multiple Rule, and the Sum and Difference Rule to that end. You also learned alternative ways of representing the derivative.

4.2 The Product and Quotient Rules

Objectives
- Use the Product Rule and Quotient Rule to find the derivative of a function

In this section, you learned how to use the Product and Quotient Rules to find the derivative of a function that is written as a product or quotient of factors.

4.3 The Chain Rule

Objectives
- Use the Chain Rule to find the derivative of a function

A composition of two functions occurs when the outputs of one function are the inputs of a second function. In this section you learned how to compose two functions. You also discovered how to use the Chain Rule to find the derivative of a composition of functions.

4.4 Exponential and Logarithmic Rules

Objectives
- Use the Exponential Rule and Logarithmic Rule to find the derivative of a function

In this section, you learned how to use the Exponential Rule and the Logarithmic Rule for derivatives. You discovered that these rules allow you to calculate the exact derivative of exponential and logarithmic functions instead of having to rely on a numerical estimate.

4.5 Implicit Differentiation

Objectives
- Use implicit differentiation to differentiate functions and nonfunctions

In this section, you learned how to use implicit differentiation to find the derivative of functions and nonfunctions. Knowing how to do implicit differentiation will help you understand related rates in the Chapter 5.

Keywords

Leibniz notation	Representing the derivative in the form $\dfrac{dy}{dx}$
Differentiate	To find the derivative.
Differentiation	The process of finding the derivative.
Differentiable function	A function whose derivative exists for all values in its domain.
Implicit differentiation	The process of finding the derivative of a function (or relation) not in the form $y = f(x)$.

Rules

Constant Rule

The derivative of a constant function $y = c$ is

$$y' = 0$$

Power Rule

The derivative of a function $y = x^n$, where x is a variable and n is a nonzero constant, is

$$y' = nx^{n-1}$$

Constant Multiple Rule

The derivative of a differentiable function $g(x) = kf(x)$ with constant k is given by

$$g'(x) = kf'(x)$$

Sum and Difference Rule

The derivative of a differentiable function $h(x) = f(x) \pm g(x)$ is given by

$$h'(x) = f'(x) \pm g'(x)$$

Product Rule

The derivative of a function $h(x) = f(x) \cdot g(x)$ is given by

$$h'(x) = f'(x) \cdot g(x) + g'(x) \cdot f(x)$$

Quotient Rule

The derivative of a function $h(x) = \dfrac{f(x)}{g(x)}$ is given by

$$h'(x) = \frac{f'(x) \cdot g(x) - g'(x) \cdot f(x)}{(g(x))^2}$$

Chain Rule

The derivative of a function $h(x) = f(g(x))$ is given by

$$h'(x) = f'(g(x)) \cdot g'(x)$$

Generalized Power Rule

Let $u = g(x)$. The derivative of a function $f(x) = u^n$ is given by

$$f'(x) = nu^{n-1}u'$$

Chain Rule: Alternative Form

If $y = f(x)$ and $x = g(t)$, then $y = f(g(t))$ and

$$\frac{dy}{dt} = \frac{dy}{dx} \cdot \frac{dx}{dt}$$

Exponential Rule

The derivative of a function $f(x) = b^x$ is given by

$$f'(x) = (\ln b)b^x$$

Exponential Rule: Special Case

The derivative of a function $f(x) = e^x$ is given by

$$f'(x) = e^x$$

Generalized Exponential Rule

Let $u = g(x)$. The derivative of a function $f(x) = b^u$ is given by

$$f'(x) = (\ln b)b^u u'$$

If $b = e$, the rule simplifies to

$$f'(x) = e^u u'$$

Logarithmic Rule

The derivative of a function $f(x) = \log_b x$ is given by

$$f'(x) = \frac{1}{\ln b} \cdot \frac{1}{x}$$

Logarithmic Rule: Special Case

The derivative of a function $f(x) = \ln x$ is given by

$$f'(x) = \frac{1}{x}$$

Generalized Logarithmic Rule

The derivative of a function $f(x) = \ln u$ where u is a function of x is given by

$$f'(x) = \frac{1}{u}u'$$

*Go to CourseMate for Applied CALC for additional practice, printable flashcards, and more! Access at **login.cengagebrain.com**.*

Learning Objectives

5.1 Maxima and Minima

Objectives
- Find relative and absolute extrema
- Use the First Derivative Test to find relative extrema

In this section, you learned how to find the relative and absolute extrema of a function. You discovered that the First Derivative Test is an excellent tool for finding the location of relative extrema algebraically.

5.2 Applications of Maxima and Minima

Objectives
- Analyze and interpret revenue, cost and profit functions

In this section, you learned how to find marginal revenue, marginal cost, and marginal profit. You also learned how to use the derivative to optimize specified areas and volumes.

5.3 Concavity and the Second Derivative

Objectives
- Use the Second Derivative Test to find relative extrema
- Determine the concavity of the graph of a function

In this section, you learned how to use the second derivative in discussing the graphical concepts of concavity and inflection points. You saw how the Second Derivative Test may be used as an alternative means of finding relative extrema. You also discovered that velocity is the derivative of the position function and that acceleration is the second derivative of the position function. You also learned some curve sketching techniques.

5.4 Related Rates

Objectives
- Solve related-rate problems

In this section, you learned how related rates can be used to measure how the rate of change in one variable affects the rate of change in another variable. You discovered that technology can simplify the arithmetic computations of related-rate problems.

Keywords

Extremum (extrema)	A maximum or minimum value of a function.
Relative maximum	A function value larger than all other function values in an open interval (a, b).
Relative minimum	A function value smaller than all other function values in an open interval (a, b).
Absolute maximum	A function value larger than all other values of a function.
Absolute minimum	A function value smaller than all other values of a function.
Critical value	A stationary or singular value.
Stationary value	A value c in the domain of a function f with $f'(c) = 0$.
Singular value	A value c in the domain of a function f with $f'(c)$ undefined.
Critical point	The coordinate $(c, f(c))$ for the critical value c.
Marginal revenue (cost, profit)	The approximate change in revenue (cost, profit) when one more unit is produced.
Concavity	The upward or downward curvature of a graph.
Second derivative	The derivative of the derivative.
Inflection point	The point on a graph where the concavity changes.
Point of diminishing returns	The point where the instantaneous rate of change reaches a maximum.

Tests and Techniques

First Derivative Test

Let $(c, f(c))$ be a critical point of a nonconstant, continuous function f.

- If f' changes from positive to negative at $x = c$, then a relative maximum occurs at $(c, f(c))$.
- If f' changes from negative to positive at $x = c$, then a relative minimum occurs at $(c, f(c))$.
- If f' doesn't change sign at $x = c$, then a relative extremum does not occur at $(c, f(c))$.

Constructing a Derivative Sign Chart

1. Label a number line with the critical values of the function f.
2. Write f' next to the number line.
3. Evaluate f' at a value in each of the number line intervals.
4. Record a "+" if the derivative is positive and a "–" if the derivative is negative.
5. Use the First Derivative Test to determine where relative maxima and minima occur and record the results on the chart.

The Second Derivative Test

Let f be a continuous function with $f'(c) = 0$. If

- $f''(c) > 0$, then a relative minimum of f occurs at $x = c$.
- $f''(c) < 0$, then a relative maximum of f occurs at $x = c$.
- $f''(c) = 0$, then the test is inconclusive.

Curve Sketching

To graph a function $f(x)$ on an interval $[a, b]$, complete the following steps. After each step, graph the corresponding point(s).

1. Find the x-intercepts of $f(x)$ by setting $f(x) = 0$ and solving for x.
2. Find the y-intercept of $f(x)$ by evaluating $f(0)$.
3. Find the relative extrema and increasing/decreasing behavior of $f(x)$ by constructing a sign chart for $f'(x)$.
4. Find the absolute extrema of $f(x)$ by evaluating $f(x)$ at each critical value and at the endpoints $x = a$ and $x = b$.
5. Find the inflection points and concavity of $f(x)$ by constructing a sign chart for $f''(x)$.
6. Connect the points, paying attention to the increasing/decreasing behavior and concavity of $f(x)$.

Go to CourseMate for Applied CALC for additional practice, printable flashcards, and more! Access at **login.cengagebrain.com**.

reviewcard **CHAPTER 6**
THE INTEGRAL

Learning Objectives

6.1 Indefinite Integrals

Objectives
- Use integration rules to find the antiderivative of a function

In this section you learned how to find the equation of some functions that have a given derivative through the process of integration. You also discovered how to use basic integration rules to integrate a variety of functions.

6.2 Integration by Substitution

Objectives
- Use the method of substitution to integrate functions

In this section you learned how to work with differentials. You used integration by substitution to find the antiderivative of a function that didn't initially appear to be integrable. You also learned that there are functions that we can't integrate simply by reversing the derivative rules.

6.3 Using Sums to Approximate Area

Objectives
- Estimate the area between a curve and the x-axis by using left- and right-hand sums

In this section, you learned how to use left- and right-hand sums to approximate the area between the graph of a function and the horizontal axis. A solid understanding of these concepts will help you grasp the material presented in the next section.

6.4 The Definite Integral

Objectives
- Apply definite integral properties
- Calculate the exact area between a curve and the x-axis by using definite integrals

In this section, you learned how to use summation notation. You discovered that the definite integral is the sum of the signed areas. You also learned how to manipulate definite integrals using definite integral properties.

6.5 The Fundamental Theorem of Calculus

Objectives
- Apply the Fundamental Theorem of Calculus
- Find the function for accumulated change given a rate-of-change function

In this section, you learned how to use one of the most powerful tools in calculus: the Fundamental Theorem of Calculus. You used this theorem to calculate a bounded area quickly and to determine the accumulated change of a rate-of-change function.

Keywords

Integration	The process of finding a function that has a given derivative.
Indefinite integral	The general form of a function with a given derivative.
Integrand	The function to be integrated.
Integrable function	A function that may be integrated.
Differential	The terms dy, dx, dt, etc.
Integration by substitution	The process of rewriting an integrand with a new variable to facilitate integration.
Left-hand (right-hand) sum	The sum of the areas of rectangles whose upper left-hand (right-hand) corners touch the graph of the function.
Riemann sum	A left- or right-hand sum.
Definite Integral	The expression $\int_a^b f(x)\, dx$.
Limits of integration	The subscript and superscript values attached to the integral sign.
Sum of the signed areas	The area of the enclosed regions above the x-axis minus the area of the enclosed regions below the x-axis.

Rules and Concepts

Power Rule for Integrals

Let $f(x) = x^n$, where x is a variable and n is a constant with $n \neq -1$. Then

$$\int x^n \, dx = \frac{x^{n+1}}{n+1} + C, \, n \neq -1$$

Constant Multiple Rule for Integrals

Let $f(x) = k \cdot g(x)$, where x is a variable and k is a constant. Then

$$\int (k \cdot g(x)) \, dx = k \cdot \int g(x) \, dx$$
$$= k \cdot G(x) + C$$
$$g(x) = G'(x)$$

Constant Rule for Integrals

Let k be a constant. Then

$$\int k \, dx = kx + C.$$

Sum and Difference Rule for Integrals

Let $f(x)$ and $g(x)$ be integrable functions of x. Then

$$\int (f(x) \pm g(x)) \, dx = \int f(x) \, dx \pm \int g(x) \, dx$$

Exponential Rule for Integrals

Let $a > 0$ and $a \neq 1$. Then

$$\int a^x \, dx = \frac{1}{\ln(a)} a^x + C$$

Integral Rule for $\frac{1}{x}$

If $x \neq 0$, then

$$\int \frac{1}{x} \, dx = \ln|x| + C$$

Meaning of $\int_a^b f(x) \, dx$ for a Function f

Let f be a function on $[a, b]$. The definite integral

$$\int_a^b f(x) \, dx$$

is the sum of the areas of the enclosed regions above the x-axis minus the sum of the areas of enclosed regions below the x-axis on the interval $[a, b]$.

Definite Integral Properties

For integrable functions f and g, the following properties hold:

- $\int_a^a f(x) \, dx = 0$

- $\int_a^b f(x) \, dx = -\int_b^a f(x) \, dx$

- $\int_a^b k \cdot f(x) \, dx = k \int_a^b f(x) \, dx$ for constant k

- $\int_a^b (f(x) \pm g(x)) \, dx = \int_a^b f(x) \, dx \pm \int_a^b g(x) \, dx$

- $\int_a^b f(x) \, dx = \int_a^c f(x) \, dx + \int_c^b f(x) \, dx$ for $a \leq c \leq b$

Fundamental Theorem of Calculus

Let f be a continuous function on $[a, b]$. Then

$$\int_a^b f(x) \, dx = F(b) - F(a)$$

where F is any antiderivative of f.

Accumulated Change of a Function

Let f be the rate-of-change function (derivative) of F on $[a, b]$. Then

$$\int_a^b f(x) \, dx = F(b) - F(a)$$

is the accumulated change in F over the interval $[a, b]$.

Go to CourseMate for Applied CALC for additional practice, printable flashcards, and more! Access at **login.cengagebrain.com**.

Learning Objectives

7.1 Integration by Parts

Objectives
- Use the integration by parts method to find the antiderivative of a function

In this section, you learned how to use the following method of integration by parts to integrate a function:

Detailed Steps for the Integration by Parts Method

Let u and v be differentiable functions of x. To integrate $\int u\,dv$, do the following:

1. Identify u and dv. The product of these two factors must equal the entire integrand coupled with the associated dx.
2. Differentiate u and integrate dv with respect to x.
3. Write the expression $uv - \int v\,du$ in terms of x.
4. Integrate $\int v\,du$ to eliminate the integral sign.
5. Simplify.

This technique is among the more challenging integration methods and should be used only after basic integration methods and the substitution method have failed.

7.2 Area Between Two Curves

Objectives
- Calculate the area of the region between two continuous curves

In this section, you learned how to calculate the area of the bounded region between two graphs. You also learned that the definite integral of the difference between two rate-of-change functions is equal to the difference between the accumulated changes in each of the individual functions.

7.3 Differential Equations and Applications

Objectives
- Solve introductory-level differential equations
- Apply Newton's Law of Heating and Cooling
- Analyze and interpret the real-life meaning of differential equations

In this section, you learned how to use integral calculus to solve first-order differential equations. You also learned how to apply differential equations in Newton's Law of Heating and Cooling and in the continuous compound interest formula.

7.4 Differential Equations: Limited Growth and Logistic Growth Models

Objectives
- Find the equation of a limited growth model
- Find the equation of a logistic growth model
- Analyze and interpret the real-life meaning of differential equations

This section covered how to use differential equations to find limited growth and logistic growth models. You also practiced interpreting the results of differential equations. Both types of models approach a constant value as the domain values grow large; however, near the origin the models behave quite differently. A limited growth model grows very rapidly at first, whereas a logistic model initially grows very slowly.

Keywords

First-order differential equation	An equation that relates an unknown function and its first derivative.
Separable first-order differential equation	A differential equation that may be written in the form $dy/dx = f(x)/g(y)$.

Rules and Concepts

Area Between Two Curves

If the graph of f lies above the graph of g on an interval $[a, b]$, then the area of the region between the two graphs from $x = a$ to $x = b$ is given by

$$\int_a^b (f(x) - g(x))\, dx$$

Difference of Two Accumulated Changes

If f and g are continuous rate-of-change functions defined on the interval $[a, b]$, then the difference of the accumulated change of f from $x = a$ to $x = b$ and the accumulated change of g from $x = a$ to $x = b$ is given by

$$\int_a^b (f(x) - g(x))\, dx$$

Solving Separable First-Order Differential Equations

The solution to the first-order differential equation $\dfrac{dy}{dx} = \dfrac{f(x)}{g(y)}$ is obtained by moving the x and y variables to opposite sides of the equal sign and integrating. That is,

$$\int g(y)\, dy = \int f(x)\, dx.$$

Continuous Compound Interest

The value of an investment account earning continuous compound interest is

$$A = Pe^{kt} \text{ dollars}$$

where A is the value of the investment after t years, P is the initial value of the investment (in dollars), and k is the continuous interest rate.

The rate of change in the value of the investment is given by

$$\frac{dA}{dt} = kA \frac{\text{dollars}}{\text{year}}$$

Newton's Law of Heating and Cooling

Let T be the temperature of an object at time t and A be the temperature of the environment surrounding the object (ambient temperature). Then

$$\frac{dT}{dt} = k(T - A)$$

where k is a constant that varies depending on the physical properties of the object.

Limited Growth

Assume that the rate of growth of a function y with maximum value M is proportional to the difference between the present value of y and M. That is,

$$\frac{dy}{dt} = k(M - y)$$

The solution to the differential equation with initial condition

$$y(0) = 0$$

is given by

$$y = M(1 - e^{-kt}).$$

Logistic Growth

Assume that the rate of growth of a function y with maximum value M is proportional to the product of the present value of y and the difference between the present value of y and M. That is,

$$\frac{dy}{dt} = ky(M - y)$$

The solution to this differential equation with initial condition

$$y(0) = \frac{M}{1 + S}$$

is given by

$$y = \frac{M}{1 + Se^{-kMt}}$$

Go to CourseMate for Applied CALC for additional practice, printable flashcards, and more! Access at **login.cengagebrain.com**.

Learning Objectives

8.1 Multivariable Functions

Objectives
- Analyze and interpret multivariable mathematical models
- Evaluate multivariable functions

In this section, you learned how to evaluate multivariable functions and saw how functions of two variables may be represented graphically. You also discovered some applications of multivariable functions.

8.2 Partial Derivatives

Objectives
- Analyze and interpret multivariable mathematical models
- Calculate first- and second-order partial derivatives of multivariable functions
- Find the equation of a cross section of a graph of a surface
- Determine the meaning of partial rates of change

In this section, you learned how to calculate partial derivatives. You saw that these partial rates of change allow you to determine the effect on the output of a change in one of the input variables.

8.3 Multivariable Maxima and Minima

Objectives
- Analyze and interpret multivariable mathematical models
- Find critical points of multivariable functions
- Locate relative extrema and saddle points of two-variable function graphs

This section examined relative extrema in the context of multivariable functions. Three-dimensional views help when visualizing extrema. In addition to extrema, three-dimensional surfaces may also have a saddle point. The section reexamined critical points and second derivative tests, this time for two-variable functions, and covered various techniques used to find relative extrema and saddle points of two-variable functions.

8.4 Constrained Maxima and Minima and Applications

Objectives
- Analyze and interpret multivariable mathematical models
- Use the Lagrange Multiplier Method to find constrained maxima and minima

This section covered how to use the Lagrange Multiplier Method to solve constrained multivariable optimization problems. Even though graphing utilities are helpful, they are not foolproof in identifying the location of constrained extrema.

Keywords

Multivariable functions	A function with two or more independent variables.
Surface	A three-dimensional shape representing a function of two variables.
Independent variable	The input variable of a function.
Dependent variable	The output variable of a function.
Domain	The set of all inputs.
Range	The set of all outputs.
Cross section	The two-dimensional face that results from slicing a three-dimensional surface.
Saddle point	A point (a, b, c) that is a relative maximum of one cross-section graph and a relative minimum of another cross-section graph.

Rules and Concepts

Function of Two Variables

A **function f of two variables** is a rule that associates each ordered pair of inputs (x, y) with a single output $z = f(x, y)$. The graph of a function of two variables is a three-dimensional **surface**.

Cobb-Douglas Production Function

$$f(L, C) = kL^m C^n$$

Partial Derivatives

The **partial derivatives** of a function $f(x, y)$ are given by

$\dfrac{\partial f}{\partial x}$ (read "the partial of f with respect to x")

$\dfrac{\partial f}{\partial y}$ (read "the partial of f with respect to y")

To calculate $\dfrac{\partial f}{\partial x}$, differentiate the function, treating the y variable as a constant.

To calculate $\dfrac{\partial f}{\partial y}$, differentiate the function, treating the x variable as a constant.

Partial Derivatives (Alternative Notation)

The **partial derivatives** of a function $f(x, y)$ may alternatively be written as

f_x (read "the partial of f with respect to x")

f_y (read "the partial of f with respect to y")

Geometric Interpretation of Partial Derivatives

The partial derivatives of a function $f(x, y)$ represent rates of change of the graph of the surface.

- $f_x(x, y)$ is the rate of change of the graph of the function $f(x, y)$ when y is held constant. It is referred to as the *slope in the x-direction*.
- $f_y(x, y)$ is the rate of change of the graph of the function $f(x, y)$ when x is held constant. It is referred to as the *slope in the y-direction*.

Second-Order Partial Derivatives

The **second-order partial derivatives** of a function $f(x, y)$ may be written as

$f_{xx} = \dfrac{\partial}{\partial x}\left(\dfrac{\partial f}{\partial x}\right)$ (read "the second partial of f with respect to x")

$f_{yy} = \dfrac{\partial}{\partial y}\left(\dfrac{\partial f}{\partial y}\right)$ (read "the second partial of f with respect to y")

$f_{xy} = \dfrac{\partial}{\partial y}\left(\dfrac{\partial f}{\partial x}\right)$ (read "the mixed second partial derivative of f with respect to x then y")

$f_{yx} = \dfrac{\partial}{\partial x}\left(\dfrac{\partial f}{\partial y}\right)$ (read "the mixed second partial derivative of f with respect to y then x")

Relative Extrema of Multivariable Functions

Consider the graph of a surface with equation $z = f(x, y)$. The value $f(a, b)$ is

- a relative minimum if $f(a, b) \leq f(x, y)$ for all (x, y) in some region surrounding (a, b).
- a relative maximum if $f(a, b) \geq f(x, y)$ for all (x, y) in some region surrounding (a, b).

Critical Point of a Two-Variable Function

An ordered pair (c, d) in the domain of f is a **critical point** of f if

$$f_x(c, d) = 0 \text{ and } f_y(c, d) = 0$$

(Both statements must be true in order for (c, d) to be a critical point.)

When we refer to the critical point (c, d), we mean that (c, d) is the pair of input values whose output is $f(c, d)$.

Second Derivative Test for Two-Variable Functions (D-Test)

To determine if a relative extremum of a function f occurs at a critical point (c, d), first calculate

$$D(c, d) = f_{xx}(c, d) \cdot f_{yy}(c, d) - (f_{xy}(c, d))^2$$

Then

- f has a **relative maximum** at (c, d) if $D(c, d) > 0$ and $f_{xx}(c, d) < 0$.
- f has a **relative minimum** at (c, d) if $D(c, d) > 0$ and $f_{xx}(c, d) > 0$.
- f has a **saddle point** at (c, d) if $D(c, d) < 0$.
- The test is inconclusive if $D(c, d) = 0$.

Lagrange Multiplier Method for Solving Constrained Optimization Problems

To optimize an objective function $f(x, y)$ subject to the constraint $g(x, y) = 0$, do the following:

1. Define $F(x, y, \lambda) = f(x, y) + \lambda g(x, y)$.
2. Set the partial derivatives of F equal to zero:

$$F_x = 0 \qquad F_y = 0 \qquad F_\lambda = 0$$

and solve to find the critical points.
3. The solution to the constrained optimization problem (if it exists) will occur at one of these critical points.

Go to CourseMate for Applied CALC for additional practice, printable flashcards, and more! Access at **login.cengagebrain.com**.

Graphing a Function

Graphing calculators, such as the TI-84 Plus, can quickly draw the graph of a function.

1. Bring up the graphing list by pressing the (Y =) button.

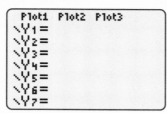

2. Type in the function using the (X, T, θ, n) button for the variable and the (^) button to place an expression in an exponent. Make sure you use parentheses as needed.

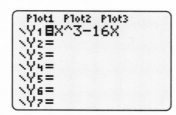

3. Specify the size of the viewing window by pressing the (WINDOW) button and editing the parameters. Xmin is the minimum x-value, Xmax is the maximum x-value, Ymin is the minimum y-value, and Ymax is the maximum y-value. Xscl and Yscl are used to specify the spacing of the tick marks on the graph.

```
WINDOW
 Xmin=-5
 Xmax=5
 Xscl=1
 Ymin=-45
 Ymax=45
 Yscl=5
 Xres=1
```

4. Draw the graph by pressing the (GRAPH) button.

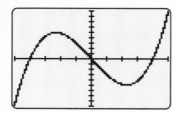

Entering a Table of Data

1. Bring up the Statistics Menu by pressing the (STAT) button.

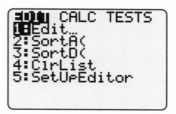

2. Bring up the List Editor by selecting 1:Edit...and pressing (ENTER).

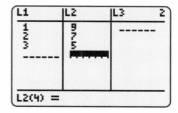

3. Clear the lists. If there are data in the list, use the arrows to move the cursor to the list heading L1. Press the (CLEAR) button and press (ENTER). This clears all of the list data. Repeat for each list with data. (*Warning:* Be sure to use (CLEAR) instead of (DELETE), which removes the entire column.)

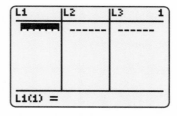

4. Enter the numeric values of the *inputs* in list L1, pressing (ENTER) after each entry.

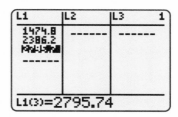

5. Enter the numeric values of the *outputs* in list L2, pressing ENTER after each entry.

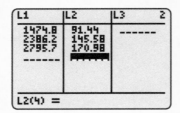

Linear Regression

The TI-84 Plus calculator has a linear regression feature that does the necessary calculations automatically.

1. Enter the data into L1 and L2.

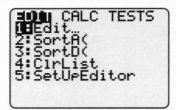

2. Return to the Statistics Menu by pressing the STAT button.

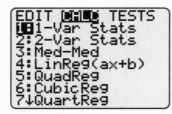

3. Bring up the Calculate Menu by using the arrows to select CALC.

```
EDIT CALC TESTS
1 1-Var Stats
2:2-Var Stats
3:Med-Med
4:LinReg(ax+b)
5:QuadReg
6:CubicReg
7↓QuartReg
```

4. Calculate the linear equation of the model by selecting 4:LinReg(ax+b) and pressing ENTER twice. The line of best fit is $y = 0.06008x + 2.6928$ and has correlation coefficient $r = 0.9999$.

```
LinReg
 y=ax+b
 a=.060078251
 b=2.692769342
 r²=.9998949304
 r=.9999474638
```

5. If the correlation coefficient r does not appear, perform this step then repeat the regression. Press 2nd then 0, scroll to DiagnosticOn, and press ENTER twice. This will ensure that the correlation coefficient r and the coefficient of determination r^2 will appear.

```
CATALOG
 DependAuto
 det<
 DiagnosticOff
▶DiagnosticOn
 dim<
 Disp
 DispGraph
```

Graphing a Regression Equation

1. Press the Y= button to bring up the Graphing List, immediately after calculating the regression equation.

```
Plot1 Plot2 Plot3
\Y1=
\Y2=
\Y3=
\Y4=
\Y5=
\Y6=
\Y7=
```

2. Bring up the Variables Menu by pressing the VARS key.

```
VARS Y-VARS
1:Window…
2:Zoom…
3:GDB…
4:Picture…
5 Statistics…
6:Table…
7:String…
```

3. Bring up the Variables: Statistics Menu by selecting `5:Statistics...` and pressing (ENTER).

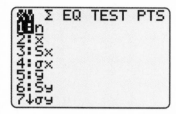

4. Bring up the Equation Menu by moving the cursor to EQ.

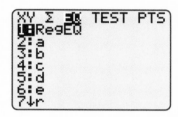

5. Select the regression equation by choosing `1:RegEQ` and pressing (ENTER). This pastes the regression equation into the Graphing List.

6. Graph the function by pressing the (GRAPH) button. You may need to adjust the viewing window to see the graph.

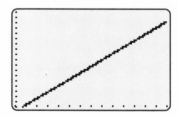

*Technology guides are available on CourseMate for Applied CALC. Access at **login.cengagebrain.com**.*

Drawing a Scatter Plot

The graph of a set of discrete data is called a scatter plot. We can draw a scatter plot by hand, using graph paper, or using a calculator.

1. Bring up the Statistics Plot Menu by pressing the (2nd) button, then the (Y =) button.

2. Open Plot1 by pressing (ENTER).

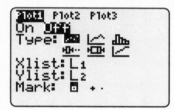

3. Turn on Plot1 by moving the cursor to On and pressing (ENTER). Confirm that the other menu entries are as shown.

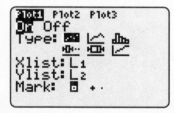

4. Graph the scatter plot by pressing (ZOOM) and scrolling to `9:ZoomStat`. Press (ENTER). This will graph the entire scatter plot along with any functions in the Graphing List. The ZoomStat feature automatically adjusts the viewing window so that all of the data points are visible.

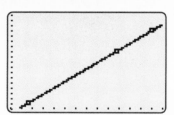

The graph shows the data points along with the linear model. We see that the linear model fits the data well.

Quadratic Regression

1. Enter the data using the Statistics Menu List Editor. (Refer to Entering a Table of Data if you've forgotten how to do this.)

L1	L2	L3	3
0	10280	------	
1	10273		
2	10592		
3	10611		
4	9871		
5	11213		
6	11804		

L3(1) =

2. Bring up the Statistics Menu Calculate feature by pressing (STAT) and using the blue arrows to move to the CALC menu. Then select item 5:QuadReg and press (ENTER).

```
EDIT CALC TESTS
1:1-Var Stats
2:2-Var Stats
3:Med-Med
4:LinReg(ax+b)
5QuadReg
6:CubicReg
7↓QuartReg
```

3. If you want to automatically paste the regression equation into the (Y=) editor, press the key sequence (VARS) Y-VARS; 1:Function; 1:Y1 and press (ENTER). Otherwise press (ENTER).

```
QuadReg
  y=ax²+bx+c
  a=157.8485883
  b=-770.6397775
  c=10268.35154
  R²=.9979069591
```

Aligning a Data Set

1. Enter the data using the Statistics Menu List Editor. (Refer to Entering a Table of Data if you've forgotten how to do this.)

L1	L2	L3	2
0	275	------	
10	237		
20	223		
30	217		
------	------		

L2(5) =

2. Move the cursor to the top of L3. We want the entries in L3 to equal the entries in L2 minus the amount of the vertical shift (in this case 210). To do this, we must enter the equation L3=L2-210.

L1	L2	L3	3
0	275	------	
10	237		
20	223		
30	217		
------	------	------	

L3 =

3. Press (2nd) then (2) to place L2 on the equation line at the bottom of the viewing window. Then press (−) and (2)(1)(0) to subtract 210.

L1	L2	L3	3
0	275	------	
10	237		
20	223		
30	217		
------	------		

L3 =L₂-210

4. Press (ENTER) to display the list of aligned values in L3.

L1	L2	L3	3
0	275	65	
10	237	27	
20	223	13	
30	217	7	
------	------	------	

L3(5) =

Exponential Regression

1. Enter the data using the Statistics Menu List Editor. (Refer to Entering a Table of Data if you've forgotten how to do this.)

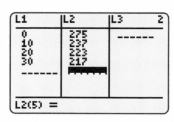

L3(5) =

2. Bring up the Statistics Menu Calculate feature by pressing (STAT) and using the blue arrows to move to the CALC menu. Then select item Ø: ExpReg and press (ENTER).

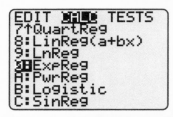

3. If the data to be evaluated are in L1 and L2, press (ENTER). Otherwise, go to step 4. (If you want to automatically paste the regression equation into the (Y =) editor, press the key sequence (VARS) *Y*-VARS; 1:Function; 1:Y1 before pressing (ENTER).)

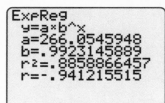

4. If the data to be evaluated are in L1 and L3, press the key sequence (2nd) (1) (,) (2nd) (3) to place the entries L1 and L3 on the home screen.

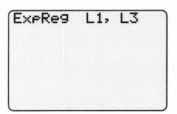

5. Press (ENTER) to display the exponential model. (If you want to automatically paste the regression equation into the (Y =) editor, press the key sequence (VARS) *Y*-VARS; 1:Function; 1:Y1 before pressing (ENTER).)

```
ExpReg
 y=a*b^x
 a=60.8078408
 b=.9285201575
 r²=.9938920452
 r=-.9969413449
```

---

*Technology guides are available on CourseMate for Applied CALC. Access at **login.cengagebrain.com**.*

Calculating the Difference Quotient for Different Values of *h*

1. Enter the function $S(t)$ as Y1 by pressing the (Y=) button and typing the equation.

```
Plot1 Plot2 Plot3
\Y1■.00713X^3-.1
11X^2+1.42X+31.6
\Y2=
\Y3=
\Y4=
\Y5=
```

2. Press the (MATH) button, scroll to Ø: Solver . . ., and press (ENTER). (*Hint:* It is quicker to scroll up than to scroll down.)

```
MATH NUM CPX PRB
4↑3√(
5:×√(
6:fMin(
7:fMax(
8:nDeriv(
9:fnInt(
0:Solver…
```

3. If the Solver already contains an equation, you may see something similar to the graphic shown. Press the blue up arrow and then press (CLEAR) to delete the equation.

```
X^2+3*A+G=0
  X=2
  A=46
  G=7
  bound={-1E99,1…
```

4. Notice that the equation is set equal to zero. We may rewrite the difference quotient

$$D = \frac{Y_1(x + h) - Y_1(x)}{h} \text{ as } 0 = \frac{Y_1(x + h) - Y_1(x)}{h} - D.$$

Type in the second equation using the (VARS);

Y-VARS; 1:Function menu sequence to enter the function Y1. To enter the variable H, press the (ALPHA) (^) key sequence. To enter the variable D, press the (ALPHA) (X^{-1}) key sequence. Press (ENTER).

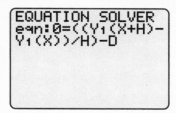

```
EQUATION SOLVER
eqn:0=((Y1(X+H)-
Y1(X))/H)-D
```

5. Enter the value of *x* where you want to evaluate the difference quotient and enter a small positive value of *H*. Move your cursor to the *D* variable and press the (ALPHA) (ENTER) key sequence to solve for the difference quotient *D*. Repeat for various values of *H*.

```
((Y1(X+H)-Y1(…=0
  X=9
  H=.001
■D=1.154671517
  bound={-1E99,1…
■left-rt=0
```

---

*Technology guides are available on CourseMate for Applied CALC. Access at **login.cengagebrain.com**.*

Calculating a Derivative with *nDeriv(*

1. Enter the function equation using the $\boxed{Y =}$ editor. (In this case, we entered *C* in Y1, *P* in Y2, and *S* in Y3. We input the variables Y1 and Y2 into equation Y3 by using the $\boxed{\text{VARS}}$ *Y-VARS; 1:Function* key sequence and selecting the respective functions.)

2. Press $\boxed{\text{2nd}}$ $\boxed{\text{MODE}}$ to quit and return to the home screen. Then press $\boxed{\text{MATH}}$ and select 8:nDeriv(. Then press $\boxed{\text{ENTER}}$.

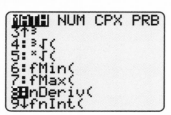

3. The nDeriv(function requires three input values: the function name, the variable of differentiation, and the value of the variable. Enter the appropriate values.

```
nDeriv(Y₃,X,8
```

4. Press $\boxed{\text{ENTER}}$ to display the result.

```
nDeriv(Y₃,X,8)
          -39895.7604
```

---

*Technology guides are available on CourseMate for Applied CALC. Access at **login.cengagebrain.com**.*

Evaluating the Derivative at a Point

1. Enter the function $f(x)$ as Y1 by pressing the (Y =) button and typing the equation.

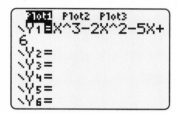

2. Graph the function over the specified domain. (You may need to press (WINDOW) and adjust the Xmin and Xmax settings. Press (ZOOM) and 0:ZoomFit then (ENTER) to automatically adjust the y-values so that the entire graph will appear on screen.)

3. Press (2nd) then (TRACE) to bring up the CALCULATE menu. Select item 6:dy/dx and press (ENTER).

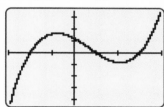

4. Type in the x-value of the point where you want to evaluate the derivative or use the blue arrows to select the point graphically. Then press (ENTER).

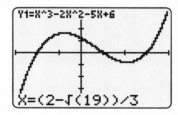

5. The value of the derivative is displayed at the bottom of the screen. (You can assume that a value like "1E-6" is equal to zero. 1E-6 = 1×10^{-6} = 0.000001. The error occurs because the calculator uses numerical methods to differentiate.)

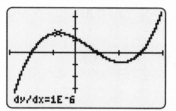

Finding the x-Intercepts of a Function

1. Enter the function $f(x)$ as Y1 by pressing the (Y =) button and typing the equation.

2. Graph the function over the specified domain. (Press (WINDOW) and adjust the Xmin and Xmax settings. Press (ZOOM) then 0:ZoomFit then (ENTER) to automatically adjust the y-values so that the entire graph will fit on the screen.)

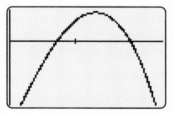

3. Press (2nd) then (TRACE) to bring up the CALCULATE menu. Select item 2:zero and press (ENTER).

4. The calculator asks `Left Bound?`. Enter an *x*-value that is smaller than the *x*-intercept. Or, if you prefer, use the blue arrows to select a point to the left of the *x*-intercept visually.

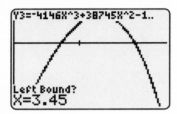

5. The calculator asks `Right Bound?`. Enter an *x*-value that is larger than the *x*-intercept. Or, if you prefer, use the blue arrows to select a point to the right of the *x*-intercept visually.

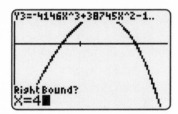

6. The calculator asks `Guess?`. You may enter a guess or simply press `ENTER`. The calculator returns the coordinates of the *x*-intercept. Repeat Steps 3 through 6 to find additional *x*-intercepts.

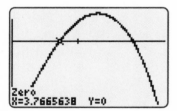

Creating a Table of Values

1. Use the `Y =` editor to enter the function(s) to be evaluated. In this case we enter a function *W* and its first and second derivatives, *W'* and *W"*.

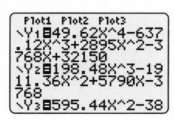

2. Press `2nd` `WINDOW` to open Table Setup. `TblStart=` is the *x*-value where we want the table to begin, and Δ`Tbl` is the distance between consecutive *x*-values. We will start the table at $x = 0$ and space values one unit apart.

3. Press `2nd` `GRAPH` to display the table.

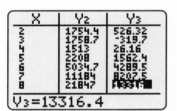

4. To see additional function values, use the blue arrows to navigate between rows and columns.

5. If you prefer to select the *x*-values of the function(s) manually, press `2nd` `WINDOW` and select `Indpnt:Ask`.

6. Press `2nd` `GRAPH` to bring up the table, then enter the desired *x*-values.

Using the Equation Solver

1. Press (MATH) `0:Solver...` to bring up the Equation Solver. Immediately press the blue up arrow. Press (CLEAR) to delete any equation that may be visible.

```
EQUATION SOLVER
eqn:0=
```

2. Enter the equation you need to solve, using the alpha keys for variables. We use the equation from Section 5.4, Example 4. Since the calculator requires the equation to be set equal to zero, we must subtract $\dfrac{dF}{dt}$ from both sides of our equation. For our equation, we let $\dfrac{dF}{dt} = A$, $\dfrac{dT}{dt} = B$, and $\dfrac{dw}{dt} = C$.

```
EQUATION SOLVER
eqn:0=(.020425*C
-.5*(.303107*W^-
.5)*C)*(91.4-T)+
B*(.474677-.0204
25W+.303107*W^.5
)-A
```

3. Press (ENTER) to display the variable input menu. Edit the variables to reflect the known values from Section 5.4, Example 4.

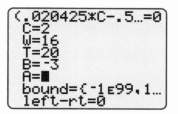

```
(.020425*C-.5...=0
C=2
W=16
T=20
B=-3
A=■
bound={-1ε99,1...
left-rt=0
```

4. Move the cursor to the line containing A= and press (ALPHA) (ENTER) to solve the equation for A.

```
(.020425*C-.5...=0
C=2
W=16
T=20
B=-3
▪A=-6.57468495
bound={-1ε99,1...
▪left-rt=0
```

---

*Technology guides are available on CourseMate for Applied CALC. Access at **login.cengagebrain.com**.*

---

tech card CHAPTER 6
THE INTEGRAL

Finding Riemann Sums with Riemann.8xp

1. Press the (PRGM) key to bring up the list of programs. Select the program RIEMANN. Press (ENTER) twice.

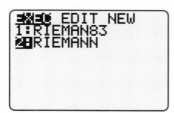

2. The introduction and credit screen is displayed. Press (ENTER).

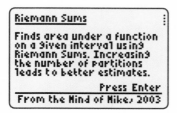

3. To find a Riemann sum of a function f on an interval [a, b] with n rectangles, enter the following:

   f(x): (Enter the function f.)
   Lower Bound: a
   Upper Bound: b
   Partitions: n

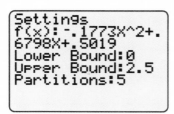

4. The calculator draws the graph of f and draws vertical lines at x = a and x = b. Press (ENTER).

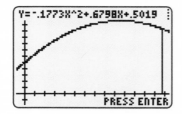

5. The calculator displays an option screen. To calculate the left-hand sum, select 2:LEFT SUM and press (ENTER).

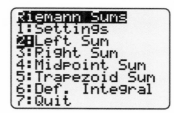

6. The calculator graphs the function, draws the left-hand rectangles, and displays the value of the left-hand sum. Press (ENTER) to return to the option screen.

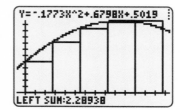

7. Select 3:RIGHT SUM. The calculator graphs the function, draws the right-hand rectangles, and displays the value of the right-hand sum. Press (ENTER) to return to the option screen.

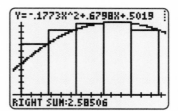

8. Select 5:TRAPEZOID SUM to calculate the average of the left- and right-hand sums. Press (ENTER) to return to the option screen.

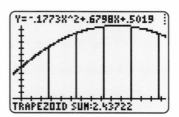

9. Select 1:Settings to change the function, interval, or number of rectangles. Otherwise, select 7:QUIT to quit the program.

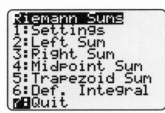

Finding a Definite Integral Graphically

When working definite integral problems, it is often helpful to use a graphing calculator to verify the accuracy of our results. The following steps describe how to find a definite integral graphically.

1. Enter the function $f(x)$ as Y1 by pressing the (Y =) button and typing the equation.

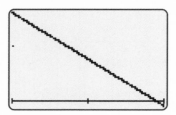

2. Graph the function over the specified domain. (You may need to press (WINDOW) and adjust the Xmin and Xmax settings. Press (ZOOM) then (0) to automatically adjust the y-values so that the entire graph will appear on the screen.)

3. Press (2nd) then (TRACE) to bring up the CALCULATE menu. Select item 7: $\int f(x) dx$ and press (ENTER).

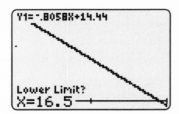

4. The calculator asks, Lower Limit? Enter the value of the lower limit of integration. Then press (ENTER). (*Warning:* The lower limit must be within the domain of the viewing rectangle.)

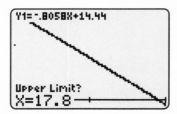

5. The calculator asks, Upper Limit? Enter the value of the upper limit of integration. Then press (ENTER). (*Warning:* The upper limit must be within the domain of the viewing rectangle.)

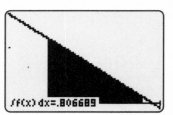

6. The calculator shades the region between the graph of the function and the x-axis and then displays the value of the definite integral. In this case, we found

$$\int_{16.5}^{17.8} (-0.8058s + 14.44)\, ds = 0.8067.$$

Calculating a Definite Integral with *fnInt*

In the previous tech tip, we showed how to use the TI-84 Plus to determine the value of a definite integral graphically. In these steps, we will demonstrate an alternative way to calculate the definite integral.

1. Press (MATH) to bring up the MATH menu. Select 9:fnInt. Press (ENTER).

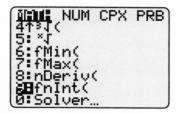

2. The calculator program *fnInt* requires four values: the function, the variable of integration, the lower limit, and the upper limit. Enter each of the values

and press (ENTER). To calculate $\int_1^3 \frac{\ln(x)}{x}\,dx$, we enter *fnInt* $(\ln(x)/x, x, 1, 3)$.

```
fnInt(ln(X)/X,X,
1,3)
```

3. The calculator displays the value of the definite integral.

```
fnInt(ln(X)/X,X,
1,3)
           .6034744804
```

---

*Technology guides are available on CourseMate for Applied CALC. Access at **login.cengagebrain.com**.*

Logistic Regression

In practice, it is easiest to create a logistic model by using logistic regression. This card demonstrates how to find a logistic model for a set of data.

1. Enter the data into L1 and L2 using the Statistics Menu List Editor. (Refer to Entering a Table of Data if you've forgotten how to do this.) If the lower horizontal asymptote of the data set appears to be something other than $y = 0$, you will need to vertically align the data in L2 before moving to Step 2. That is, subtract the estimated value of the lower horizontal asymptote from each value in L2.

```
L1    |L2   |L3      3
    0 | 1   |▬▬▬▬▬
    1 | 2   |
    2 | 4   |
    3 | 8   |
    4 | 10  |
    5 | 11  |
------|-----|
L3(1) =
```

2. Bring up the Statistics CALC menu, select item B:Logistic, and press ENTER.

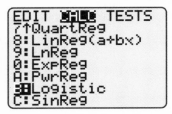

```
EDIT CALC TESTS
7↑QuartReg
8:LinReg(a+bx)
9:LnReg
0:ExpReg
A:PwrReg
B:Logistic
C:SinReg
```

3. If you want to automatically paste the regression equation into the Y= editor, press the key sequence VARS Y-VARS; 1:Function; 1:Y1 and press ENTER. Otherwise press ENTER. If you vertically aligned the data in Step 1, you will need to manually add the value of the lower horizontal asymptote to the regression equation.

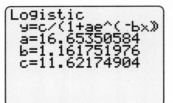

```
Logistic
 y=c/(1+ae^(-bx))
 a=16.65350584
 b=1.161751976
 c=11.62174904
```

*Technology guides are available on CourseMate for Applied CALC. Access at **login.cengagebrain.com**.*